**Gettysburg College Library**
GETTYSBURG, PA

Dr. Kenneth L. Smoke

Chairman - Dept. Psych.

Memorial Trust Fund

Cognition and Neural Development

# COGNITION AND NEURAL DEVELOPMENT

Don M. Tucker and Phan Luu

# OXFORD
UNIVERSITY PRESS

Oxford University Press is a department of the University of Oxford.
It furthers the University's objective of excellence in research, scholarship,
and education by publishing worldwide.

Oxford   New York
Auckland   Cape Town   Dar es Salaam   Hong Kong   Karachi
Kuala Lumpur   Madrid   Melbourne   Mexico City   Nairobi
New Delhi   Shanghai   Taipei   Toronto

With offices in
Argentina   Austria   Brazil   Chile   Czech Republic   France   Greece
Guatemala   Hungary   Italy   Japan   Poland   Portugal   Singapore
South Korea   Switzerland   Thailand   Turkey   Ukraine   Vietnam

Oxford is a registered trademark of Oxford University Press in the UK and certain other
countries.

Published in the United States of America by
Oxford University Press
198 Madison Avenue, New York, NY 10016

© Oxford University Press 2012

All rights reserved. No part of this publication may be reproduced, stored in a
retrieval system, or transmitted, in any form or by any means, without the prior
permission in writing of Oxford University Press, or as expressly permitted by law,
by license, or under terms agreed with the appropriate reproduction rights organization.
Inquiries concerning reproduction outside the scope of the above should be sent to the Rights
Department, Oxford University Press, at the address above.

You must not circulate this work in any other form
and you must impose this same condition on any acquirer.

Library of Congress Cataloging-in-Publication Data
Tucker, Don M.
  Cognition and neural development / Don M. Tucker and Phan Luu.
    p. cm.
  Includes bibliographical references and index.
  ISBN 978-0-19-983852-3 (hardback: alk. paper)   1. Neurophysiology.
  2. Cognition.   3. Brain.   4. Memory—Physiological aspects.
  I. Luu, Phan, 1968-   II. Title.
  QP406.T885 2012
  612.8—dc23
             2011051636

9 8 7 6 5 4 3 2 1
Printed in the United States of America
on acid-free paper

# Preface

In this book, we begin with what we think is an insight. The mind is the brain. The mind is not just brain *function*, as if function could be separated from structure. Rather it is the very substance of the brain. As the mind operates, it reflects the ongoing neurodevelopmental process, continuing the self-organization of neural connections that began in the embryonic differentiation of the neural tube and that continues throughout life.

We think this insight is implicit in today's scientific research, but it is not yet fully apparent. Once it does become apparent, everyone will think "Of course we knew this all along." We might say this insight is still preconscious in the minds of psychologists and neuroscientists, and even those who should bridge these disciplines, cognitive neuroscientists. The progress in science prepares us for this insight, yet it is still difficult to realize.

One problem is that mind naturally seems different from brain. The objectivity required for flexible scientific analysis is difficult when we have personal awareness of mental processes (more or less) but have no personal awareness of the neural mechanisms. From the naive view of the subjective mind, we then experience psychological processes as fundamentally separate from the bodily processes that generate them. It is natural, then, even for scientists to think that the brain function that creates our familiar psychological phenomena can arise from a brain whose basic structure, its anatomy, is fixed. Rather, the insight is that the phenomena of mind are due to the ongoing developmental changes in neuroanatomical differentiation. Every thought reorganizes the tissue.

Another problem is that the insight requires developmental reasoning. In modern training in both psychology and neuroscience, development is seen as a specialized topic, dealing with the minds of babies or the brains of embryos. If development is not your specialty, you never think about it. This is an important mistake. Development is the process of life. Just as biology cannot be understood except in the light of evolution, psychology cannot be understood except in the light of brain development. Psychology is, indeed, in each moment, brain development.

These impediments should be overcome soon enough. The theoretical progress that makes it possible to understand the neurodevelopmental nature of human cognition has been slow over the last couple centuries, but it is now well advanced. Perhaps most importantly, we have learned how neural connectivity implies cognitive function. This has come largely through connectionist computational modeling, showing how complex and naturalistic representation of concepts can be organized in the pattern of connections among simple neuronal elements. In addition, we have a lot of concrete evidence of the isomorphism (parallel forms) of psychological function and brain activity. This is being gathered by powerful methods of observing the brain's activity in controlled cognitive experiments, first with hemodynamic (functional magnetic resonance imaging or fMRI) measures, and now with the temporal dynamics shown by dense array electroencephalography (dEEG). Both of these are being augmented by increasingly accurate measurement of network structure with nerve fiber tractography.

Now, the structure-function illusion still remains in this research. Cognitive neuroscientists still think the neuroanatomical structure is fixed, and the hemodynamic or electrophysiologic measures reflect the activity of the tissue that somehow gives rise to psychological function without changing its structure. But activity becomes structure. The advances in technology and measurement are preparing us for this insight. We will soon appreciate that the cognitive mechanisms of learning and memory are achieved through developmental neural plasticity. Neural structure is developing continuously, and this development is brain function. As a result, our familiar psychological experience throughout life is the direct reflection of ongoing neural ontogenesis.

## Overview

The task of this book is to help the insight along, by showing how specific neurodevelopmental mechanisms not only explain the brain's continuing growth and differentiation, but at the same time explain important features of human cognition. At the core of cognition is adaptive memory consolidation, through which learning produces structural changes in the brain and the self. We begin in Chapter 1 with an overview of the embryonic mechanisms of synaptogenesis and activity-dependent pruning that shape the brain's anatomy in utero and then are continued in postnatal neural development to form the neural substrate for learning. A theoretical understanding of mammalian learning is particularly important to explaining the neural substrate of cognition, because mammals are found to learn not through passive reflex associations (as was thought in traditional learning theory), but through active expectancies for future events. A neurophysiological account of mammalian learning therefore must explain the elemental cognitive processes that allow mammals to represent adaptive goals, anticipate actions in relation to these goals, and adapt both expectancies and behavior in the face of unexpected consequences. Through continued neural ontogenesis, mammals continue to grow and differentiate their neural networks throughout life. Neural growth is continuous, including during sleep, as the neurophysiology of memory consolidation patiently and incessantly integrates each day's experience within the brain's connectional patterns. Learning is then a literal neurodevelopmental process, shaping the neural growth and differentiation as it continually anticipates future events. Just as cognition must be understood as a neurodevelopmental process, mammalian brain development must be understood as a cognitive process.

Once neural development is considered in this way, a basic question becomes the motive control of neural activity. With activity-dependent specification of neural connections, the control of neural activity controls both the growth of neural networks and the ongoing cognitive process. Although the self-regulation of neural activity is a complex issue, we assert that even a simple model of neural cybernetics (control systems) can be useful. We propose that there are two fundamental forms of neuromodulation, described as *redundancy* and *habituation*. These shape the continuity of neural activity, and therefore memory consolidation, over time. A key theme throughout the book is understanding how these primitive controls alter not only activity-dependent plasticity of neural networks, but the structure of ongoing working memory. Memory structure is biased between focused (redundancy) and expansive (habituation) modes of cognitive representation. By altering ongoing neural activity in time, these controls on redundancy and habituation shape both the consolidation of memory and the connections that form the dynamic architecture of the developing brain.

Chapter 1 thus outlines a theory in which learning and the growth of the brain are the same thing. And they are shaped by the same neural control systems.

In Chapter 2, we consider more specifically how neural control systems achieve memory consolidation by changing the brain's structure. Within the mammalian brain, memory consolidation negotiates between two boundaries: the interface with the environment, as represented by the *somatic* (sensory and motor) nervous system, and the interface with the internal requirements for adaptive homeostasis, as represented by the *visceral* (internal milieu) nervous system. The somatic nervous system has its basis of representation in sensory and motor cortices. The visceral nervous system has its basis of representation in limbic cortices. The organization of the mammalian neocortex then reflects an ordered set of networks (including heteromodal and unimodal association cortices) spanning between the visceral limbic *core* of the hemisphere and the somatic sensorimotor networks forming the outer *shell* of the brain, interfacing with the world. In order to understand the neural mechanisms through which motive expectancies regulate mammalian learning, we complete Chapter 2 by considering specific phenomena of neurophysiological excitement, such as limbic kindling and thalamic recruiting responses, that may provide clues to the adaptive mechanisms of memory consolidation that communicate between the brain's visceral core and its somatic shell. These are the boundaries of the mind.

In Chapter 3, we consider these neurophysiological mechanisms of learning in relation to specific requirements of motivated (viscerally charged) sensory and motor (somatic) operations. Cognition is nothing more than motivated sensation and action, extended by memory. We call this *the theory of action regulation*. We will see how even complex features of human executive control of cognition could emerge from the basic mammalian frontolimbic networks for action regulation. There are two of these frontolimbic networks: one dorsal and one ventral. The dorsal control system provides *feedforward* control of behavior from the anterior cingulate and mediodorsal frontal networks. The ventral control system provides *feedback* control from the insular cortex, extended amygdala, and orbital and ventrolateral frontal networks. Chapter 3 will explain how the dorsal frontolimbic control system operates with the habituation bias, such that the feedforward control of cognition (as well as action) is impulsive, expansive, and well suited to grasping the holistic context for action. In contrast, the ventral frontolimbic control system operates with the redundancy bias, supporting the sustained focus on external constraints that is critical to the feedback guidance of action. This same sustained focus on external constraints provides essential cybernetics to allow cognition to emerge from action regulation. It allows the differentiation of discrete semantic objects not only to guide actions, but to structure perception and language.

Thus we see how cognition is emergent from elementary motivational controls on sensation and action. These motivational controls are the same neural mechanisms that shape the activity of the embryo's nascent sensory and motor networks, and they continue to shape cognition (neural development) throughout life.

Chapter 4 then illustrates, in broad outline, some of the general psychological implications of this theory of motivated learning. In psychological terms, the idea is that the corticolimbic networks simultaneously control dual functions: motivation and memory. Although these functions have been isolated in modern neuroscience and neuropsychology, they were considered inseparable in Freud's early psychoanalysis, as captured by the term *motive-memory*. We propose that there are in fact *two* modes of strategic control of the motive-memory, arising from what we now recognize as dual limbic systems. We describe these modes of control in psychological terms: they are the *impetus* (impulse) and the *artus* (constraint). These two concepts provide concise theoretical tools for analyzing the adaptive control of neural, and psychological, development.

The goal of Chapter 4 is psychological theory, even though we frame this theory in neurocybernetic terms. As clinical psychologists, we have learned about psychology through working with real people in real lives, where 25% of the population suffers from some mental disability. As students of the brain, we see this clinical pragmatism as an advantage for theoretical work, causing us to be disinterested in narrow experimental paradigms and demanding of explanatory concepts that account for the remarkable range of human developmental outcomes. The concepts of the impetus and the artus are broad, and somewhat vague, but they have the requisite theoretical scope to describe everyday psychology.

After Chapter 4, the theoretical argument changes in its basic form. For theoreticians, this is where the fun begins. Up to this point, we have drawn on evidence on how learning occurs in the brain to formulate an integrated description of psychological and neural self-organization. This description is admittedly simplistic, and yet once we get the picture, the obvious question arises: how did these dual limbic systems get here in the first place? How did it happen that the dorsal limbic system is regulated by habituation, whereas the ventral limbic system is regulated by redundancy?

The answer, of course, is evolution. And the ontogenesis that is its mechanism. What makes the theoretical work so fun at this point is that evolution has left its marks on each phase of mammalian neuroembryological development. Phylogeny is history, and now long gone. But it is recapitulated, in major residuals of form, by ontogeny. The story of mammalian cognition can be told, at least in its essential outline, simply by reading the developmental residuals of the evolved mammalian neural architecture as they unfold in the embryo's remarkable journey of self-organization. Once we know what to look for, even nonexperts (such as psychologists) can read the story of neuroembryology to see the outlines of the remarkable events of mammalian neural evolution. It was these events that created the transformational cognitive capacities of dual limbic systems.

Chapter 5 sets the stage for the neuroembryological analysis, by taking the neuropsychological theory from Chapter 4 back to specific neural mechanisms of the impetus and the artus. These mechanisms arise from differences between the dorsal and ventral divisions of the hemisphere that involve both anatomical connectivity and neurophysiological activity. These specific neurocognitive mechanisms are what need to be explained by an evolutionary-developmental analysis of the neurodevelopmental process.

After this discipline of studying the specific neural mechanisms in Chapter 5, Chapter 6 gets to the most interesting part. Because we now realize that cognition is a continuing neurodevelopmental process, we can make the evolutionary-development analysis work not just by looking at the psychological development of the child but by looking at the morphogenetic development of the embryo. The evolutionary-developmental analysis of mammalian neuroembryology gives a remarkably clear explanation for how and why unique neurocognitive cybernetics emerge from the dorsal and ventral divisions of neocortex. In Chapter 6 we begin with the basics of mammalian neuroembryology, and then consider recent evidence on the patterns of gene expression in embryonic development. This evidence implies that the dual limbic systems, and the related dorsal and ventral divisions of neocortex, reflect the residuals of a major mutational event in mammalian evolution, what we describe as the *collapse of the reptilian telencephalon*. This is neuroarchaeology, looking at the pieces and seeing the story of the historical progression. Through what appear to have been a set of developmental mutations, the 3-layered pallium and the subpallial (basal ganglia) architectures of the vertebrate (amphibian and reptilian) telencephalon seem to have collapsed, catastrophically. In this collapse of their morphogenetic boundaries, they confused their neuronal substrates irrevocably, and thereby fused in a new morphogenetic configuration, the 6-layered mammalian neocortex. Both the structural and the functional differentiations of the dorsal and ventral divisions of the mammalian neocortex, with their unique and intricate patterns of connectivity with the subcortical neuraxis, may be traced to unique mechanistic biases emergent from this major speciation event. Even as the pallial and subpallial structures were combined in the 6-layered cortex, their differential cybernetic algorithms appear to have been retained, now manifested in the specialized cognitive and neurodevelopmental biases of the dorsal and ventral limbic-neocortical divisions.

With the basic theoretical outline framed at this point, we conclude and take stock of the implications in Chapter 7. Because cognition is the neurodevelopmental process, the evolutionary-developmental analysis provides a new perspective on the structure of the mental process. For example, we can see how both classical and recent evidence on the unique features of the human brain, such as hemispheric specialization or the enlarged frontal poles, must be understood in the light of the evolved developmental strategies of expectant mammalian consolidation. We humans are just mammals that never quite mature. As a result we retain a kind of embryonic transience throughout life, our neuroanatomy changing (irreversibly) through each experience and behavioral choice. As a novel way to understand the specifics of this transience, we can look to the primitive roots of embryonic network differentiation, in pallial and subpallial neurocybernetic mechanisms. The roots are primitive, yet they are retained as the functioning postnatal ontogenetic mechanisms of cognition and memory. Because of our radical neoteny, the developmental mechanisms are freed from the constraints of maturation and allowed to elaborate neural plasticity to form highly idiosyncratic patterns of intelligence and personality. The mechanisms of human cognition are highly individualized, and highly sensitive to the specific cultural context. Yet at the same time they are rooted in primitive control systems that can be seen to emerge from their reptilian foundations in embryogenesis, shape the uniquely mammalian forms of the child's corticolimbic plasticity, and then continue to regulate memory consolidation and psychological self-organization throughout the life span.

# Contents

1. Neurodevelopmental Mechanisms of Learning — 1
2. Consolidating Memory — 24
3. Regulating Action — 59
4. Opponent Complementarity in Psychological Function — 90
5. Structural Clues to Dorsal-Ventral Specialization — 139
6. The Evolved Structure of Mammalian Memory — 151
7. Self-Organizing Ontogenesis on the Phyletic Frame — 190

References — 219

Index — 259

# 1
# Neurodevelopmental Mechanisms of Learning

Human learning is a remarkable phenomenon, allowing a neotenous juvenile primate to acquire the skills of a complex modern culture. Working memory may allow active self-regulation of the learning process, as multiple cognitive processes are integrated in service of a goal (Coolidge & Wynn, 2005; Hamidi, Tononi, & Postle, 2008; Lenartowicz & McIntosh, 2005; B. T. Miller, Deouell, Dam, Knight, & D'Esposito, 2008; Miller & Cohen, 2001; Muller & Knight, 2006; O'Reilly & Frank, 2006; Postle et al., 2006). Although learning has often been considered to be a psychological process, and merely supported by the necessary biological mechanisms, we propose that modern learning theory and research on neural plasticity, taken together, suggest a more literal interpretation. The process of learning can be understood as the continuation of the embryonic neurodevelopmental process. Learning is neural morphogenesis, continued into postnatal development. The mechanisms of neural differentiation sculpt the general structure of the embryonic brain through multiple genetic and epigenetic chemotrophic and neurotrophic mechanisms, and then achieve the fine structure of cerebral architecture through overabundant production of synapses (neuronal connections) and activity-dependent specification (removal of unused connections). Learning is then the continuation of neural differentiation throughout development, simultaneously differentiating the child's cortical network architecture and her psychological capacities in self-regulation.

The general outline of the evolutionary order of mammalian neural systems can be seen in the progressive stages of neural embryogenesis in Figure 1-1. The telencephalon, including the basal ganglia, limbic circuits, and cortex, becomes increasingly important in later stages of neuroembryogenesis, just as it has become increasingly important in mammalian evolution.

The unique forms of active learning that organize human working memory may continue to be motivated by primitive morphogenetic mechanisms that

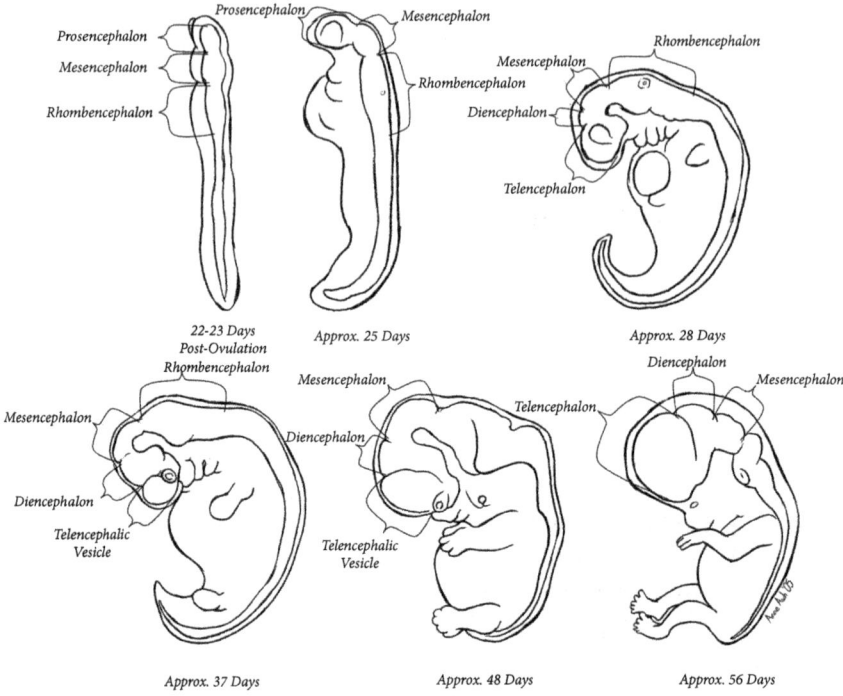

Figure 1-1. Development of the human brain with progressive elaboration of the forebrain or prosencephalon (comprising telencephalon and diencephalon) in later developmental stages.

remain embryonic in their basic form. As a result, these mechanisms require an evolutionary-developmental analysis, considering both their phyletic origins and the ontogenetic process that actualizes the individual brain. Humans seem to have developed not only a massive neocortex to support activity-dependent synaptic specification throughout development, but certain mutations of embryonic morphogenetic control that allow new cognitive capacities to be actualized through continuous regulation of synaptic differentiation over two or more decades of juvenile immaturity.

## Learning and Synaptic Plasticity

Research into neural plasticity has suggested the importance of a developmental analysis for many years (Greenough, 1975; West & Greenough, 1972). The specification of the fine structure of neural connections is activity-dependent, meaning that only synapses that are engaged by ongoing brain activity are retained (Black & Greenough, 1986; Greenough & Black, 1992). Early in mammalian cerebral development there is a process of overabundant synaptogenesis, followed by pruning and elimination of unused connections.

To differentiate between genetic and learning mechanisms of plasticity, Black and Greenough suggested that plasticity common to the species could be described as *experience-expectant*, in that the mechanisms of plasticity evolved because all individuals encounter the appropriate environmental stimulation (such as visual patterns). In contrast, plasticity unique to the individual could be described as *experience-dependent* (Black & Greenough, 1986) because it reflects the specific environmental adaptation of that individual. Although Black and Greenough point out that there is no known mechanistic distinction between species-specific and individual-specific plasticity, they suggest that critical periods of plasticity may be more important to species-specific plasticity.

Although research on neural plasticity has continued to demonstrate the sensitivity of multiple neural mechanisms to environmental exposure (Markham & Greenough, 2004), and ontogenesis clearly involves critical periods during which environmental experience is required for a neurodevelopmental process (such as the formation of ocular dominance columns), the distinction between species- and individual-specific plasticity is not well supported by evidence. In fact, the more interesting theoretical interpretation may be that neural ontogenesis in mammals has maintained embryonic mechanisms of plasticity well into postnatal growth and development, such that the juvenile period itself functions as an extended critical period for learning.

A general principle in the evolution of intelligence may be that there is increasing delay interposed between stimulus and response (Herrick, 1948; Tucker, Derryberry, & Luu, 2000). In simple vertebrates, stimulus-response connections are reflexive. In mammals they have become mediated by increasingly complex mechanisms of memory. Because simply being slower to respond is not adaptive, the mechanisms of mammalian memory became *expectant*, preparing the animal to respond rapidly to future events (Tucker et al., 2000). Within the neurodevelopmental theory of cognition, we hypothesize that the embryonic mechanisms of neural plasticity have evolved to manage cognitive expectancy to support postnatal learning. As we will see later in this section, this is not only the species-specific experience-expectant ontogenesis of Black and Greenough (1986), but a control of ongoing expectancy for hedonic and aversive events that shapes the mammalian learning process. Just as embryonic neural plasticity is by its nature expectant for activity-dependent synaptic differentiation, mammalian learning appears to extend and continually modify this expectancy as a strategy for flexible behavioral adaptation before, not just after, critical environmental events.

Through this dynamic control of expectant learning, the structure of mammalian neural connections is *self-organizing*, as the developing organism's spontaneous behavior determines the sensory, motor, and motivational events that shape neural traffic and thus synaptic differentiation. In the early embryonic stages, the spontaneous activity is spasmodic and disorganized. However, as originally observed by studies of the chick embryo (Hamburger, Balaban, Oppenheim, & Wenger, 1965), this early self-generated activity appears to be the mechanism for organizing functional neural circuitry, through creating coordinated (and thus retained) patterns of sensation and action. In the course of early postnatal human development, the spontaneous acts of self-organization become increasingly less influenced by innate rhythms of arousal, sleep, feeding, and elimination, and more influenced by the brain's systemic integration of states of attention, memory consolidation, and motivated social interaction. Juvenile mammals exercise learning capacity through a unique behavioral adaptation: play. For human infants, the embryonic mechanisms of neural differentiation are still active, but they now begin to operate within the child's psychological self-determination within the social context.

## Plasticity and Network Structure in Neurodevelopmental Theory

To formulate the neuronal basis for learning in terms of the modern understanding of distributed representations in neural networks, it is important to go beyond the notion that connections are strengthened by use, and explain the changes in network connectivity that are caused by particular patterns of use. The modern analysis of functional connectivity of neural networks is often traced to Donald Hebb's model of the *cell assembly*, formed because synapses are strengthened by synchronous activity of the communicating neurons (Hebb, 1949). Although ignored for many years, Hebb's model eventually proved to be essential for connectionist models of distributed representations in neuronal networks. These included not only those models expressed in generic computational simulations (Rumelhart & McClelland, 1986) but also those formulated in realistic neuroanatomical architectures (Grossberg, 1970; Sejnowski, 2002).

Half a century before Hebb, the neurologist and neuroanatomist Sigmund Freud had attempted to explain learning and memory within a model of self-organizing connections within neuronal networks (Freud, 1895; Pribram & Gill, 1976). In his *Project for a Scientific Psychology*, Freud proposed that memory could be explained as a strengthening of neural connections through use (Pribram & Gill, 1976). As the process of learning modifies neuronal connections, Freud recognized that neuronal networks face an inherent dilemma of maintaining stability or allowing plasticity. He proposed that certain phi neurons are specialized for perception (which is kept stable) and other psi neurons are specialized for memory (which is plastic). Even in these early neuronal models, Freud saw that the motive control for learning is integral to regulating plasticity. So much so that he came to describe it as the *motive-memory*.

An important limitation of Freud's reasoning about stability and plasticity was that he considered memory a property of the discharge tendency of the neuron itself. It would be half a century later, with the work of McCulloch and Pitts in the 1940s, before there was initial theoretical insight into the distributed nature of cognitive representation in nerve nets (Heims, 1991; McCulloch & Pitts, 1990). Hebb's concept of the cell assembly captured this notion of distributed representations strengthened through use (Hebb, 1949). It was several decades more before the nature of stability, plasticity, and catastrophic interference were fully understood in distributed representations (Grossberg, 1980). Once the neurodevelopmental evidence on activity-dependent plasticity (Hamburger et al., 1965) became apparent as a mechanism for cortical development (Rakic & Singer, 1988), the way was paved for the fundamental insight that mammalian learning is the extension of the neurodevelopmental process (Singer, 1987; von de Malsburg & Singer, 1988).

The challenge that we address in this chapter is how to progress from generic notions of neural plasticity to the specific nature of self-organization of adaptive learning, not only in mammalian behavioral development but in human cognition. The modern literature on animal learning provides the necessary theory. Whereas it has long been assumed that animal learning proceeds through a passive association of stimuli and responses, the modern evidence shows that mammals learn, and thus organize their cerebral networks, through primitive, yet active and expectant, processes of cognition. What we learn through a careful analysis of animal learning becomes informative for basic mechanisms of human cognition.

## Modern Learning Theory

The elementary physiological mechanisms of learning were conceived by Pavlov to be organized like a reflex. *Conditioning* is the process through which a new stimulus is inserted, through repeated association, into an innate, "unconditioned" or reflexive stimulus-response association. Although many psychologists and neuroscientists continue to think of learning theory in this way, the modern evidence has led to a different realization. Whereas passive associations may define learning in simple vertebrates (Tennant & Bitterman, 1975), mammals learn new behavioral patterns through what may be described as elemental cognition: actively expecting future events and then systematically modifying those expectancies based on outcomes.

Even at the height of behaviorism, in academic psychology in the mid-20th century, there were cognitive explanations for animal behavior. Krechevsky (1932) took issue with the notion that an animal approaches new learning situations with random behavior. Drawing from the Gestalt theory of learning, in which learning is a process of gradual articulation and differentiation from a holistic perceptual structure, Krechevsky provided a cognitive interpretation of the behavioral biases that animals show in a new learning situation.

When presented with a discrimination learning problem, rats exhibit systematic biases in their responses, such as choosing a rightward direction in a maze. Furthermore, the animal shifts from one type of bias (e.g., predominantly going right) to another, rather than acting randomly, until the correct response is achieved. These biases, Krechevsky argued, are products of the animal's experience applied to the new learning situation. He called these biases *hypotheses*. For Krechevsky, hypotheses lead to behavior that is systematic, purposive, abstract, and does not require environmental influence for its initiation. Krechevsky's analysis proposed that animals approach new learning situations as active agents of exploration, rather than passive random emitters of behavior.

An example of modern evidence documenting the active cognition of mammals is the *blocking* effect, repeatedly demonstrated in learning studies with rats (Rescorla, 1980). A stimulus, such as a tone, can be shown to be perfectly adequate as a conditioned stimulus. This means that if it is paired with an innate (unconditioned) stimulus (such as food), this new tone stimulus will elicit a response (such as a lever press), even when presented without the food. However, if the response is first trained to a different stimulus, such as a light, then presenting the tone together with the light does not lead to conditioning of the tone. Presenting the tone alone then fails to elicit the response, even after many pairings.

A passive, reflexive association theory of learning cannot explain these results. It becomes obvious that the rat fails to learn the association of the tone with the food because the light is already an effective predictor of the food event. There is then no cognitive representation of the tone in the animal's expectancy model. Given these results, it may not be overly anthropomorphic to say that, in the presence of the adequate expectancy cue provided by the light, the tone is not remembered because it fails to receive attention.

Although modern learning theory is expressed in technical and mathematical terms, thus avoiding the appearance of anthropomorphic explanations, an influential recognition of the role of expectancies in modern learning theory was expressed by Kamin when he commented that his rats didn't seem to learn unless they were surprised (Kamin, 1965). The more formal description of this insight appearing in the literature emphasized that a cue (conditioned stimulus) is learned because of a discrepancy in the rat's existing predictions, such that the new cue is now recognized as predictive of the desired outcome (Rescorla & Wagner, 1972). Essentially, Kamin's observation remains accurate: learning of new cues is indeed most effective in the case of surprise, particularly an aversive surprise, when important outcomes are discrepant with old cues and habits. Learning thus emerges from the need to change. As in the blocking example, new cues are not learned if the old cues are still effective predictors. Learning is not a reflex that is impressed upon by passive repeated associations, but the outcome of an elementary expectancy for motivationally significant outcomes.

## Control and Communication in the Animal and the Machine

Theoretical principles of animal learning have important parallels with principles of machine learning. Although animal learning theory adopted what were essentially cognitive models in the later 20th century, there was little use of the powerful theoretical tools of control theory that had been developed decades before. In Wiener's cybernetic analysis, first published in 1948, control theory applies mechanisms of feedback and continuous control to both biological and machine learning (Wiener, 1961). The cybernetic approach of the middle 20th century had widespread influences on technical, scientific, and popular thinking (Heims, 1991), with the notable exception of academic psychology. Whereas behaviorists maintained that animal learning is a matter of passive associations, cognitive psychologists of the later 20th century found little use for either control theory or motivational influences, and they focused instead on information processing capacities in their experimental studies.

Yet studies of machine learning quickly integrated both cognitive representation and control theory into their theoretical models. Beginning with the work on the *perceptron*, a simple 1-layer connectionist model (Rosenblatt, 1958), connectionist or artificial neural network simulations provided highly instructive theoretical models for how complex concepts could be trained within distributed neural networks, such as those of the mammalian cortex (Rumelhart & McClelland, 1986). Control theory also brought important insights to robotics, where it had failed to do so in cognitive psychology. In attempting to program robots to navigate through complex territories, it became clear that there were differential advantages to specific control strategies, depending on the robot's representational stores (Hendler, 1995). In the case where the robot holds extensive prior knowledge of the environment, such as of the distances and locations of objects, navigation can proceed through *feedforward* control: the robot has all the information required for expecting the environmental features at each turn. However, in the face of changing circumstances, *feedback* control becomes necessary, in which sensory data must be used to recognize and adapt to unexpected objects in order to incorporate them into the continually modified behavioral plan (Hendler, 1995). Because feedback control is well suited to adaptation in the face of uncertainty, it readily gained recognition as the primary cybernetic control in engineering.

As we examine the complex and sometimes apparently arbitrary mechanisms of evolutionary adaptation that have allowed mammalian learning, it is important to keep these generic requirements for cybernetic effectiveness in mind. Evolution operated on varied mutations of the vertebrate plan, but the requirements for self-regulation, through variants of feedforward and feedback control, were continuous constraints on the emerging mechanisms. Both feedforward and feedback strategies are integral to the learning process, in both animals and machines.

In the modern scientific analysis of learning, at the same time that behavioral observations in animals showed that new learning results from the need to change, neural network models of learning were showing that new learning may be costly, specifically through disrupting old learning by modifying the connections weights of the old learning (Grossberg, 1980). The *stability-plasticity dilemma* faced by neural networks in the presence of new input is a fundamental principle that underscores the importance of regulating learning, not only to add new information, but to protect the old. Change is thus not inevitable in development; it has a cost that must be managed through explicit mechanisms that assign value to stability versus plasticity (Grossberg, 1984; Tucker & Desmond, 1998).

The stability-plasticity dilemma in machine learning is an important perspective to bring to a neurodevelopmental theory of learning. The mechanisms of embryonic morphogenesis must maintain the structural integrity of the emerging neocortical architecture, at the same time as achieving openness to adaptation through activity-dependent specification. In the continuation of the mechanisms of embryogenesis throughout life span development, mammalian learning must maintain the stability of the developing neural organization at the same time as adapting this organization to changing environmental input. When development is understood as structural organization, the balance in neural as well as psychological structural development may therefore require negotiation between *assimilation*, adapting new input that is consistent with the internal structure, and *accommodation*, adapting the internal structure to the new input (Piaget, 1936/1992).

## Neural Control of Expectancy

Expectancy may be the neurodevelopmental algorithm for bridging assimilation and accommodation, through preparing the internal representational structure for significant (and possibly disruptive) new information. Remarkably, the processes of assimilation and accommodation seem to be regulated by differing mechanisms of neurophysiological consolidation, mediated by specific limbic and thalamic circuits. This realization leads us to search out the neurophysiological nature of these neural mechanisms, in order to understand their roles in structural development.

The thalamus and hypothalamus (Figure 1-2) of the diencephalon have been described as the *interbrain* because they are positioned between the telencephalon and the brainstem (Swanson, 2003). As we will see, circuits of the diencephalon are integral to the *limbic system* that includes the telencephalic structures of the amygdala, hippocampus, and the cortical areas of the limbus or border of the hemisphere surrounding the interbrain and brainstem (Figure 1-3).

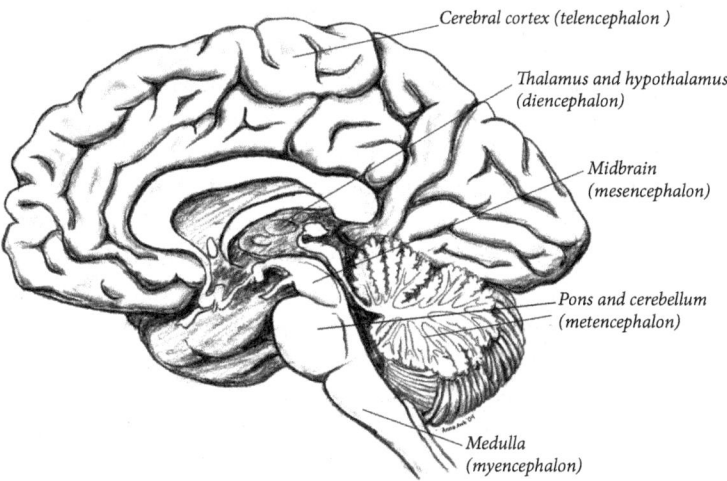

Figure 1-2. In the human brain, the telencephalon is composed of the cerebral cortex, basal ganglia including pallidum, and limbic system. The diencephalon or interbrain is situated at the top of the brainstem and includes the thalamus and hypothalamus. The more primitive brainstem structures (mesencephalon, metencephalon, and myelencephalon) remain integral to arousal and regulatory functions of the forebrain.

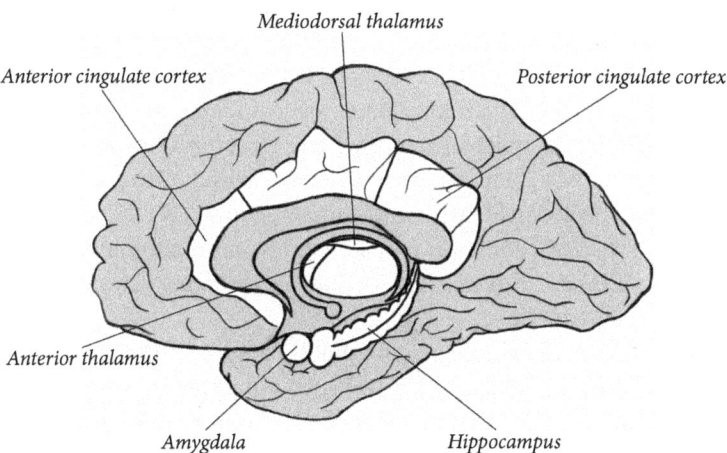

Figure 1-3. Within the human limbic system, a dorsal or top division includes the cingulate regions of limbic cortex (anterior, middle, and posterior cingulate areas) that are strongly connected to the hippocampus and are linked through the anterior nuclei of the thalamus to the hypothalamus. A ventral or bottom division of the limbic system links ventral limbic cortex (including the insula and orbital frontal lobe) and the amygdala, with thalamic circuits through the mediodorsal nucleus.

Studies of the neurophysiological mechanisms of animal learning (Gabriel, 1990; Gabriel, Sparenborg, & Kubota, 1989) have been consistent in important ways with modern cognitive models of learning, in which the animal holds an expectancy for events, and learns new behavior in relation to that expectancy. The neurophysiological model of learning emerging from this work may also be consistent with the computational and control theories that emphasize the balance between stability and change. Certain neural mechanisms appear specialized for holding expectancies for the valued events in the environmental context. Other neural mechanisms appear specialized for disrupting and modifying expectancies when they are discrepant with events. We think these specializations can be interpreted to support the differing cybernetic requirements for feedforward and feedback modes of control, respectively.

In discrimination learning paradigms, an animal is presented with a stimulus (such as an auditory tone) that predicts (CS+) the delivery of a reward or punishment, or it is presented with a stimulus that does not (CS–). The animal must make a conditioned response (CR) to the CS+ in order to obtain a reward or avoid the punishment. Through training, the animal learns to discriminate between the CS+ and CS–. Gabriel and Sparenborg (1986) have studied this paradigm with microelectrode recordings of neuronal activity implanted into the brain circuits that are particularly important to memory in rabbits.

The challenge for the rabbits was to learn which response to make to avoid punishment. Gabriel and Sparenborg looked for neurons that coded this learning, with increases in what they called *training-induced activity*. Neurons were identified to respond differently to CS+ and CS– as a result of learning in cingulothalamic networks integrating both cortical (cingulate) and subcortical (limbic nuclei and thalamic) structures. These structures include the anterior cingulate cortex (ACC) and posterior cingulate cortex (PCC), hippocampus, amygdala, and mediodorsal (MD) and anterior ventral (AV) nuclei of the thalamus. The most significant effect was an increase in training-induced activity in response to CS+ after learning, rather than by a decrease in training-induced activity in response to the CS–.

Particularly striking was the finding that the responsive neurons are linked within different cingulothalamic networks, and that these networks are active at different stages of the learning process. The network comprising the ACC, amygdala, and MD thalamus is particularly important in the *early stages* of learning, when learning requires effort to override habitual or automated responses that appear obligatory to the conditioned stimuli. Because the rabbits were reacting to the changes made by the experimenter, this early learning required considerable feedback control. In contrast, the network comprising the PCC, hippocampus, and AV thalamus (the major components of the Papez circuit) controls what Gabriel and Sparenborg term the *late stages* of learning. In these late stages, the Papez circuit codes a stable and regular context for action.

This is a feedforward mode of control, because the expectancy (the context model) works well to predict environmental events.

Lesions to the early learning network thus affect rapid acquisition of new responses, whereas lesions of the late learning network affect the ability to organize asymptotic/automated performance within a well-learned context model of the environment. Although functioning differentially, these cingulothalamic networks do show some overlap in the different stages of learning (Gabriel and Sparenborg, 1986). For example, even though the hippocampus is clearly part of the late learning system because of its critical role in contextual coding, it also contributes to new learning by inhibiting actions that are no longer appropriate in new learning conditions. At the same time, the specificity of the hippocampus and Papez circuit is clear; the hippocampal neurons do not show the training-induced differentiation of CS+ from CS- until the later stage of learning.

The Gabriel and Sparenborg neurophysiological model describes rabbit learning, and it is not dependent on cognitive constructs such as attention or data derived from human studies. Nonetheless, we think there are important parallels to cognitive models of the stages of human learning. In general, the role of the PCC (and related parietal cortex) in maintaining the global context model is similar to that seen for the *posterior attention system* in both primates (Mesulam, 1981) and humans (Posner & Petersen, 1990). Furthermore, the importance of the ACC in supporting rapid shifts of attention is well known in the human cognitive neuroscience literature (Fair et al., 2007; Luu & Pederson, 2004; Posner & Rothbart, 1998).

New learning situations require executive control processes in order to adapt new patterns of performance (Chein & Schneider, 2005; Schneider & Chein, 2003). In contrast, once automated perceptual and action schema are organized, performance is relatively effortless, in a stable mode that could be described as the context model of asymptotic learning for rabbits or as the flow experience for humans (Csikszentmihalyi & Rathunde, 1993). Moreover, if the neurophysiological learning model can be extended to human cognition, it accounts for executive control processes without having to posit an executive homunculus (Posner, 1978). For rabbits, executive control is simply the process of focused associative learning in the early stages, triggered by interruption of ongoing behavior when familiar expectancies fail (Gabriel, Taylor, & Burhans, 2003). For humans as well as rabbits, repeated practice allows new regularities to be extracted to form the new stable expectancy, the new context for action (Tucker & Luu, 2007).

## *Motivated Learning*

There may be intrinsic motivational biases, linked to primitive forms of cognitive expectancy, associated with the differing learning mechanisms for the stable context versus discrepancy detection. In research on human cognition, motivation is often assumed. However, motivation becomes a more practical and therefore apparent factor in animal research, such that the arbitrary separation of motives and memories is less tenable. The initial studies by Gabriel and associates focused on discriminations in avoidance learning, and this may have been important to the strong role of the amygdala and ventral limbic networks (with direct projections to the ventral ACC) in feedback guidance of the learning process under threat. For example, when the rabbit needed to perform an avoidance action in response to a CS+ to avoid a shock, Poremba and Gabriel (1997) found that the amygdala is critically important in the early stage of learning. If the amygdala is destroyed or inactivated, the animal shows profound deficits for learning within the first day of training compared to controls. Furthermore, with amygdala lesions, the animal fails to show the increased neuronal responses to CS+ in either the ACC or MD that signals the engagement of the early learning system, implying that the amygdala's motivational input is integral to the memory consolidation of this cingulothalamic circuit. However, the rabbit with an amygdala lesion can still be trained to performance criterion, suggesting that the slow (Papez circuit, feedforward) learning system is still intact. Once the animal was highly trained, a subsequent amygdalar lesion did not then affect performance.

In contrast to avoidance discrimination learning, for which the ACC-MD-amygdala early learning circuit is essential, approach discrimination learning does not depend on the amygdala and only indirectly depends on the early learning system. In approach learning, the animal learns to perform an action in response to a stimulus in order to receive a reward. Smith et al. (2002) showed that in control animals approach discrimination learning during the early stages was not associated with differential neuronal encoding of CS+ and CS- in structures such as the ACC, as it is with avoidance discrimination learning. Nevertheless, these animals demonstrated the normal learning curve. That is, they were quickly able to learn to approach the reward stimulus (e.g., drinking spout) when presented with the CS+, implying that feedback control from the amygdala-MD-ACC circuit was not required.

At the same time, when structures of this circuit, such as the MD, were lesioned, the animals showed profound deficits in early learning, although they were able to learn to criterion levels after extended training. This early learning deficit was characterized as a failure to inhibit approach responses to CS-. Smith et al. interpreted these findings to indicate the importance of the fast learning network (and we would say feedback control) in associating the CS- with no reward, so that approach behavior can be inhibited.

Taken together, this series of studies points to an amygdala-centric system that is sensitive to negative feedback. In avoidance learning, this system codes information that guides avoidance discriminations. In approach learning, the amygdala and ventral limbic system are not essential: the gradual learning system and context model supported by the cingulate cortex appears quite able to mediate the approach responses.

The ACC has a pivotal and complex role in the control of learning. With strong inputs to the ventral (subgenual) ACC from the amygdala, insular, and anterior temporal regions (Price, 1999), the ACC appears integral to linking the ventral limbic input, with its sensitivity to avoidance cues, to the developing action plan. At the same time, the ACC is a primary cingulate substrate for the dorsal corticolimbic contribution to planning and motor control (Paus, 2001; Vogt, Rosene, & Pandya, 1979), providing the route through which the context model of the PCC and parietal lobe, apparently with an integral approach motivation, shape the expectancy guiding the unfolding action plan (Luu & Tucker, 2003; Tucker & Luu, 2007). Together, these complementary learning circuits permit rapid early learning, perhaps by forming positive expectancies to frame actions (with PCC input to ACC) and then monitoring for negative discrepancies from the expectations (with amygdalar and ventral limbic input to the ACC). Once a viable context model (with implicit behavioral response biases) is established, the frontal and ventral limbic control is less important, as stable adaptation is maintained under posterior, perceptual, corticolimbic control.

Thus, we conclude that by relating specific capacities in learning to specific neural mechanisms, the research by Gabriel and associates provides an important way of understanding the inherent motive biases underlying mammalian learning. Learning in mammals appears to require an elemental form of expectancy, requiring both a form of primitive cognition, describing the animal's capacity for representing significant expectancies for future events, and control theory, describing the cybernetic requirements for regulating the balance between a valued context model (which is biased against change) and the disruptive effects of discrepancy with unique control properties to focus attention and effect rapid new learning, particularly under conditions of threat (Tucker & Luu, 2007). The modern animal learning evidence shows that the primary signal for new learning is often disruptive: the disruptive event signals that current expectations are wrong and that change is required (Kamin, 1965; Rescorla & Wagner, 1972). At this point, the constraints from feedback control mechanisms are important. In contrast, other evidence suggests that the hedonic (approach) value of the animal's current expectancy shapes the learning process, very likely through the positive value inherent to the familiar context model. This seems to reflect the operation of feedforward cybernetics, anticipating events that are consistent with the hedonic expectancy.

An example of the operation of hedonic expectancy in learning is shown by the *successive negative contrast effect*. This is illustrated when a normally adequate reward (one that easily motivates acquisition of a new habit) proves inadequate when it is substituted for an established preferred reward (Papini, 2003). As with other mammalian learning effects, the explanation can only be cognitive: the animal must be evaluating the environmental event (the new reward) in relation to its current expectancy. Once the hedonic charge of the expectancy has been set with the preferred reward, the normally adequate reward is no longer adequate. Although the working memory of a rat or rabbit may be primitive, it appears to be charged with value. The inadequacy of the new stimulus arises not because of an automatic, passive reward mechanism, but through a motivated cognitive evaluation of the stimulus in relation to the current hedonic expectancy.

## Vertical Integration of Learning

The requirement for expectancy to guide learning appears to be unique to mammals. Fish and reptiles appear capable of forming elementary stimulus-response associations, but they are not capable of directing behavior with expectancies in the way that mammals do (Amsel & Stanton, 1980; Bitterman, 1968; Papini, 2003). A neurodevelopmental theory of learning may suggest how elemental embryonic mechanisms preparing for neural plasticity continue to prepare for postnatal behavioral adaptations. In humans, the neocortex is extensively elaborated, yet it is based on the venerable mammalian plan. The mechanisms of working memory and adaptive expectancy appear to be integral to regulating this neocortical architecture, such that human cognition may require the specific mechanisms mediating between stability of a context model and the change required to adapt to discrepant input.

At the same time, the mammalian neocortex is not an isolated organ. It depends on the evolved architecture of the neuraxis. Just as the process of embryonic morphogenesis builds the new organism through a kind of successive metamorphosis of the evolved mechanisms of the rhombencephalic, mesencephalic, diencephalic, and telencephalic levels of organization (Figures 1-1 and 1-2), the learning process engages control systems across each of these levels. As a result, effective learning requires what can be described as *vertical integration*. Even as we emphasize that mammalian learning is inherently cognitive, requiring the neocortical consolidation of sensory experiences with appropriate behavioral strategies, we must recognize that even the most complex cognition of mammals does not occur without neurophysiological support at all levels of the evolved brain. This support includes arousal control from brainstem neurotransmitter projection systems, motivational control from hypothalamic and limbic circuits, regulation of cortical integration from thalamic nuclei, and cognitive as well as motor initiative, differentiation, and sequencing from the basal ganglia.

In embryonic neural development, vertical integration of the neuraxis is required to organize the arousal and activity controls across multiple levels in order to organize activity-dependent specification of neural architecture, including the extensive cortex of the developing human fetus (Figure 1-1).

In the neurodevelopmental process of learning, the same vertical integration may be required in each learning instance. By preparing the juvenile mammal's brain for learning, the embryonic neurodevelopmental mechanisms establish a kind of expectancy to guide postnatal synaptic differentiation. As these neurodevelopmental mechanisms continue to guide the learning process, they not only tune the arousal of the neocortex adaptively, but they extend memory capacity to dynamically regulate hedonic and aversive expectancies. The theoretical challenge for a neurodevelopmental theory of cognition is therefore to explain the elementary neural mechanisms through which motives shape brain activation and arousal to support both hedonic and aversive expectancies. How can the control of neural activity be related to the memory processes that are integral to expectant learning?

One perspective on this question comes from examining the neuromodulator systems that control both embryonic neural activity and the arousal and activation of the cognitive process.

## Self-Regulation Through Activation and Arousal

Following the discovery of the reticular activating system of the brainstem (Moruzzi & Magoun, 1949), there was an important line of research and theory on how elementary mechanisms of activation and arousal could be integral to both animal learning and human motivation (Lindsley, 1957, 1960; Malmo, 1959). In psychology, the concept of *arousal* was often taken to mean autonomic arousal (Schacter & Singer, 1962), with the longstanding assumption by psychologists that physiological arousal influences behavior primarily as it is interpreted cognitively (James, 1884). A key scientific challenge of 20th-century psychology was understanding how the self-regulation of arousal could be understood in relation to basic brain mechanisms of motivation (Cannon, 1915, 1927). This challenge was largely unmet. The concepts of psychological self-regulation through brainstem activating systems promised to provide a neurophysiological basis for psychological and psychiatric theory (Lindsley, 1957, 1960; Malmo, 1959). Yet these concepts were not adopted by psychological theorists of the latter half of the century (who limited themselves largely to explanations in terms of cognitive representations). The result has been a cognitive neuroscience that has categorized cognitive faculties in relation to hemodynamic localization of brain activity, largely without reference to neurophysiological mechanisms.

The increasing evidence that brainstem activating systems operate through specific neurotransmitter release (Carlsson, 1988; Descarries & Lapierre, 1973; Schildkraut, 1965) was clearly relevant to the effects of drugs on emotional arousal, including both street and psychiatric drugs (Cooper, Bloom, & Roth, 1974) (Figure 1-4). The widespread projections of

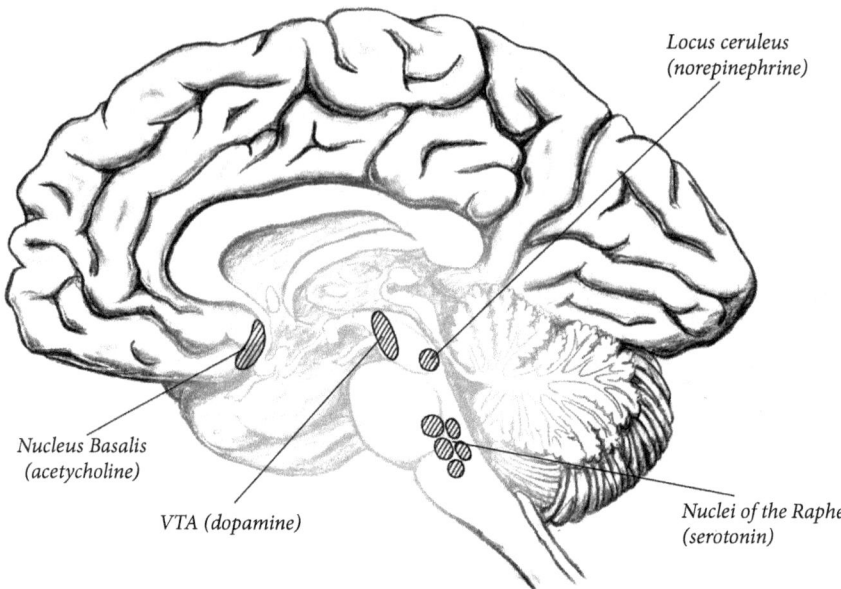

Figure 1-4. A few cell groups in the brainstem, together with the nucleus basalis in the forebrain, extend widespread and neurochemically specific projections throughout the brain. These neuromodulator systems regulate sleep, arousal, and alertness through specific actions on their target neurons.

these neuromodulator systems (Figure 1-5), emanating from focal nuclei of origin, provides an obvious anatomical basis for motivational control of neural activity. Although a considerable body of animal research examined behavioral and learning effects of drug and lesion manipulation of these neuromodulator systems, it seemed difficult through many decades of research to develop a lasting theoretical framework to explain how the organism self-regulates cerebral activation and arousal in a way that is relevant to motivational control.

An important theoretical account of how elementary neurophysiological arousal mechanisms could influence attention and cognition was provided by Pribram and McGuinness (1975). Certain brain mechanisms, including those targeting the basal ganglia, appear important to *tonic* neural activation in support of motor readiness. Other mechanisms operate in a more *phasic* fashion, and yet also provide important controls on attention. Taking up these concepts of primitive neural controls, Tucker and Williamson examined the literature on specific brainstem neuromodulator systems that could provide explanatory mechanisms (Tucker & Williamson, 1984). The control of tonic activation and motor readiness seems to rely on dopamine (DA) and acetylcholine (ACh) projection systems, with projections to limbic and frontal cortical networks as well as the basal ganglia. In contrast, the control of perceptual orienting seems to

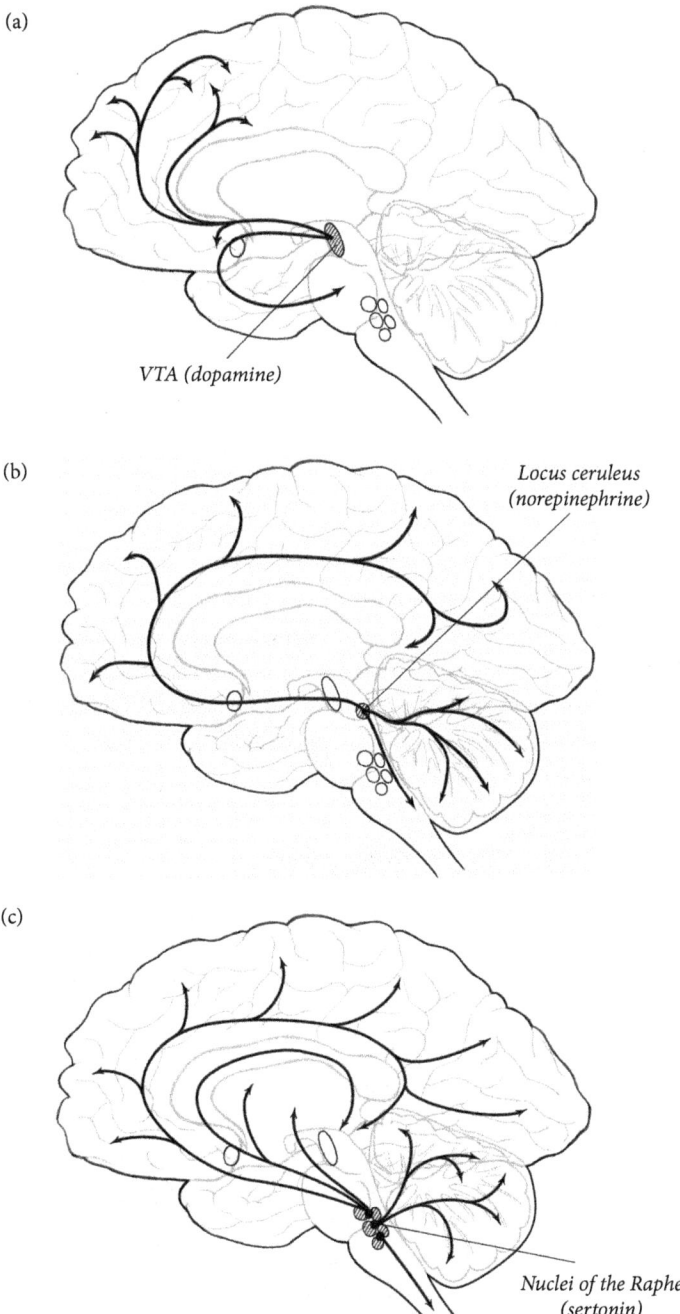

Figure 1-5. Each of the major neuromodulator projection systems controls widespread regions of the brain. In the following chapters, we will consider whether certain of these systems may be differentially specialized for regulating dorsal versus ventral corticolimbic territories.

(d)

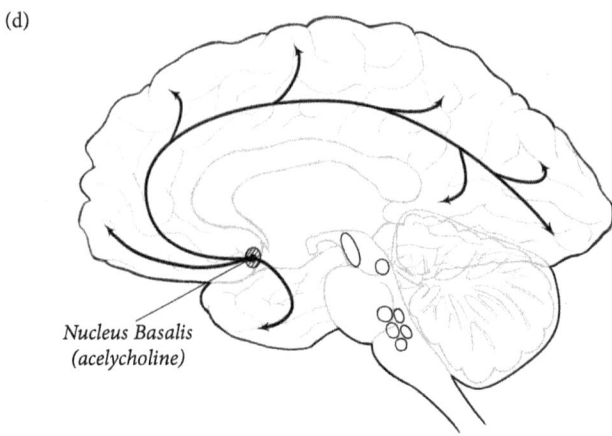

Figure 1-5. (Continued).

rely on a phasic arousal system engaging norepinephrine (NE) and serotonin (5-HT) brainstem projection systems.

Given the important roles of these neuromodulators in regulating synaptic differentiation in embryonic neural morphogenesis (Marin-Padilla, 1998; Rakic, 2009; Trevarthen, 1985), a model of motivated self-regulation through operation of these neuromodulator systems could form the basis for a neurodevelopmental theory of cognition.

## *Motive Restriction of Working Memory*

Based on the restriction of the range of behavior observed in animals given dopaminergic agonists, with higher doses leading to repetitive and stereotyped actions, Tucker and Williamson (1984) theorized that the tonic activation system produces a sensitization or *redundancy bias* on working memory, restricting the range of information in the current store at the same time that it increases the sustained representation of that information. There is thus a structural effect on working memory from this primitive redundancy control on neural activity: the current cognitive representation becomes focused and sustained.

A specific cybernetic influence such as this may be relevant to the fast learning mechanism examined by Gabriel et al. (Gabriel, Kubota, Sparenborg, Straube, & Vogt, 1991) that allows the animal to adapt to changes in the environment when the context model is inadequate. The sustained neural activity, supporting focused attention on the discrepant event, is a suitable mechanism for this learning process (Tucker & Luu, 2007). The motivational quality of tonic activation, described as vigilance and anxiety by Tucker and Williamson, is also consistent with the engagement of the fast learning system under aversive conditions (Gabriel et al., 1991).

This formulation of anxiety as the motivational bias of the mesolimbic dopamine system was at variance with the widespread view in neuroscience that dopamine projections are a primary mechanism of the brain's reward system (Fibiger & Phillips, 1986; Wise & Rompre, 1989), and thus more important to positive rather than negative affect. More recently, the analysis of the behavioral role of the mesolimbic dopamine projections has pointed to their importance to the incentive salience of environmental events, what Berridge and his associates have described as *wanting* as opposed to *liking* (Berridge, 2009; Kringelbach & Berridge, 2009, 2010). In the incentive (rather than aversive) context, this Berridge et al. formulation would be consistent with the sustained neural activity and focused attention in strong need states (such as hunger) that would support the fast learning mechanism. It would also be consistent with the affective quality of *anxiety* associated with the redundancy bias of tonic activation in humans, which may be engaged not only for threats, but also for strong need states (Tucker & Williamson, 1984).

### Motive Expansion of Working Memory

A contrasting, and indeed opposite, influence on neural activity may be provided by the phasic arousal system. Based on the increased alertness, yet paradoxically the rapid habituation of orienting, in animals given norepinephrine agonists, Tucker and Williamson (1984) theorized that a *habituation bias* could structure attention and working memory in qualitatively unique ways. Because any specific content rapidly habituates, working memory capacity is expanded by the habituation bias, even as any given representation is shallow and ephemeral. This strategy for ongoing representational control may be suited to the slow learning mode of Gabriel and associates, in which the context model is largely predictive of ongoing events, and only minor alterations are required for continuing effective cognitive predictions (Tucker & Luu, 2007). Based on the traditional catecholamine hypothesis of the affective disorders in psychiatry (Schildkraut, 1965), Tucker and Williamson proposed that the strong norepinephrine modulation of the habituation bias is associated with the elation of mania. This affective quality would be consistent with the animal learning evidence that there is a hedonic charge to the expectancy for future events that guides the learning process (Papini, 2002, 2003).

At least in broad outline, the habituation bias and elation could be aligned with the *liking* element of motivational control characterized in the Berridge et al. model (Berridge, 2009; Kringelbach & Berridge, 2009, 2010). The recent neuroscience evidence points to the importance of circuitry in the nucleus accumbens and ventral pallidum in the motivational control of liking; understanding how this evidence relates to the limbic control of phasic arousal, and to the hypothesized norepinephrine/serotonin mediation of the habituation bias, is an important challenge for neurodevelopmental theory. A key point is that these neuromodulator systems shape the activity-dependent control of neuroembryogenesis, and then continue, with qualitatively similar forms of neuromodulation, as control systems in mammalian behavioral and cognitive self-regulation.

## *Spatiotemporal Trade-Offs in Activity-Dependent Plasticity*

The idea of motivational control of neural activation and arousal, while involving both complex neural mechanisms and controversial neuroscience literatures, is the foundation for our neurodevelopmental theory of learning and cognition. The concepts of redundancy and habituation provide elements for a theory of *qualitative* controls on neural activity, explaining not just the amount of activity, but how it is controlled over time. Admittedly, these two concepts are only a crude theoretical approximation to the complex set of controls applied by the neuromodulator systems of the mammalian brain. Nonetheless, they may provide a theoretical basis for reasoning from elementary controls on neural activity to the organization of plasticity in neural networks.

Because the spatial connectivity and thus representation in the brain is achieved through physiological processes that are extended in time, there are inherent trade-offs between spatial connectivity and the temporal control of neural activity. A Hebbian network modulated by a redundancy bias emphasizes the dominant, currently active cell assembly. Activity that is ongoing is maintained, and so are the connections among the active neurons. The pattern of connections is constant or tonic in time.

But this redundancy of the present cell assembly limits the connectivity of other neurons that could have been engaged in the same interval of time. The effect on the spatial structure of connectivity is then to restrict the spatial scope of synchronized connections over time. Temporal constancy, and thus increased strength of dominant connections, is therefore achieved at the cost of reducing scope in representational space.

In a similar but opposite fashion, a cell assembly modulated by a habituation bias is inherently ephemeral. It is limited in time. Within the overall synaptic activity network, however, the effect of habituating activity in any dominant cell assembly is to broaden the spatial pattern of connections engaged over a given interval of time in the entire network, thus expanding representational scope and including more remote connections.

With these simple theoretical elements, we can formulate a neurodevelopmental theory of cognition, in which the mechanisms regulating neural activity over time in embryological activity-dependent specification are conserved in the extended (and increasingly parentally protected) ontogenesis of mammals. Even before considering the specific network architectures of the cortex (which we will see are indeed important and specific to these neuromodulator influences), we can characterize the effects of habituation and redundancy biases in allocating the resources of representational space, and working memory, in neural connections. By expanding the scope of connections kept active over time, the habituation bias maintains synaptic density broadly within the network. By restricting the scope of connections kept active over time, the redundancy bias prunes the synaptic connectivity to retain only the most active cell assemblies.

For each of these cybernetic modes of tuning network structure, the animal learning evidence suggests that the specific motive control engages a primitive form of cognitive expectancy. These are the adaptive modes of the motive-memory. Elation (with its inherent phasic arousal and habituation) engages expectancy for pleasure. Network representations (concepts) are then expansive. At the same time, they are transient, leading attention to be allocated rapidly to a diverse array of changing events. In contrast, anxiety (with tonic activation and the integral redundancy bias) engages expectancy for threat, or it focuses expectancy in high need states. Network representations are restricted, and so is working memory (Tucker & Luu, 2006; Tucker & Williamson, 1984).

## Summary

Table 1-1 presents a list of terms that describe properties of the dorsal and ventral divisions of the neocortex, and what we will describe as dorsal and ventral limbic systems. Many of these terms will not be introduced until Chapter 3 or later, but it may be useful to see the multiple dichotomies that will be explored in the neurodevelopmental analysis.

Table 1-1. Summary of terms describing properties of the dorsal and ventral divisions of the neocortex and limbic system

| Dorsal | Ventral |
|---|---|
| Projectional | Reactive |
| Feedforward | Feedback |
| Impetus | Artus |
| Internality | Externality |
| Intention | Attention |
| Pragmatic | Semantic |
| Limbifugal | Limbipetal |
| Egocentric | Allocentric |
| Visceromotor | Viscerosensory |
| Assimilation | Accommodation |
| Habituation | Redundancy |

We begin this book with the insight that the activity-dependent organization of neural networks in embryonic ontogenesis continues throughout life as the mechanism of learning (Singer, 1987). To bring this insight to neuropsychological theory, we must recognize that mammalian learning is not a passive association of stimulus and response, but an active process of cognitive expectancy for future events, both hedonic and aversive. The expectancy sets the frame for evaluating events, and consolidating them in memory. Cognition, in this functional analysis, is invariably motivated, in simple mammals as well as humans. The question then becomes how the brain can regulate the neural activity of its key networks in ways that support motivated cognition.

Research into the neurophysiological mechanisms of approach and avoidance, such as in the Gabriel et al. studies of rabbits, has shown that limbic circuits apply specific control biases on primitive cognitive expectancies to support learning. Considering this neurophysiological evidence in relation to modern cognitive theories of animal learning provides clues to the adaptive roles of limbic and thalamic circuits in controlling cognition. Considering this evidence in relation to control theory in engineering provides clues to the specific cybernetic biases of the neurophysiological mechanisms. Maintaining the hedonic expectancy of the context model operates as a feedforward bias for expectant learning. Disrupting the context model in the face of discrepancy or threat operates as a feedback bias for rapid, focused, adaptive learning. We will develop this model of neural control more fully in Chapter 3, but even at this point it should be clear that learning is invariably directed by motive controls.

The challenge for a neurodevelopmental theory of psychology is understanding how these specific control systems could arise from neurophysiological mechanisms for activity-dependent organization of neural networks. The brain's vertical integration, across the multiple evolved levels of network organization, shows that the fundamental platform for neural activity rests on elementary neuromodulator systems of the brainstem, midbrain, and ventral striatum. These systems remain poorly understood, yet it is clear that they change neural arousal, and emotional state, not only quantitatively, but qualitatively. We concluded Chapter 1 with concepts of qualitative neural control, redundancy, and habituation, that may suggest how primitive biases on neural activity shape both its temporal course and the resultant spatial scope of network connectivity to organize cognitive expectancy and the consolidation of memory.

To build a workable neurodevelopmental theory of psychological function, we must understand the operation of these primitive motive-cognitive biases within the specific architectures of human neuroanatomy. The human brain is a complex elaboration of the mammalian brain, which in turn has evolved corticolimbic networks for cognitive organization of both perception and action to support learning. Furthermore, although the neocortex appears to be essential for the representational capacity that allows mammals their unique forms of expectant learning, it operates only through continual recruitment of multiple subcortical neural controls, including neuromodulator systems, required for effective vertical integration. To examine these issues, Chapter 2 begins with principles of cognitive representation within distributed neural networks. It then examines the unique connectional architecture of the mammalian brain, an architecture that links the networks of the neocortex to their adaptive substrates in brainstem, striatal, and limbic circuitry. Finally, Chapter 2 considers the neurophysiological mechanisms of consolidation and network excitement that shape neural activity to achieve lasting adaptive connectivity.

# 2

## Consolidating Memory

To be integrated within the structure of mammalian memory, sensory information and motor habits must be *consolidated* within the neocortex. This means the information must be processed through neurophysiological mechanisms that involve some form of neural communication between the sensory and motor networks at the lateral surface of the neocortex, the limbic networks at the inner core of the hemisphere, and the "association" cortices interposed between these boundaries. If they are to affect memory, controls on the dynamics of neural activity (perhaps such as redundancy and habituation biases) must shape the communications within the corticolimbic architecture. In the neurodevelopmental theory of learning, the mechanisms of consolidation emerge from the mechanisms of embryonic neural plasticity. Consolidation in learning is thus an extension of the embryonic mechanisms for self-organizing the architecture of cortical networks in fetal growth. As it operates continuously, in the background of waking cognition and in sleep, consolidation may be the neurophysiological mechanism for ongoing neural ontogenesis.

The role of limbic networks in consolidation is shown by the empirical evidence on mammalian learning. Once it is recognized, this role raises basic questions about mammalian cortical evolution. The limbic networks are mostly 6-layered cortical networks on the limbus or border of the medial hemisphere. They are adjacent to the more primitive justa-allocortex and peri-allocortex, and they are closely regulated by the primitive limbic circuits including the hippocampus, a 3-layered cortical rudiment (Cajal, 1899), and the amygdala, a complex structure that is often considered as a protocortical rudiment (Butler & Molnar, 2002). Although memory consolidation is most often spontaneous and ongoing, rather than conscious and volitional, it is perhaps the most important component process of cognition. The result of consolidation is a cognitive representation, a *concept*. We will see that the evidence from anatomy and from studies of consolidation show that the concept is distributed across the multiple linked networks of the cortex. To the extent that it is regulated actively—by ongoing motives in relation to an adaptive goal—consolidation generates expectant working memory.

Whereas intact communication between multiple neocortical and limbic regions is necessary to consolidate memory, the representation, once established, may be functional in guiding experience and behavior within the neocortex without ongoing limbic support (Squire, 1986a). The spontaneous and ongoing nature of consolidation, as well as the requirement for continuous neurophysiological activity during the consolidation process, are well illustrated by the human clinical evidence on memory disorders. Amnesia following brain damage is *retrograde*: the greater the severity of the insult, the farther the amnesia extends into the past (Squire, 1987). This fact implies that consolidation, disrupted by the injury, must be ongoing in the background of normal neurophysiology.

Remarkably, the consolidation process appears to extend not just over hours, or days, but over years. Patients with memory impairment caused by repeated electroconvulsive treatment were examined in relation to the loss of memory for events in years before the treatment by testing their recall of televisions shows (Squire, 1986b). The patients showed a loss of memory that was increasingly retrograde over years of past history as a function of the number of electroconvulsive treatments they received (Squire, 1986b). The more shock treatments they received, the further in the past their memory was impaired. During the course of memory consolidation a secondary process, known as reconsolidation, wherein memory traces are reactivated and in an amenable state, can strengthen or alter the already stable memory traces (Nader & Einarsson, 2010).

From this evidence, we discover that memory is a fragile, active resource of the mind, and that consolidation and reconsolidation is an ongoing life-span ontogenetic process. The neurodevelopmental process of cognition is both cumulative and ongoing. It organizes experience within the fine connectional architecture of the brain's networks, but the connections alone are not adequate to retain memory. Rather, there is an ongoing neurophysiological process sustaining personal developmental memory. If this continuous process is disrupted, from brain injury or electroshock, there is a graded (retrograde) loss of access to the past. The structure of the mind is dynamic, and fundamentally transient.

At the same time that it is spontaneous and ongoing in daily experience, and supported by the essential mechanisms of sleep (Walker, 2009), memory consolidation is continuously motivated. The limbic networks of the brain not only control consolidation; they also control motivation and emotion. Although largely ignored in neuropsychology and neuroscience, this dual function of limbic networks is no accident. The representations in limbic networks are closely linked to visceral responses through hypothalamic and autonomic circuits, and these circuits are integral to memory consolidation. As a result, a theoretical analysis of the neurodevelopmental process of cognition requires a careful study of the specific motivational mechanisms, and biases, emergent from specific subcortical control systems.

In this chapter, we examine the structure of human cerebral networks with the modern insight that the structure of mind can be read from connectional anatomy. At the same time, we search for the dynamic controls on activity within cerebral networks to understand how the connectional anatomy develops, through continuous motivation, over the life span.

## Principles of Distributed Representation

To understand the mechanisms of activity-dependent control of connectivity within cortical networks, we can gain considerable insight from modern computational models of distributed representation in artificial neural networks. Continuing the classical insights into nerve nets reviewed in the last chapter, we have come to understand how connectional anatomy can be related to psychological function largely through the computational simulations of parallel distributed processing models (Grossberg, 1980; Rumelhart & McClelland, 1986). These are often called *connectionist* computational models. One principle of this work is that information is represented within the *pattern of connections* among neurons (processing elements), while the neurons themselves may have only elementary response properties (*dumb neurons*). Another principle is that connected networks gain important properties of learning patterns in the world (through various modes of training the connectional weights) when they are separated into different *representational levels*. When all of the neurons are at the same level, the representational pattern is subject to *catastrophic interference*, as the pattern of weights used for previous learning is restructured and therefore degraded in order to accommodate the new learning. This is because the representations are distributed throughout the network: the connections modified by the new learning are the same ones that store the old learning.

In the general sense, the problem of maintaining old learning in the process of acquiring new learning is described as the *stability-plasticity dilemma* (Grossberg, 1980). Whereas in Freud's Project this problem was recognized in relation to whether individual synapses would be modified or not (Freud, 1895), in the modern parallel computational models the stability-plasticity dilemma takes on a more specific form, as modifying weights for new learning leads to disruption not just of individual weights, but also of the distributed patterns of old learning. When separate network levels (*hidden units*) are segregated, then the representational pattern may become organized hierarchically, allowing greater plasticity of the input level, while retaining the continuity of prior learning within the network as a whole (Rumelhart &

McClelland, 1986). The hierarchic organization of representational levels has continued to be an important principle in recent generative models of neural networks, in both engineering applications and in analysis of primate vision (Hinton, 2010).

Even a brief summary of connectionist principles provides an important framework for understanding the significance of cortical network organization for psychological function. At least in principle, we can see that there may be a way to read the patterns of connectivity from anatomy and thereby interpret the structure of cognition.

Because of the distributed nature of cognitive representations in the mammalian brain, the management of learning appears to require strategies for managing the stability-plasticity dilemma. Some strategies seem to be architectural, with different functions segregated to different representational levels, as in connectionist computational models. Not unlike Freud's phi neurons for stable perception (Freud, 1895), other evolved strategies for mammalian neural plasticity seem to be developmental, modifying the maturation of neurophysiological mechanisms of plasticity in certain cortical networks to maintain stability in those networks after the early juvenile plasticity is complete. For both architectural and developmental strategies, a crucial theoretical question becomes the neurophysiological control systems (for functions such as arousal, motivation, and memory consolidation) that organize the neural plasticity of learning within these general strategies. As we will see, the discovery of dual (dorsal and ventral) subcortical engines of consolidation implies that there are dual strategies in the motivational control of neurodevelopmental plasticity and cognition.

## Representation in Corticolimbic Networks

Studies of the connection density of the primate neocortex have clarified a general architectural pattern that was not apparent when the cortex was seen as a collection of lobes and major fiber tracts. As shown in the series of studies by Pandya and associates (Barbas & Pandya, 1984, 1987; Pandya & Barnes, 1987; Pandya & Seltzer, 1982; Pandya & Yeterian, 1984, 1985; Yeterian & Pandya, 1988), each sensory area is linked through a series of four adjacent networks (primary sensory, unimodal association, heteromodal association, limbic) to the limbic core of the hemisphere (Figures 2-1 and 2-2). With important exceptions, a more or less reverse pattern of connectivity is observed for the pathways from limbic areas through the frontal lobe to motor cortex (Figures 2-1 and 2-2) (Barbas & Pandya, 1987; Pandya & Barnes, 1987). Quantitative analysis showed that the intrapathway connections, between adjacent networks in each sensory or motor pathway, make up the majority of corticocortical connections.

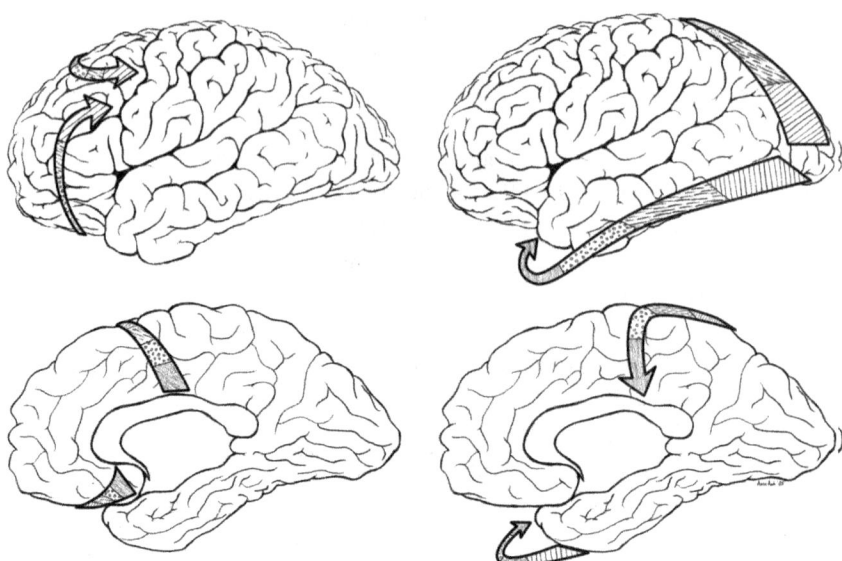

Figure 2-1. The primate cortex shows an organized pattern of networks around the limbic core of the hemisphere, including both motor function in the anterior brain (left) and sensory functions in the posterior brain (with the visual pathway shown at right). The direction of processing is often thought to proceed from primary sensory cortex (striped; vision shown here) to unimodal association cortex (dashed) to heteromodal association cortex (stippled) to the limbic networks, at the limbus or border of the hemisphere (shaded). The organization of action would then proceed primarily from limbic regions through a similar level of adjacent association networks in the frontal lobe toward motor cortex. For both sensory (except auditory) and motor cortices, there are both dorsal or archicortical and ventral or paleocortical divisions. After Pandya and Yeterian (1984).

The limbic (border) cortical networks at the core of the hemisphere are described by Mesulam as *paralimbic* because they are adjacent to the limbic structures, the hippocampus and amygdala (Mesulam, 2000). Because they form the limbus or border of the medial hemispheric wall, we retain the traditional term *limbic cortex*. These limbic networks, although comprising 6-layered neocortex for the most part (with layer 4 strangely absent in much of the anterior cingulate cortex) are the least differentiated cytoarchitectonically of any cortical region. Each further adjacent "ring" of neocortex (heteromodal, unimodal, and primary sensory or motor cortex) shows increasing cytoarchitectonic differentiation (Barbas & Pandya, 1987; Barbas & Rempel-Clower, 1997).

As suggested in the cartoon in Figure 2-3, the density of regional interconnections shows an opposite progression, with increasingly fewer connections, and thus greater local isolation, for the next (farther from limbic) increasingly differentiated neocortex. In this progression, the most local connections are observed for the primary sensory or motor cortex within the pathway (Barbas &

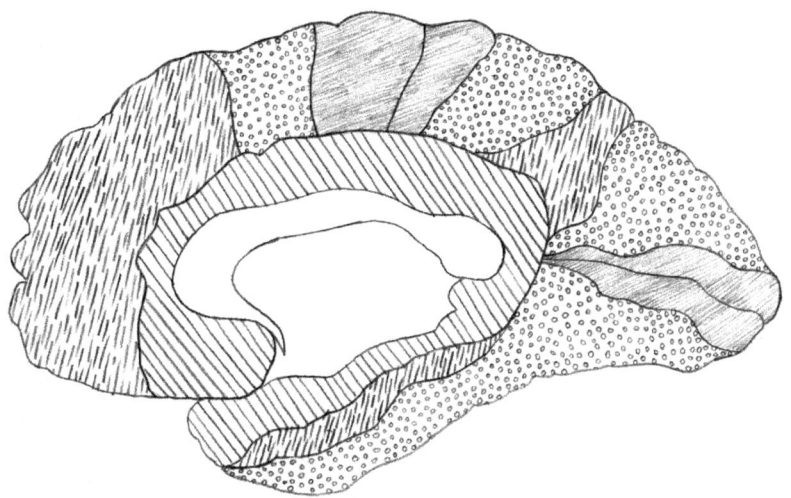

Figure 2-2. Medial wall of the right hemisphere, showing approximate positions of limbic (shaded), heteromodal (stippled), unimodal (dashed), and primary sensory or motor cortex (shaded), with the same markings as in Figure 2-6. After Mesulam (2000).

Pandya, 1989). Furthermore, the interpathway connections (for example, from visual to auditory) are found to respect pathway network levels. For example, heteromodal-to-heteromodal are common but heteromodal-to-unimodal connections are rare (Barbas & Pandya, 1989).

### *Viscerosomatic Consolidation*

Thus the embodiment of human cognition can be understood in relation to the bodily functions of the neural networks that effect cognition. There are *visceral* constraints at the limbic core of the hemisphere, providing motivational control. There are *somatic* constraints at the sensory and motor cortices, providing a kind of reality testing as concepts are mapped to the environmental requirements for perceptions or actions. The connectional architecture of the cortex then exhibits a highly organized 4-level (limbic, heteromodal, unimodal, primary cortex) structure (Figures 2-1, 2-2, and 2-3) that is framed by these visceral (limbic) and somatic (primary sensory or motor cortex) requirements. As a result, memory consolidation must arbitrate between the sensory and motor requirements at the somatic shell and the emotional and motivational requirements at the visceral core.

Embodiment is not an obvious property of the mind, at least to academics. Reading the literature in today's cognitive neuroscience, it may seem as if there are specialized cognitive modules within the cortex, with little connection to bodily function. It is then a curious but unexplained fact in this

Figure 2-3. Cartoon unfolding of the cortex from Figure 2-2. For both sensory (posterior) and motor (anterior) networks, the limbic cortex at the core of the hemisphere links to the adjacent heteromodal association cortices, which link to the adjacent unimodal (or premotor) association cortices, which link to adjacent primary sensory or motor areas. The interpathway connections (circular lines between each pathway level) are most dense at the limbic core, and decrease toward sensory and motor cortices, which, with the exception of motor-somatosensory projections, are relative islands in the cortex, linked only to their adjacent unimodal (or premotor) cortex. In this cartoon of the human brain, large cortical pathways are shown for the dorsal and ventral visual areas (right) and dorsal and ventral motor areas (left), with smaller areas for somatosensory (middle top) and auditory (middle bottom) pathways.

literature that highly sophisticated capacities, such as self-monitoring or empathy, are attributed to primitive limbic regions, such as the anterior cingulate cortex or the insula. These were the first rudimentary cortices in primitive mammals, and before the neuroimaging results neurologists and psychologists assumed these limbic regions were only concerned with visceral and motivational functions.

However, through simple inspection of the cortical anatomy underlying the neuroimaging results, we find that the brain's architecture is organized for distributed representation (rather than arbitrary cognitive modules), in clearly defined pathways that support the visceral-limbic motive control of somatic-cortical sensory and motor functions. Every pathway between neocortex and the limbic core of the hemisphere is associated with a sensory modality, or with motor control. Cognition and memory are not disembodied mental spirits.

Rather they must arise in each instance from the arbitration between visceral and sensorimotor constraints.

### An Evolutionary-Developmental Order

When it was first recognized (Sanides, 1970), the increasing cytoarchitectonic differentiation of the cortex across the four levels (limbic, heteromodal, unimodal, primary) appeared to support an evolutionary interpretation of "growth rings" reflecting the evolution of increasing complexity of the mammalian neocortex (Sanides, 1970). In the Sanides model, the limbic cortex reflects the earliest evolved form, and each stage of evolutionary differentiation added adjacent network levels, progressing through heteromodal association cortex, then unimodal association cortex, and finally primary sensory (and motor) cortex. With the advent of quantitative studies of primate cortical anatomy, the connectivity seemed to support the Sanides hypothesis of growth rings (Pandya & Yeterian, 1984).

However, this evidence and interpretation has been rejected by most neuroscientists because it is incompatible with the traditional assumption that the sensory and motor areas are primitive and common to all mammals, whereas it is the association areas that have evolved in more complex mammals such as humans (Kaas, 1988, 1989; Northcutt & Kaas, 1995). In the growth ring view, the primary sensory and motor cortices would not be common to all mammals, but would be the more recent additions in more complex mammals.

Whether an evolutionary order of cortical differentiation is accepted, the connectional architecture of the mammalian neocortex has been characterized in detail by the quantitative tracing studies by Pandya and his associates (Barbas & Pandya, 1984, 1987; Pandya & Barnes, 1987; Pandya & Seltzer, 1982; Pandya & Yeterian, 1984, 1985; Yeterian & Pandya, 1988). It is interesting to consider the evidence on consolidation within the framework of this connectional architecture. For sensory memories to be consolidated, some interaction is required within the pathway between the more differentiated neocortical (primary or association) and less differentiated limbic networks (Desimone & Ungerleider, 1989; Squire, 1987). The fact that amnesias are often modality-specific, caused by lesions of a specific sensory pathway, supports the pathway (modality) specificity of consolidation. Thus visual memory, and imagination, are organized (and embodied) within the visual pathway. A similar form of pathway-specific consolidation within frontolimbic networks is necessary for motor memory as well, and, as we will see, for working memory particularly (Goldman-Rakic, Bates, & Chafee, 1992; Wilson, O Scalaidhe, & Goldman-Rakic, 1993). With only these sensory and motor networks for its physical instantiation, we see how the mind must emerge from a sensorimotor substrate (Jackson, 1931). With the modern evidence on cortical connectivity organized around the limbic base, we can also see how the mind must frame even abstract concepts in relation to their visceral significance (Tucker, 2001, 2002).

If we assume that the result of consolidation is some form of synaptic modification within the networks of the pathway, resulting in new functional links across adjacent networks, connectionist reasoning would imply that the neurodevelopmental process of consolidation must be critical to cortical organization, at first in genetic outline in the spontaneous self-organization of the fetus, and then in the process of self-actualization in the particular environment of childhood. In the ongoing cognitive process, consolidation must create a four-level concept, with unique properties at each level. It seems apparent that the limbic level must capture the concept's motivational significance and general syncretic properties (given the dense interconnectivity among limbic regions); the heteromodal level must reflect a second generic, semantic representation (but now once-removed from visceral embodiment, such as in Broca's area in the motor control of language); and the unimodal and primary cortical representations must reflect increasingly sensory-specific or motor-specific properties that are aligned with the interface with the environment effected through that sensory receptor/motor effector surface.

### *Levels of Sensory and Memory Representation and the Direction of Feedforward Control*

Anatomical studies have thus detailed the network structure of the connections between cortical representational levels. This structure implies the order of the processing that must support consolidation. There is a regular *laminar* pattern of connectivity (referring to cortical layers 1–6) between the adjacent network *levels* (limbic, heteromodal, unimodal, primary) of a given pathway (Pandya & Yeterian, 1984). This pattern is illustrated in Figure 2-4. Neuroscientists typically have considered the cortex's processing to begin with the sensory (thalamic) input to primary cortex (at the right of Figure 2-4). This is a natural perspective of a stimulus-response experiment that begins with the sensory stimulation. The researchers therefore describe the information stream moving from primary to association networks (in the limbipetal or toward-limbic direction) to reflect *feedforward* processing. The opposite (limbifugal) direction has been traditionally assumed to reflect *feedback* processing (Figure 2-4).

The conventions in neuroscience are reasonable from the perspective of the experimenter, in which the brain processing in the experiment starts with passive reception of the sensory input. From the perspective of the organism, however, actively organizing perception and behavior, the opposite direction of control may be more important. Recognizing the role of expectancy in perception (Shepard, 1984), and considering the logic of control in both animal and machine learning (Hendler, 1995; Kamin, 1965), we will adopt the organism's

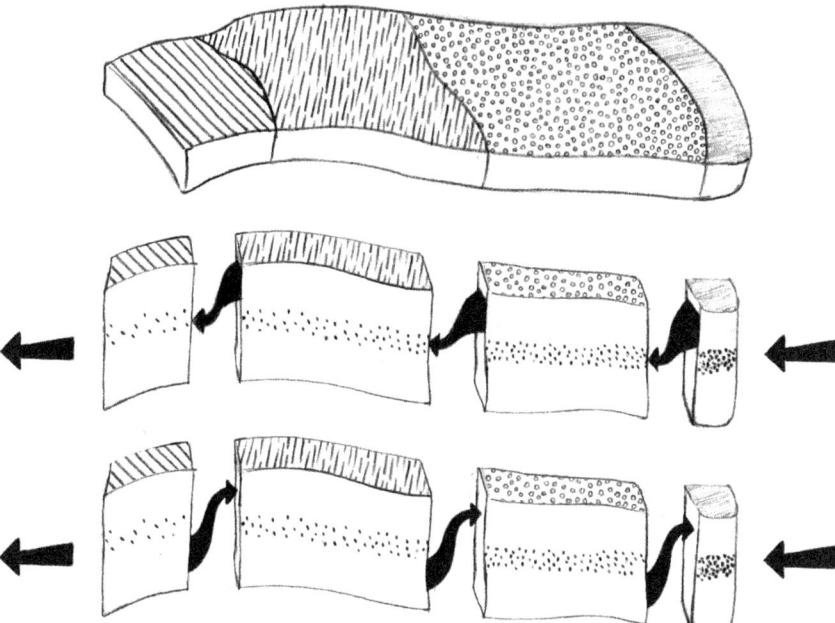

Figure 2-4. For each corticolimbic pathway (as seen in Figure 2-3) such as vision (top), there is a regular but not exactly reciprocal pattern of laminar projections (between layers of the 6-layered neocortex) (Pandya &Yeterian, 1984). The visual-to-limbic or *limbipetal* direction (middle), typically described as the *feedforward* pathway in the literature, involves projections from supragranular to granular layers. The limbic-to-visual or *limbifugal* direction (bottom), typically described as the *feedback* pathway, involves projections from deep (cortical output) layers to superficial areas.

perspective and describe the limbifugal processing as *feedforward*. Thus, expectancies arise from memory to shape the processing of perceptual detail. In Shepard's terms, perception is "hallucination constrained by the sensory data" (Shepard, 1984).

From this perspective, the "input" projections from sensory cortex would provide *feedback* articulation of those expectancies, based on the constraints of the sensory data (Shepard, 1984; Tucker & Luu, 2006). Of course both directions are important. Because consolidation operates over extended intervals of time, it would seem to require recursive and reentrant interactions across the multiple networks of each pathway. Reasoning from the regular and patterned connectivity, we can consider how local cortical processing contributes to the four-level concept described above. What are typically termed *back projections*, from less differentiated (closer to limbic) toward more differentiated (closer to primary cortex) networks, arise in infragranular (5, 6) layers of the cortex (Figure 2-4). This suggests that these are *output* projections, perhaps more fully processed by the cortical columns of the originating (closer to limbic) cortical area. From the organismic perspective of perception through active expectancy, this inferior-to-superior laminar projection is the *feedforward* direction of intercortical communication, starting from the representations of motive-memories in the relatively-closer-to-limbic network and ending in the representations of reality constraints in the relatively-closer-to-sensory network.

In the reverse, limbipetal or sensory-to-association/limbic direction, the layer 4 thalamic input to primary cortex is relayed to secondary or unimodal cortex (to its layer 4), after some form of local processing (Figure 2-4). The fact that this relay emanates from supragranular (2, 3) layers to layer 4 suggests it is a transmission of the input pattern as it is seen by that network, after some initial processing. The transmission is received in layer 4 of the association cortex, which we must assume is still an input layer. From the organismic perspective, of perception beginning through expectancies held in memory (Shepard, 1984), this is the *sensory feedback* direction of intercortical communication.

As with other features of cortical connectivity that we have examined, the interlayer connection pattern for adjacent cortical regions would seem to support the Sanides growth ring interpretation. As each more differentiated cortical region evolved, it became interposed between the thalamic (granular layer 4) input and its evolutionary predecessor (closer to limbic) network. It is as if that new more differentiated cortical region presented a new more thoroughly processed version of the thalamic input to its predecessor cortical region (Tucker, 1992).

Confident interpretations of these processing steps will of course require a clearer understanding of the functional mechanisms of cortical processing. Furthermore, as we will see in the next section, the thalamic communication with each of these cortical layers is critical to both the intracortical communication and the subcortical control of the consolidation process. However, even a rough outline of the architecture of the corticolimbic sensory pathways suggests there may be important structure to the perceptual process that can be understood from studying the laminar specificity of intrapathway connections. Sensation and perception appear to be organized in relation to visceral significance through four network levels. The greatest differentiation and modularity is seen at the sensory cortex, whereas the greatest integration and fusion of multimodal with motivational influences is found at the hemisphere's visceral limbic core. The arbitration between these boundaries, consolidating the structure of memory, proceeds through regular patterns of laminar connectivity. This is the ordered corticolimbic architecture to be achieved by the neurodevelopmental organization of the mammalian telencephalon, in broad outline before birth and in fine structure through the experiences of life.

## The Vectoral Cybernetics of Action

At least in broad outline (as in the cartoon of Figure 2-3) the development of motivated actions follows a similar processing architecture as for perception (Barbas & Pandya, 1987), with limbifugal processing in frontal networks (from limbic toward motor) shaping the feedforward direction in the control of actions. Limbipetal processing provides feedback to the motive-action plan from the motor effectors in primary motor cortex. As it does, it appears to condition or guide the action organization and consolidation process according to the requirements for motor specification. In addition, the direct somatosensory and kinesthetic input from primary somatosensory cortex to primary motor cortex appears to provide additional constraint on the process of motor articulation. Although these connections are genetically determined in broad outline, we can expect that their exercise through activity-dependent specification is a critical process of embryonic organization of the cortex, continuing to shape the development of motor skills in childhood.

The organismic perspective on feedforward and feedback is particularly useful for motor control networks of the anterior brain. The pattern of connections between adjacent premotor association and motor networks in the anterior brain is generally similar to that between adjacent sensory association and sensory networks in the posterior brain (Pandya, & Seltzer, 1982; Pandya & Yeterian, 1985). Recognizing this, the conventional neuroscience terms of feedforward and feedback projections from sensory areas (mostly applied to vision) obviously cannot apply to motor control in frontal areas (Shipp, 2005). The conventional "feedforward" or limbipetal (layer 3 to 4) connections, as typically labeled for sensory areas, could not provide feedforward control in the motor areas. Rather, from the organismic perspective these limbipetal (layer 3 to 4) connections can be seen to provide *feedback* control, allowing actions to integrate feedback constraints from the requirements for motor patterns in primary motor cortex, and from kinesthetic and somatosensory feedback from the linked postcentral somatosensory cortex. Similarly, what are described conventionally as "back-projections" or "feedback," the limbifugal (5 to 2-3) connections in the sensory pathways must be seen to provide the primary outgoing initiative for action, *feedforward* control emanating from limbic motives, in the motor pathways (Shipp, 2005).

The cybernetics of feedforward and feedback control—of the representations of both sensory and motor cognition—are thus vectoral or directional, with specific laminar connections mediating the direction of information flow in the interpathway networks. We think that these connections can be interpreted functionally and literally. A key point is that each pathway, for a sensory modality or for motor control, is bounded by visceral functions (need states) at the limbic core, just as it is bounded by the requirements for interface with the world (reality testing) at the primary sensory and motor cortices. Because cognition must emerge from the neurophysiology of these linked limbic, association, and primary sensory and motor cortices, it is clear that the visceral and somatic (sensorimotor) functions at the boundaries of each cortical pathway are not simply related activities of the brain. They are integral to each cognitive representation.

Cognition is thus embodied in the process of viscerosomatic consolidation. The classical insight into the significance of the brain's mediation between visceral and somatic domains was Yakovlev's notion of *motility control* and the process of *exteriorization* (Yakovlev, 1948). The brain's regulation begins with the interior milieu, at what Yakovlev called the "sphere of visceration." Behavior, or motility, is organized progressively to actualize the visceral needs through contact with the world in the process of exteriorization. The model in action regulation is progressive specification. It is exemplified in motor control by the global organization of posture and action preparation through the trunk muscles. At this stage action regulation is close to the sphere of visceration, in what the developmental psychologist Heinz Werner called the *postural-affective matrix* (Werner, 1957). Action then continues to develop through specifying finer appendicular motor control, through limb movements. It is finally articulated in fine oral and digital movements. This is the template not only for motor control but cognition in Brown's microgenetic theory (J. Brown, 1979, 1988, 1989, 2011)..

The connectivity of the primate cortex, and with some modifications the human cortex, fits with this traditional emphasis on the visceral, limbic basis of behavior (Figure 2-5). The greatest density of interconnection is limbic, with progressively more specialized and isolated network patterns in heteromodal, unimodal, and primary sensory cortices (Mesulam, 2000). This same architecture is found with the high connection density of the limbic networks at the base of the frontal lobe (anterior cingulate and orbital frontal), with progressive differentiation and regional modularity for prefrontal (mediodorsal and ventrolateral frontal lobe) and both dorsal and ventral divisions of primary motor cortex.

## *Dorsal and Ventral Divisions of the Cortex*

As the connectivity of the primate frontal lobe was characterized with quantitative tracer studies (Barbas & Pandya, 1989), it became clear that the cytoarchitectonics as well as connection patterns differ importantly between the dorsal (mediodorsal) networks (with the hippocampus and anterior cingulate cortex forming their limbic base), and the ventral (ventrolateral) networks (with the amygdala, insula, anterior temporal, and orbital frontal networks as their limbic base) (Barbas & Pandya, 1987, 1989). As described by Shipp (2005), the mediodorsal frontal lobe (including the anterior cingulate cortex) is not only

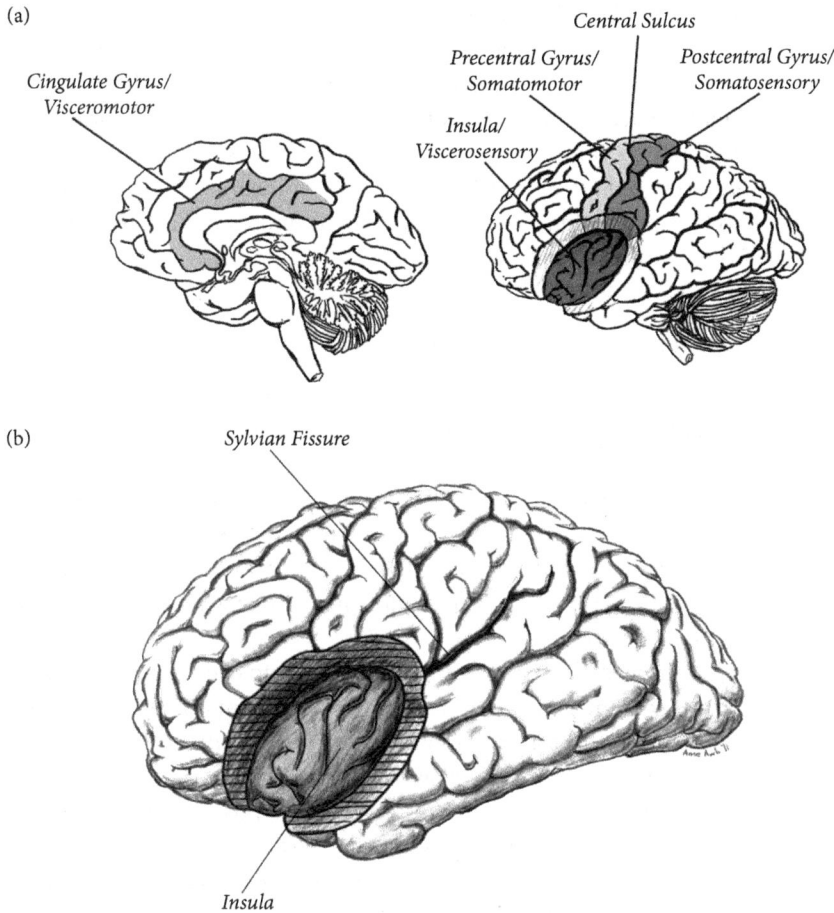

Figure 2-5. The *somatic* sensory and motor regions of the cortex are well known, here shown on the lateral surface of the hemisphere (right), divided by the central sulcus. In addition, there is a division of *visceral* sensory and motor function in limbic cortex, with the visceromotor function represented in the anterior cingulate cortex (left) and viscerosensory function represented in the insular cortex (right). The insular cortex is interposed between the frontal and temporal lobes, covered by lateral cortex (bottom right).

dominated by a pyramidal local cellular architecture, but it is also missing the granular layer (layer 4). This dorsal-ventral separation was an important foundation for the Sanides theory, in which the differentiation of the neocortex proceeded from dual limbic origins, from the hippocampus and archicortex on the dorsal surface and the amygdala, insula, and piriform cortex on the ventral surface (Pandya & Yeterian, 1984; Sanides, 1975).

The separation of mediodorsal and ventrolateral networks of the frontal lobe has long been apparent in relation to differing routes of connectivity to limbic regions (Nauta, 1971). With the more detailed studies of connectivity in primate cortex, this separation now includes recognition of differing patterns of local network architecture for the agranular and granular frontal networks. The functional implications of these different architectures are intriguing: there appear to be differing engines of consolidation in the dorsal and ventral limbic circuits that exert unique organizing influences on memory and cognition in action as well as perception. This is a key issue in understanding the adaptive control of the neurodevelopmental process.

## Anterior-Posterior Complementarity

For both dorsal and ventral divisions of the cortex, the general pattern of neocortical differentiation observed in the posterior brain applies in the frontal lobe as well. The limbic base of the cortex is characterized by highly interconnected and relatively undifferentiated cortical architecture, and each adjacent level of network (frontal cortex paralleling heteromodal cortex, with premotor paralleling unimodal sensory, and primary motor cortex paralleling sensory cortex) shows both increasing cellular differentiation and increasing local (rather than widespread) interconnectivity.

Because the greatest density of connections is within each sensory or motor pathway that links limbic cortex to association to primary sensory or motor cortex, the entire brain is in a sense modality-specific (Sereno, 1998).

At the same time, there is extensive connectivity to support interpathway integration, including connectivity within the posterior brain's representational networks, within the dorsal and ventral frontolimbic pathways, and between posterior association and anterior association cortices. There are important examples in the current literature of reciprocal connections between sensory and motor association networks that are necessary for both perception and action. Motor control, for example depends not only on frontal organization of action plans, but on reentrant connectivity with parietal lobe representations for planning and guidance (Denny-Brown, 1966; Fleming & Crosby, 1955; Mountcastle, Lynch, Gorgopoulos, Sakata, & Acuna, 1975). Similarly, interpreting actions appears to engage the same regions of premotor cortex, including *mirror neurons*, that are involved in preparing those actions (Rizzolatti & Fadiga, 1998; Rizzolatti, Fadiga, Fogassi, & Gallese, 1999). Organizing actions thus requires reciprocal control from perceptual representations, and perceiving often engages covert actions, in a kind of analysis by synthesis. A major dimension of the brain's organization, forming the anterior-posterior axis of the cortex, thus separates somatic motor from somatic sensory functions, and yet it links these functions in a kind of balanced, reciprocal, sensorimotor complementarity. This complementarity must be integral to the differentiation of the cortex, achieved through the ongoing consolidation of cognition.

## The Asymmetric Structure of Memory

For both sensory and motor functions, given the requirement for communication between limbic and neocortical networks in the consolidation process, the architecture of the cortex places clear constraints on both the representational form and the neurophysiological mechanism of consolidation. The fact that the cortex is organized in relation to sensory and motor pathways implies that memory consolidation proceeds within this architecture. Memories are structured not in some abstract semantic space, but in relation to specific embodiments of sensation and action.

Moreover, the limbic base of each pathway, sensory or motor, comprises the greatest connection density of the networks in that pathway, implying that the most integrative representations must be formed at this limbic base. The visceral influences on limbic networks, from hypothalamic, limbic thalamic, brainstem, and ventral striatal circuits, must provide continuing motive control over the adaptive significance of limbic conceptual representations. Consolidation thus mediates between a holistic, syncretic motive base at the core of the hemisphere and a differentiated, localized, and articulated sensorimotor interface with the environment formed in primary neocortical networks (Tucker, 1992, 2001). To the extent that abstract cognition is embodied in neural networks, we must find it organized within this architecture, weaving representational structures between the syncretic motive core and the articulated environmental interface (Tucker, 2007; Tucker & Luu, 2007).

### *Hemispheric Specialization for the Visceral Versus Somatic Boundaries*

Understanding the bodily constraints of memory organization may provide an important perspective on several aspects of neurodevelopmental self-organization. For example, the hemispheric structure of core-association-shell organization describes both sides of the brain. Yet in humans there seems to be an asymmetry in the relative development of these networks, with greater elaboration of limbic and related association networks in the right hemisphere, versus greater elaboration of primary sensory and motor cortices in the left hemisphere (Goldberg & Costa, 1981). Based on anatomical evidence of larger premotor networks in the left hemisphere, as well as a larger area devoted to auditory cortex (Galaburda, Sanides, & Geschwind, 1978), Goldberg and Costa reasoned that the left hemisphere may be specialized for unimodal processing. Specialization of primary motor cortex in the left hemisphere may be consistent with the typical right-hand motor dominance. Because the right hemisphere actually has a wider frontal lobe, and larger superior parietal region shown by the angle of the sylvian fissure (Galaburda, LeMay, Kemper, & Geschwind, 1978), these features would imply a corresponding elaboration of limbic and heteromodal association networks in the right hemisphere. A greater elaboration of limbic and association cortices in the right hemisphere may be consistent with its apparent specialization for emotional communication and its skills in holistic integration of spatial information (Borod, 1993, 2000).

If human hemispheric specialization does involve differential development of core versus shell networks within the hemisphere, there may be a corresponding emphasis on the direction of corticolimbic pathway processing, linking visceral to somatic networks. With greater development of core limbic and heteromodal networks in the right hemisphere, we might expect a bias toward feedforward control, as motive influences from the limbic core shape the ongoing consolidation across the linked networks. This is a processing vector proceeding from the visceral domain at the core of the self toward the somatic interface with the world. This direction of processing would support holistic cognition within the right hemisphere organized by personal need and desire (Tucker, 1981).

In contrast, a specialization for unimodal somatic networks in the left hemisphere may allow that side of the brain to develop articulated models of the interface with the environment. The implication for cognition would be a more differentiated representational capacity in the left hemisphere, relatively separated from the densely connected limbic core, that is closely tuned to the realities of sensory and motor contact with the world.

### The Excitement of Consolidation

Although there may be an inherent asymmetry in the influence of limbic constraints, for both hemispheres the memory evidence (Squire, 1986a, 1987) shows that the consolidation of cognition requires interaction, very likely reentrant, between both limbic and neocortical networks. Important clues to the neurophysiological mechanisms of this interaction are provided by observations on the electrophysiological excitability of limbic networks. This excitability appears integral to the functional influence of limbic networks on memory consolidation.

In the phenomenon of *kindling*, electrical stimulation of any region of neocortex elicits an increasing or *augmenting* response that is particularly strong in limbic regions (Adamec, 1993, 2000). This may explain why epilepsies often begin with extratemporal foci in children, but progress to become temporal lobe epilepsy in adolescents and adults if unchecked: epileptic activity within the hemisphere recruits an ongoing, chronic kindling of limbic regions. The relevance of this electrophysiological responsiveness for understanding memory consolidation is shown by evidence that the limbic kindling response, and seizures, can be *classically conditioned*, meaning that with training the kindling responds to a previously neutral sensory cue (Janowsky, Laxer, & Rushmer, 1980; Myslobodsky, Mintz, Lerner, & Mostofsky, 1983). Through a mechanism analogous to kindling, limbic-thalamic-hypothalamic circuits may be expected to influence cortical function, and the consolidation process, in the normal, nonepileptic brain through their excitability in response to motivationally significant (and thus limbic responsive) events (Harkness & Tucker, 2000).

As we have seen from the behavioral evidence in Chapter 1, controlling the processing of significant events in consolidation is not a matter of simple reflex circuits; this control must at the same time engage representational processes that create expectancy for adaptively significant objects and their environmental contexts. The theoretical question then becomes how physiological reactivity contributes to representational processes in limbic networks that could then shape recurrent consolidation processes across the linked corticolimbic pathway.

A machine learning model for how to regulate the adaptive significance of representations has been suggested by Grossberg in his theory of *adaptive resonance* (Grossberg, 1980). In this model, the cognitive process involves a kind of negotiation between expected patterns held by the brain's memory systems and the incoming patterns in sensory networks mirroring the sensory data. An adaptive resonance is then established by the coincidence of motivated expectancy with actualized sensation. In the case of a mismatch, control signals are engaged that trigger and support the continued search for a match (Grossberg, 1984; Grossberg & Versace, 2008).

From the evidence on mammalian learning, such as the negative contrast effect (Papini, 2003), the representations of expectancy that shape learning include an integral hedonic charge. This would imply that the regulation of adaptive resonance has a base in control signals from visceral homeostatic circuits (and the encephalization of these signals within limbic networks). It may be instructive, therefore, to relate the control signals for resonance and discrepancy detection to the evidence on the importance of expectancy confirmation versus discrepancy in animal learning. Considering the connectional architecture of the cortex that frames the search for coincident patterns and adaptive resonance, it is interesting to recognize that the representational process is formed not in some abstract space, but in a network architecture bounded by concrete representational networks for sensation and action (Damasio, 1989). Furthermore, the connectionist analysis of cortical networks suggests where we can find executive control of the entire hemispheric architecture: not in some specialized module of association cortex, but in the hemisphere's densely interconnected and adaptively reactive limbic core, including anterior cingulate, medial temporal, and orbital frontal networks (Tucker & Luu, 2006).

## *Cortical Lesions and Limbic Disinhibition*

Understanding the neural structure of human memory thus requires a model of connectional architecture, best articulated by primate studies. It also requires appreciating the unique elaborations in humans, which includes hemispheric specialization for differing patterns of dominance of limbic (right hemisphere) and neocortical (left hemisphere) networks. The classical observations on effects of cortical lesions on emotional reactivity may be relevant here. With lesions of the neocortex, a patient may lose the capacity for voluntary control over movements contralateral to the lesion. When asked to make a smile, for example, the facial movements are largely one-sided, with a pathological weakness opposite to the lesion. However, in the same patient a spontaneous emotional response, such as smiling in response to a joke, produces normal and symmetric facial movements (Monrad-Krohn, 1924), apparently due to the influence of undamaged limbic networks.

Consistent with the apparent right hemisphere specialization for limbic control, normal facial expressions are stronger on the left side of the face (Borod, 2000; Sackeim, Gur, & Saucy, 1978), even in infants (Rothbart, Taylor, & Tucker, 1989). The implication is that right hemisphere specialization is important to limbic control of behavior early in development, and that the right hemisphere's sensitivity to emotional communication (Borod, 2000) may emerge from its specialization for elaborating limbic representations throughout that side of the brain.

## Subcortical Regulation of the Consolidation Process

Although the elaboration and expansion of the mammalian neocortex is particularly notable in humans, it is not possible to explain memory and cognition solely in terms of cortical, or even corticolimbic, function. The neurophysiological and therefore psychological function of the cortex is highly dependent on the related activity in subcortical networks and circuits, including the subcortical telencephalon (amygdala, hippocampus, and basal ganglia), the diencephalon (thalamus and hypothalamus), and multiple brainstem structures (including the cerebellum, colliculi, reticular formation, and brainstem neuromodulator projection systems). Within the limited scope of the present theoretical analysis, we can at least point to the importance of multiple subcortical influences, and then attempt to illustrate how certain of these may shape the consolidation process. This is complex material, and we recognize it is challenging to read in the terse, dense format presented here. But a basic consideration of subcortical controls will prove essential as we consider the motivational controls on the neurodevelopmental process in the chapters ahead.

Perhaps the fundamental question is whether there are physiological mechanisms, such as cholinergic support of thalamic circuitry regulating the cortex, that may explain the kindling of limbic excitement in the consolidation process. Both recent anatomical evidence and computational modeling suggest that the circuitry of the thalamus and cortex provides control over corticocortical resonance and coincidence detection that may provide the representational mechanism for consolidating sensory and motor patterns in relation to adaptive, motivational controls. Furthermore, the brainstem projections of the classical reticular activating system are clearly important to consolidation, with highly specific influences associated with both states of consciousness and stages of sleep. Finally, striatothalamic circuits are typically not considered as mechanisms of consolidation, but their support of recruitment responses may be important regenerative mechanisms for the neurophysiological control of widespread thalamocortical networks. These circuits also have motivational properties that are important to the adaptive control of memory.

For each of these subcortical mechanisms regulating cortical networks, there are suggestions that the regulation may be regionally specific. For example, there is some evidence that the major neuromodulator systems may be relevant to hemispheric asymmetry (Glick, Ross, & Hough, 1982; Tucker & Williamson, 1984). For another example, we will address the issue of whether subcortical circuits may be differentially involved in the control of dorsal versus ventral divisions of the cortex. Most control systems influence the entire cortex, but there may be important dorsal/ventral regional specialization that proves critical in tuning the different forms of learning and consolidation in the archicortical (dorsal; hippocampal) and paleocortical (ventral; insular and amygdalar) bases of the consolidation processes within each cerebral hemisphere. Because subcortical control is so integral to cerebral activity, it becomes a major clue that certain mechanisms of subcortical control serve as unique mechanisms of memory consolidation for the dorsal and ventral divisions of the cerebral hemisphere.

### *Thalamic Control of Cortical Traffic*

It has long been known that differentiated thalamic projections define the general topography of the cortex, with specific thalamic connections for sensory input, other connections for motor output (including the basal ganglia loops) and still others for association cortex. More diffuse thalamic projections from the midline and intralaminar nuclei have long been thought to provide global modulation of cortical arousal (Jones, 2007b). Although nothing in this classical view (shown in Figure 2-6) has been refuted by recent evidence, the picture is more complex, in interesting ways. There has been remarkable progress in understanding the more specific mechanisms of thalamic regulation of cortical function, both for the specific (sensory and motor) and the "nonspecific" thalamic projections (Jones, 2007a, 2007b).

An important recent demonstration of the thalamic regulation of cortical connectivity has come from electrophysiological research on the spread of cortical activity induced with transcranial magnetic stimulation (TMS) (Massimini et al., 2005). In contrast to a robust spread of cortical activity in the waking state, the same TMS stimulation when the subject was asleep produced only

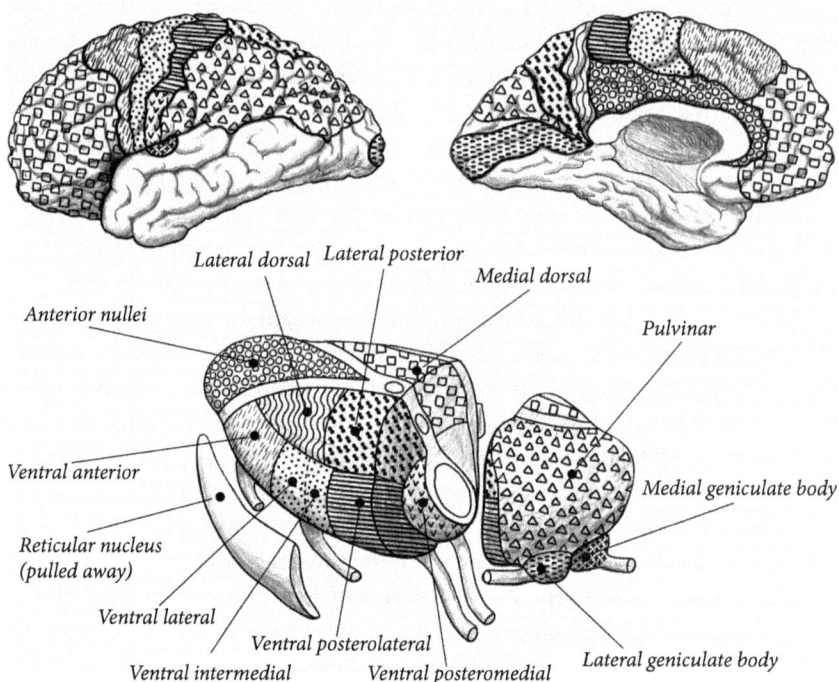

Figure 2-6. Classical view of thalamic regulation of cortex, including specialized sensory and motor nuclei, the pulvinar projecting to parietal visual and sensory integrative regions, the anterior nuclei projecting to the cingulate cortex, and the mediodorsal nucleus (MD) projecting to the frontal lobe. After F. Netter.

local neuronal responses. The difference was apparently due to thalamic gating of thalamocortical projections in sleep. These thalamocortical projections appear therefore to be critical in regulating corticocortical connections that support the spread of TMS (Massimini et al., 2005; Steriade, 2003). Very likely this differential regulation operates through the specific thalamic circuitry mediating corticocortical connectivity recently articulated by Jones (2007a, 2007b).

Unique functions have been discovered for thalamic neurons with differing calcium channels (Figure 2-7). Whereas the specific thalamic projections to granular layer 4 of sensory cortex—described by Jones (2007a) as *core* projections—are from neurons with parvalbumin calcium channels, there is a corresponding set of more widespread *matrix* thalamic projections to layer 1, including layer 1 of adjacent cortical areas, from thalamic neurons with calbindin calcium channels (Jones, 2007a, 2007b). With their more widespread interregional connectivity, the matrix projections appear to be important for regulating corticocortical traffic, providing what Jones describes as a *nonspecific binding circuit* (Jones, 2009).

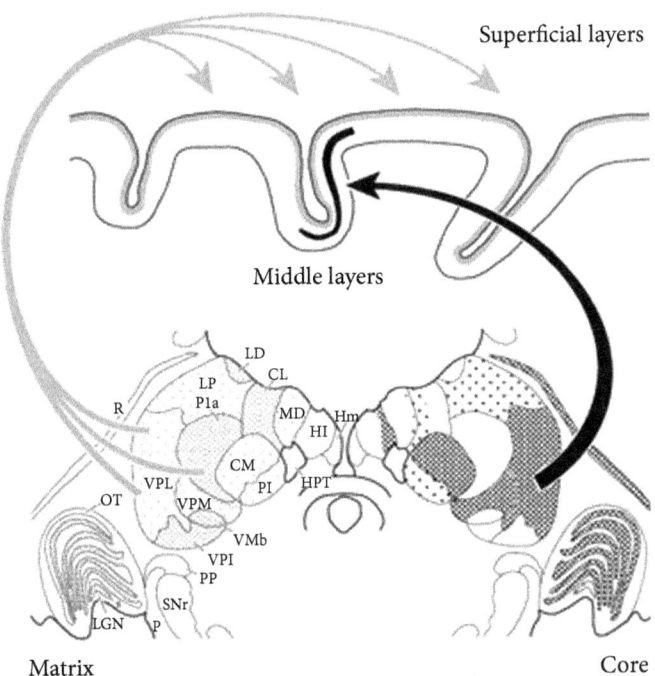

Figure 2-7. Each major nucleus of the thalamus includes core projections primarily to layer 4 of a specific cortical region, plus matrix projections that appear to target superficial cortical layers over a broader region, apparently facilitating regional communication. Reprinted from Jones, Edward G. *The Thalamus*, 2 volume set, second edition, New York: Cambridge University Press, Copyright © 2007 E.G. Jones. Reprinted with the permission of Cambridge University Press.

Complementing the matrix projections are the interregional corticothalamic projections of the layer 5 neurons of the cortex (Figure 2-8). Through responding both to the thalamic matrix projections and local cortical processing, the layer 5 cortical neurons project not to the originating sensory thalamic nucleus (for example, the lateral geniculate nucleus for vision) but to the association thalamic nucleus (for example, the pulvinar). In this way, both the matrix thalamic projections and the output from cortical columns interact to organize the corticocortical connectivity within a sensory pathway.

Thus the corticocortical connectivity shown in Figure 2-4 is not a fully functional apparatus. Rather, it seems to be complemented by resonant circuits in which the cortex uses its thalamic base to regulate its own interregional traffic.

Figure 2-9 (taken from Jones, 2009) draws from the model of Llinas and Pare (1997) to outline how the thalamocortical matrix binding circuit, together with the dispersed projections of the cortical layer 5 neurons, may interact with corticothalamic projections to create and maintain oscillatory (functional

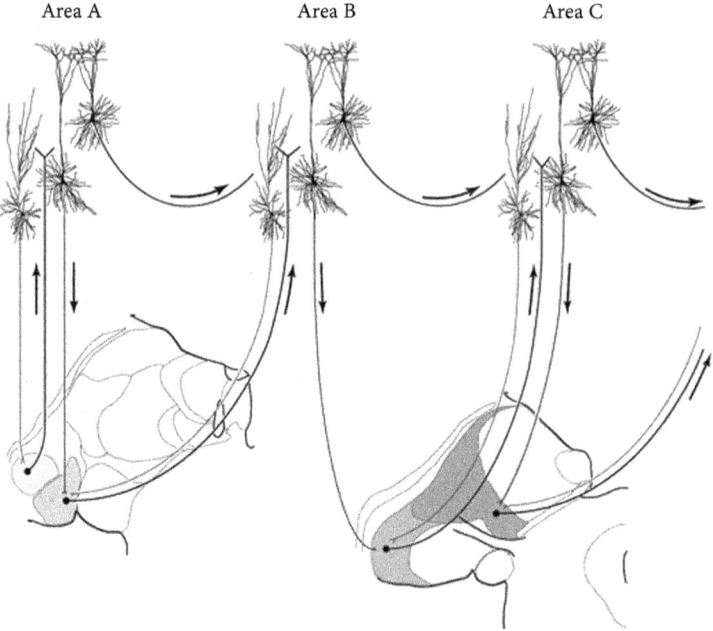

Figure 2-8. The areas of adjacent cortical networks (A, B, C), such as in primary, unimodal, and heteromodal visual cortex, not only include regular corticocortical projections shown in Figure 2-9, but ordered thalamocortical projections through which outputs from layer 5 project to regions of the pulvinar for different cortical regions, thereby creating corticothalamic loops that serve to integrate corticocortical connections. Reprinted from Jones (2007b) [with permission from Cambridge University Press—permission forthcoming; wait for revised caption/credit line].

neurophysiological) activity within a cell assembly of a corticothalamic network. A key component of this circuitry is the thalamic reticular nucleus, a thin network of inhibitory neurons surrounding the thalamus (Figure 2-6) that suppresses thalamocortical neurons and thus suppresses their influence on the cortex. The cortex appears to self-regulate in part through its inhibitory projections to the thalamic reticular nucleus, thereby disinhibiting the associated, and topographically organized, thalamocortical projections (Steriade, Jones, & Llinas, 1990).

Thus intercortical communication is mediated in part by mutual (but not exactly reciprocal) thalamocortical and corticothalamic projections, with the thalamic reticular nucleus providing a critical mechanism to support differentiated cortical control of thalamocortical synchronization (Steriade, 2003). The excitement of consolidation therefore involves not only the strong reactivity of limbic cortex and its primitive circuits (hippocampal, amygdalar), but also a linked hierarchy of limbic-association-neocortical networks with each

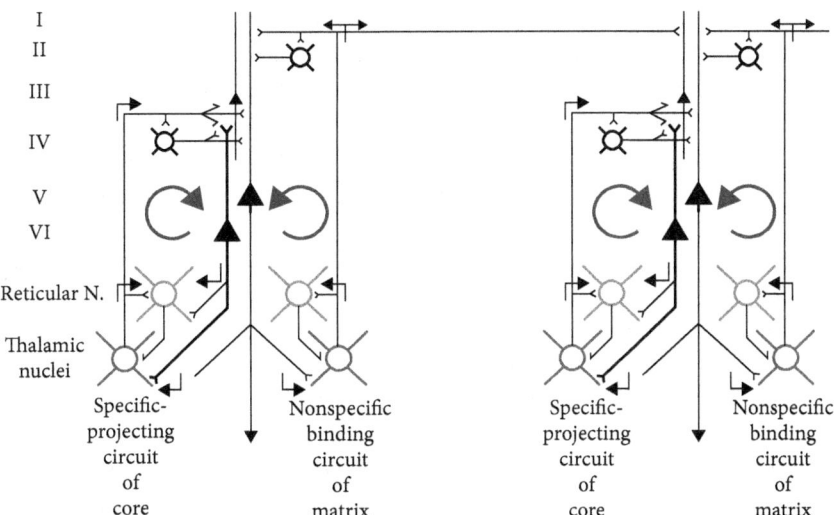

Figure 2-9. Schematic model for synchronization of thalamus and cortex. Reprinted from Jones (2009) with permission from John Wiley and Sons.

boundary of network connectivity mediated by corticothalamocortical projections. The thalamic and cortical mechanisms are themselves enveloped by primitive yet pervasive control influences from the brainstem reticular formation, including specific neurotransmitter projection systems, and by forebrain projections from the anterior hypothalamus and ventral striatum. A specific set of projections from the ventral striatum, comprising the forebrain cholinergic system, is particularly important to setting the tone of the thalamic and cortical networks, and thereby tuning important qualities of the consolidation process.

### Forebrain Cholinergic Control Over the Thalamus and Cortex

Recognizing the importance of the new evidence on the mechanisms of corticothalamocortical cell assemblies organized by Jones (2007a, 2007b), Grossberg and Versace cast the general model of adaptive resonance between cortical networks (Grossberg, 1980) into the specific mechanisms of resonant corticothalamocortical cell assemblies (Grossberg & Versace, 2008). This *simultaneous matching adaptive resonance theory* or SMART model has a particularly important feature for a theory of consolidation: a specific mechanism for controlling the arousal or activity of cerebral networks on the basis of the matching or resonance that is detected. As described above, the thalamic reticular nucleus is a pivotal mechanism for regulating thalamocortical control, and it does so under the supervision of cortical inhibition and disinhibition (Steriade et al., 1990). Thus a general, diffuse control over the thalamic reticular nucleus, apparently setting the tone of its function rather than specific topographic pattern, is provided by the forebrain cholinergic projections from the nucleus basalis (Jones, 2007a) (Figures 1-4 through 1-5). Through projections from nonspecific thalamic nuclei to the nucleus basalis, Versace and Grossberg propose that a mismatch event is able to generate a form of *reset signal*, provided by nucleus basalis cholinergic projections to the reticular nucleus, that prolongs the search for the appropriate coincidence detection.

This reset signal is interesting in light of evidence that the phase of the midline frontal theta rhythm is reset in human self-monitoring when an error is detected (Luu & Tucker, 2003a). Because the limbic theta rhythm may index the coordination of widespread corticolimbic networks in memory access (Chrobak & Buzsaki, 1998; Tucker & Luu, 2006), this reset function may be consistent with the role of cholinergic function not only in the SMART model but in physiological self-regulation of limbic networks (Mesulam & Mufson, 1984). The nucleus basalis cholinergic projections target widespread regions of the cortex in each hemisphere, with extensive cortical regulatory properties similar to the brainstem norepinephrine and serotonin projections (Mesulam, Mufson, Levey, & Wainer, 1983). Similar to the asymmetric (one-to-many but not reciprocated) projections of these brainstem activating systems, the nucleus basalis projects to the entire cortex, but there are not reciprocating cortical projections back to the nucleus basalis. Rather, the controlling input for the cholinergic nuclei is confined to limbic networks. In this way, the limbic networks determine the excitatory cholinergic regulation, an effect that may be considered as gain control (Mesulam et al., 1983), or adaptive binding (Tucker & Luu, 2006) for the corticolimbic interactions of the entire hemisphere. When combined with thalamic circuitry supporting intercortical traffic, this cholinergic regulation may be integral to limbic excitement and kindling in memory consolidation.

### *Limbic Elaboration of Hypothalamic and Brainstem Controls*

The pivotal role of limbic networks in regulating memory must be explained by the close links between these networks and the homeostatic and motivational mechanisms of the brain, particularly those of the hypothalamus (Luu & Tucker, 2003b). In higher mammals including humans, both memory and adaptive controls appear to be supported by the limbic cortex (including perirhinal, piriform, anterior temporal, and insula in the ventral limbic division and the entorhinal, posterior cingulate in the dorsal limbic division) rather than the limbic structures (amygdala and hippocampus) alone. Whereas in rats lesions of the amygdala and hippocampus produce severe memory deficits, similar lesions in primates do not unless they are extended to include adjacent limbic cortex (Mishkin, 1982; Zola-Morgan, Squire, & Mishkin, 1982). The increasing cortical elaboration of control processes has been described by the classical concept of *encephalization* (Denny-Brown, 1966). Although this concept has been rejected by most of today's neuroscientists as unsupported by direct evidence, it may be useful for an evolutionary-developmental analysis of the brain, such as in explaining why primate limbic cortex assumes functions restricted to subcortical limbic structures in rodents.

Following this line of reasoning, the *encephalization of the visceral function* may be a way to understand the highly developed capacities of human limbic networks in representing visceral functions, and thus motivational significance of concepts, in working memory. We have theorized that the cognitive elaboration of visceral regulatory influences may be important to *allostatic* self-regulation, in which there is motivational adaptation in expectation of needs, rather than reactive homeostatic regulation after a need is present (Luu & Tucker, 2003b). In this way, motive controls may be applied directly to the consolidation mechanisms of expectant learning.

There is an important specificity of visceral influence on the representations in limbic cortex, an apparent specialization for control of visceral input versus output. The anterior cingulate cortex (the limbic base of the mediodorsal frontal lobe) appears to regulate *visceromotor* functions, whereas the insula (the limbic base of the orbital and ventrolateral frontal lobe) is specialized for *viscerosensory* functions (Neafsey, 1990; Neafsey, Hurley-Gius, & Arvanitis, 1986; Neafsey, Terreberry, Hurley, Ruit, & Frysztak, 1993; Price, 1999). It seems likely that these limbic networks elaborate not just autonomic but motivational functions of the hypothalamus, organizing the encephalization of viscerosensory and visceromotor functions generally (Luu & Tucker, 2003b). Although the connectivity of limbic cortex provides extensive routes for regulation of the entire cerebral hemisphere, both autonomic and motive controls require extensive vertical integration with brainstem activating systems (Nauta & Haymaker, 1969).

Thus the adaptive control of cognition may be achieved as component concepts of needs or goals in limbic networks resonate with component concepts formed in more differentiated (heteromodal) association areas. In association areas, the component concepts would comprise both elementary sensory images, or action intentions, and more complex images of goals. The notion of *component concepts* in this formulation implies the specific embodiment of a pattern in a single cortical network representational level. The full concept is then a multinetwork (limbic, hetero-, unimodal, primary) representation. For the physiological regulation of hypothalamic influences on consolidation, support from brainstem projection systems must also be important. As described above for thalamic regulation, the brainstem (pedunculopontine) and forebrain (nucleus basalis) cholinergic systems are critical to memory. There are marked memory deficits observed with anticholinergic drugs (Gold & Zornetzer, 1983) that are clearly more specific than the effects of drugs affecting other neurotransmitter systems. It seems likely that cholinergic modulation may be important to controlling the resonant excitement that links the multiple limbic, thalamic, and cortical networks in consolidating an integrated concept.

Furthermore, whereas the monoamines, including norepinephrine and serotonin, may not be as critical as acetylcholine to memory consolidation per se, they may have important facilitating roles on arousal and cortical modulation that sculpt the form of the consolidation process and thus the scope of working memory. In sleep, hypothalamic control is central to the arousal state transition (Doran, Van Dongen, & Dinges, 2001; Minzenberg & Carter, 2008). Furthermore, the major brainstem neurotransmitter systems show highly specific patterns of activity during specific sleep stages. The norepinephrine and serotonin projections, for example, are virtually silent in REM sleep, when cholinergic control is strongly active (Steriade & McCarley, 1990).

In the working memory of the waking state, norepinephrine and serotonin may shape consolidation through qualitative influences on mood, arousal, and attention. This shaping may be consistent with the habituation bias (Tucker & Williamson, 1984) and its expansive, integrative attentional mode. The rapid habituation of working memory to any specific element guarantees that it will be broadly distributed over many elements. Consistent with the hypothesis that norepinephrine function is exaggerated in mania (Schildkraut, 1965), the habituation bias of high NE tone would be associated with the expansive mental associations that occur in an elated mood (Tucker & Williamson, 1984).

## *Ventral Striatum and Paleostriatum*

Although not typically related to memory capacities beyond motor learning, the circuitry of the striatum (caudate, putamen, nucleus accumbens) and pallidum (external and internal divisions of the globus pallidus), also figure importantly in the regenerative excitatory neurophysiological processes underlying the self-regulation of memory consolidation. Close interactions of the striatal networks with the intralaminar nuclei of the thalamus appear critical to recruiting responses specifically (Jones, 2007a,b), and perhaps the organization of memory for action generally.

The circuitry of the basal ganglia represents a curious residual of mammalian evolution (Figure 2-10). The cortex sends extensive projections from both sensory and motor/premotor areas to the striatum, including both the sensorimotor components (caudate/putamen) and what appear to be the integrative limbic components (nucleus accumbens/ventral striatum) (Alheid & Heimer, 1988; Nauta, 1986). The striatum (formerly called the neostriatum) then projects to a more primitive telencephalic center for motive and motor control, the pallidum (formerly called the paleostriatum). Whereas the striatal-pallidal output once provided primary telencephalic motor control of action in simpler vertebrates, in mammals the striatal output proceeds to the pallidum but is then routed back to the thalamus and cortex (Alheid & Heimer, 1988). The basal ganglia "loops" thus formed are important not only to sensorimotor integration,

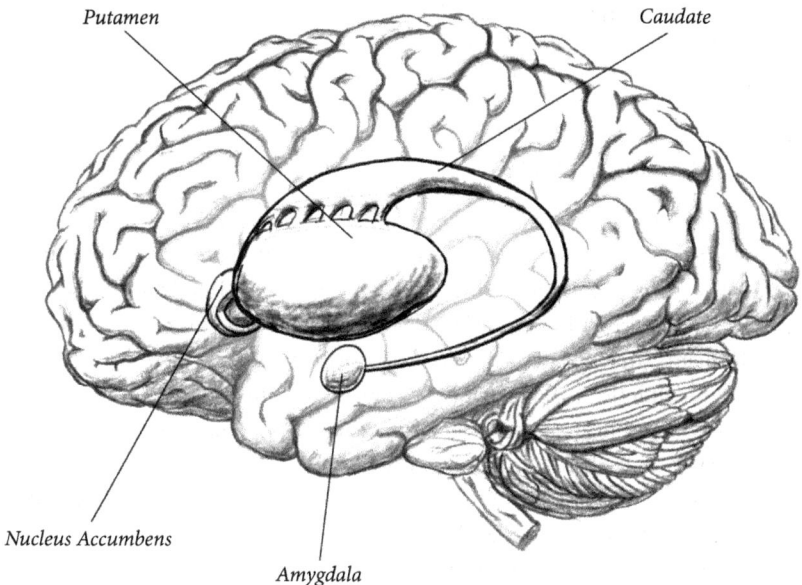

Figure 2-10. The basal ganglia (nerve bundles at the bottom) or neostriatum include dorsal striatal structures (the caudate and putamen) and ventral striatal structures (including the nucleus accumbens).

but to limbic and motivational control, including unique circuits engaging the ventral striatum and nucleus accumbens (Nauta, 1986).

As shown in many fMRI studies, for example (Knutson, Fong, Adams, Varner, & Hommer, 2001), activity in the ventral striatum is associated with perception of, and anticipation of, reward. As pointed out above, recent investigations of reward processes have differentiated between craving (wanting) and hedonic (liking) components, with strong contributions from regions of the pallidum to the hedonic component (K. C. Berridge & Kringelbach, 2008; Kringelbach & Berridge, 2010). An important realization from the recent research is that many of the earlier lesions targeting lateral hypothalamus actually damaged the centers in the pallidum (K. C. Berridge, 2009). Thus the most basic motive controls involve ventral striatal (and paleostriatal or pallidal) as well as hypothalamic and brainstem circuits.

Although there are close relations between amygdala and ventral limbic networks with the basal ganglia, the output from the hippocampus and dorsal limbic structures also targets the ventral striatum, providing an important pathway for dorsal limbic engagement of the primitive hedonics of liking. It has long been an important anatomical fact that the primary output of the hippocampus is to subcortical regions (Pribram, 1981). Whereas the postcallosal projections of the fornix are to the anterior nuclei of the thalamus (and thereby to cingulate cortex), the precallosal projections target the septum and nucleus accumbens, providing a route for dorsal limbic (Papez circuit) control of the ventral striatum and pallidum.

## Action Readiness and Inhibitory Specification in the Basal Ganglia

Motive control from ventral limbic networks appears to operate through a craving or wanting mechanism, particularly through engaging neuromodulation through the mesolimbic dopamine projections (C. W. Berridge, 2006; K. C. Berridge, 2009; Kringelbach & Berridge, 2010). As we consider specific neural mechanisms that exert widespread control over cerebral networks, and thus memory consolidation, the specialized neurotransmitter systems of the basal ganglia (dopamine and acetylcholine) are important points of integration. In addition to the nigrostriatal dopamine projections regulating motor function of the striatum, the ventral tegmental (mesolimbic) dopamine projections target ventral limbic and anterior cingulate territories in a pattern that may be particularly relevant not only to motivational craving but to the fast (avoidance) learning mechanisms in animal studies. These projections are theorized to underlie the constancy or redundancy bias: a focusing of attention to support working memory for a restricted set of perceptual objects (Tucker & Luu, 2007; Tucker & Williamson, 1984). This effect could be seen as a qualitative shift in the consolidation process, favoring focus on a few motivationally central elements at the expense of a global allocation. Such a control bias might be appropriate under conditions of anxiety, in response to either an external threat or a high internal need state (Tucker & Williamson, 1984).

In motor control, the basal ganglia circuits provide a specific form of inhibitory specification of actions through surround inhibition, in which a single action, such as of a finger, is differentiated through inhibition of competing actions, such as from adjacent fingers (Shin, Sohn, & Hallett, 2009). This influence appears to be mediated through the full basal ganglia and corticothalamic circuit, including projections from the caudate and putamen to the globus pallidus, from there to the thalamus, from which the inhibitory specification appears to be supported by projections to the cortex. The importance of basal ganglia control of the inhibitory surround in regulating fine motor control is suggested by hemispheric specialization for this mechanism, apparently providing a mechanism for the fine motor control of the dominant hand.

A more general model of inhibitory specification for the basal ganglia suggests that control from these structures exerted over thalamocortical relations allows the cortex to select certain actions while inhibiting others (Redgrave, Prescott, & Gurney, 1999). In a computational model that accounts for specific neurophysiological divisions of the basal ganglia, Redgrave et al. propose that one mechanism provides for the selection of a dominant action, whereas the second mechanism provides for ongoing control of the selected action, with this ongoing control provided by the modulation from dopamine projections. From the severe cognitive impairments observed with dysfunction of the basal ganglia in Parkinson's disease, it seems clear that the control properties such as inhibitory specification emergent from the basal ganglia, together with their dopamine modulation, are not restricted to elementary motor control, but are integral to human cognition and learning generally (Tucker, Luu, & Poulsen, 2008).

The central role of dopamine in motivational control has often been linked to the concept of "reward circuits." Although mesolimbic dopamine release has a discrete pattern of response in anticipation of predicted reward, and is suppressed under conditions of extinction (Schultz, Tremblay, & Hollerman, 1998), the response of DA and the basal ganglia may be more important to prediction and discrepancy than to the hedonic qualities of "reward." Berridge and associates have drawn the distinction between the hedonic quality of an event, or "liking," for which dopamine is not central, and the anticipation or "wanting," for which it is (K. C. Berridge, 2009; K. C. Berridge & Kringelbach, 2008). This role of wanting in anticipation and incentive may be similar to the notion of motor readiness as a cognitive control (Pribram & McGuinness, 1975), and it may help to link dopamine modulation not only to motivation globally, but to the specific controls on action regulation provided by the basal ganglia.

In addition to supporting well-known thalamocortical projections for motor control, the pallidal-thalamic projections may contribute to general cerebral arousal control through targeting the intralaminar nuclei of the thalamus. As reviewed by Jones (Jones, 2007a), the evidence that recruiting responses are mediated by the intralaminar nuclei (Morrison & Dempsey, 1942, 1943) led to widespread interest in thalamic mechanisms of neural control (Lindsley, 1951, 1957; Malmo, 1959). Over the next several decades this important research seems to have been forgotten (Jones, 2007a). If we consider that sustaining, regenerative neurophysiological phenomena such as kindling are important candidates for mechanisms of consolidation, then the evidence on thalamic recruiting responses would suggest that the classical "diffuse" thalamic projections from the intralaminar nuclei may also be important candidates for the neurophysiological control of widespread cortical networks in memory integration. Jones (2007b) proposes that the new insights into the matrix projections of thalamic nuclei, targeting the integrative superficial layer of the cortex, may provide modern clues to the traditional evidence on the cortical control provided by the intralaminar nuclei.

## Vertical Integration in the Consolidation of Experience

The theoretical challenge for explaining the mechanisms of memory consolidation in the mammalian brain is to understand vertical integration. The cortex has an ordered structure, centered on limbic networks, and the evidence on amnesia shows that corticolimbic traffic is essential for memory consolidation. But this traffic is regulated by multiple subcortical systems, and it is important to understand the contribution of each major system. In Chapter 1 we outlined the general principle of activity-dependent specification of neural connections, and we pointed to brainstem and forebrain neuromodulator projection systems as regulating neural activity in ways that shape that specification. Mediating between the corticolimbic architecture and elementary neuromodulator controls on neural activity are multiple subcortical control systems, each of which influences the cortex in specific ways, and many of these ways involve subcortical control over the brainstem neuromodulator projection systems.

The complexity of the multiple levels of neural control systems can seem overwhelming. The perspective necessary to organize this complexity may come from evolutionary-developmental reasoning, as may be gained from examining Figure 1-1. Researchers studying the human brain tend to become concerned with unique features of the human cortex. Whereas these are indeed fascinating, the general operation of the human brain is given by the venerable mammalian plan, and the unique neocortex of mammals may be understood best through appreciating its extension of the more basic self-regulatory circuitry of protomammalian telencephalon, the basal ganglia and limbic structures. These, in turn remain closely linked to regulatory control by diencephalic structures, the thalamus and hypothalamus, and it is largely through the more primitive telencephalic and diencephalic structures that the brainstem activating systems are engaged and controlled.

The evolutionary order of these vertically integrated brain systems can be seen from gross anatomy (Figure 1-1). As we examine the neuroembryological evidence in Chapter 6 we will see that the neurodevelopmental process, at least in utero, achieves vertical integration across these systems through recapitulating (more or less) the evolutionary order. The implications of vertical integration for cognition and memory are made clear by an examination of the critical circuits for memory consolidation, which involve essential contributions from multiple vertical levels. There are two distinct circuits for memory consolidation, each with an apparently unique evolutionary, and perhaps developmental, history.

## Cortical, Limbic, and Diencephalic Levels in the Dual Circuits of Memory Consolidation

We have seen that the cortical pathways involved in memory consolidation include both dorsal (archicortical) and ventral (paleocortical) divisions, and that each division has unique roots in subcortical circuits. As recognized by Mishkin (Mishkin, 1982), several lines of evidence suggest there is not one memory circuit in the mammalian brain, but two (Figure 2-11). The dorsal

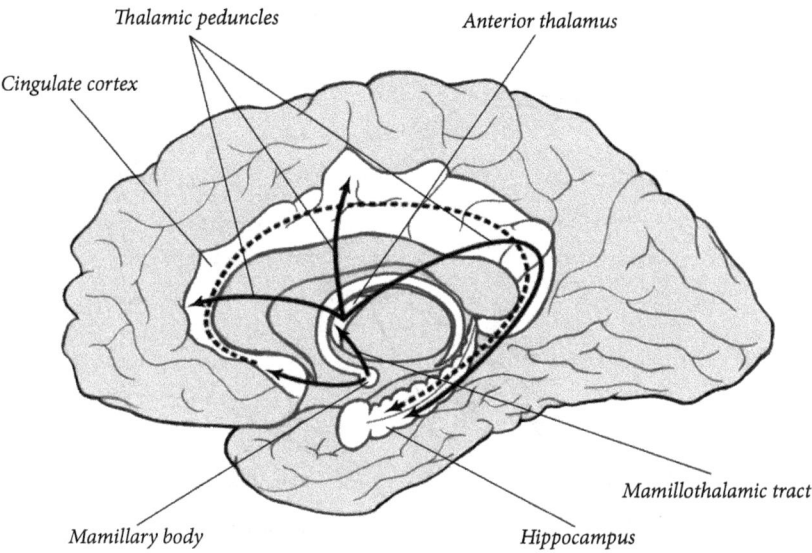

Figure 2-11. Dual memory circuits. A dorsal memory mechanism is organized through the classical Papez circuit, centered on the hippocampus, and engaging the cingulate cortex and mammillary bodies of the hypothalamus through the anterior nuclei of the thalamus. After Brand and Markowitsch (2006).

division of the cortex is strongly interconnected with the hippocampus through the cingulate gyrus. The hippocampus connects to the anterior hypothalamus and nucleus accumbens through the fimbria-fornix. The projections from the cingulate through the anterior ventral thalamus and mammillary bodies of the hypothalamus make up the classical Papez circuit, the classical cortical-diencephalic mechanism of memory consolidation (Squire, 1998; Zola-Morgan & Squire, 1993). The hippocampal phenomena such as long-term potentiation and kindling represent classical neurophysiological mechanisms of memory consolidation. As we will see in reviewing the neuroembryological evidence in Chapter 6, these mechanisms engage primitive reptilian and amphibian 3-layered network architectures that may represent formative stages in ontogenetic as well as phylogenetic development.

A separate memory circuit is centered on the amygdala (Mishkin, 1982), engaging interconnected regions of ventral limbic cortex (anterior temporal, insular, orbital frontal), with unique thalamic circuitry (through projections to the mediodorsal thalamus and from there to multiple regions of the frontal lobe; Figure 2-12). The physiological reactivity of amygdalar circuits (Adamec, 1990) appears integral to the consolidation within this pathway. The close connections of the amygdala and ventral limbic circuits with the ventral striatum suggest these circuits are particularly important to the basal ganglia networks and their role in cerebral arousal and memory integration.

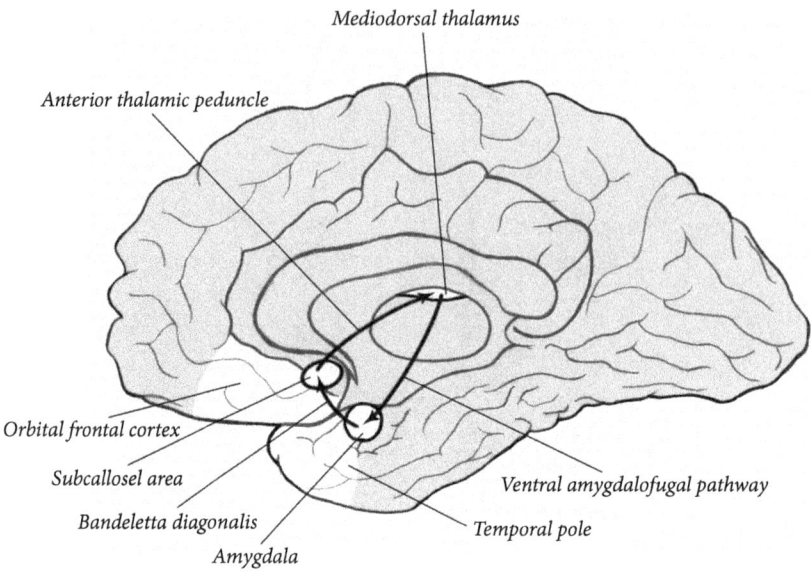

Figure 2-12. Dual memory circuits. A ventral memory mechanism is organized through a triangular circuit with projections integrating the (1) amygdala and temporal pole with (2) the subcallosal and orbital frontal area, and (3) the mediodorsal nucleus of the thalamus. After Brand and Markowitsch (2006).

The general principles of consolidation appear similar between the dual limbic memory systems. There are linked cortical networks for both, with greater density of interregional connections for limbic networks and greater modularity and specificity in the progression from limbic to heteromodal to unimodal to primary sensory or motor cortices. There is a strong reactivity of the limbic circuits, engaging thalamic and striatal circuits and hypothalamic and brainstem systems to regulate the activity of the consolidation process on the basis of motivational significance. But, given the radically different subcortical bases in the hippocampal versus amygdala circuitry, are these dual limbic circuits just two ways of motivating memory, or are they two fundamentally different forms of consolidation?

Certainly the cognitive products appear to be different. The *spatial* or *configural* memory of the hippocampal-dorsal networks seems to involve a unique and in many ways opposite conceptual structure to the *object* or *item* memory that is organized within the ventral limbic networks and their insular-amygdalar base (Aggleton & Brown, 1999; Ungerleider & Mishkin, 1982; Yonelinas, 2002). But is this opposite conceptual structure achieved through similar mechanisms of motivation and consolidation, or does it result from two fundamentally different mechanisms?

These questions about the differing mechanisms of cognition are fundamental to the present theoretical formulation. We will examine them systematically in the next chapter, working toward a neurodevelopmental analysis of dual memory systems in elementary sensory and motor learning processes.

## Summary

In this chapter we have examined the connectional architecture of the mammalian brain, and we have considered principles of distributed representation that may explain how cognition is organized in this architecture. The primary dimension of the cortex runs from visceral evaluation to somatic actualization, and this must be the structure of cognition. There are two directions for the consolidation process. One runs from the core visceral self toward the somatic interface with the world, and the other runs in the opposite (limbipetal) direction, from the realities of the sensorimotor context to constrain the operations of the core.

The neurophysiological mechanisms of regulating the activity and connectivity within this structure are complex, reflecting intrinsic excitatory and inhibitory circuits of the cortex, both focal and more distributed projections to and from the thalamus, limbic (hippocampal, insular, amygdalar) and striatal (basal ganglia) circuits of motivational control, and both forebrain and brainstem neuromodulator projection systems regulating activation, arousal, and neural tone in specific ways. Consolidation is then a vertically integrated process, with some of the most important contributions provided by the highly conserved mechanisms of the brainstem, midbrain, and diencephalon.

We began this chapter with the anatomy of the cortex, with the idea that connectivity implies function, and quickly faced the fact that the connectivity of the cortex reflects its embodiment, in the sensory and motor maps of the neocortex, and in the visceral functions of the limbic core. Both these boundaries of embodiment must frame the process of memory consolidation. There are multiple mechanisms that shape neural activity adaptively, according to sensory, motor, and motivational requirements, and each of these is based in subcortical circuits. The challenge of a neurodevelopmental theory is to explain how the mechanisms of consolidation operate within these boundaries to build the brain, continuously throughout life.

In the next chapter we will consider evidence that the major subcortical control systems for memory consolidation are organized in relation to the dual memory circuits. One is organized to support the dorsal networks of the cortex, those with archicortical (hippocampal) origins and linked to visceromotor functions. A second control system appears to have evolved to support the ventral networks of the cortex, those with paleocortical (piriform and amygdala) origins and linked to viscerosensory functions. The interesting theoretical possibility is that these two control systems may generate fundamentally different modes of consolidation.

The sensory-motor specializations of the neocortex are well known for the *somatic* networks, organized on the rostral-caudal (frontal-parietal) dimension. Running orthogonal to this, the dorsal-ventral dimension of *visceral* input-output (visceromotor and viscerosensory) specialization may be equally important to mammalian cortical development. We will begin Chapter 3 with the executive controls that are particularly important to the motive strategies of human developmental learning. The networks of the frontal lobe form representations of values, goals, and plans that are not only the active elements of expectant learning; they also have the unique connectivity to recruit the multiple control systems shaping arousal, consolidation, and the learning process throughout the neural hierarchy.

By studying dorsal and ventral corticolimbic systems of action regulation, we will see that the specific forms of expectancy seen in the animal learning literature we examined in Chapter 1 may be understood as specific neurodevelopmental mechanisms, which after birth organize both neural architecture and the ongoing process of cognition. One form of expectant cognition is biased toward maintaining a context representation with hedonic goals. Another form engages sustained attention to modify the model in the face of aversive or discrepant events. The interesting theoretical notion is that these mechanisms self-regulate experience and behavior through integrating multiple levels of the neuraxis in much the same way that they engage vertical integration to self-regulate the embryonic process of neural morphogenesis.

# 3
## Regulating Action

Much of memory consolidation operates automatically, in the background of waking and sleep, integrating experiences in relation to their motivational significance. This is the effective unconscious. At the same time, the evidence on animal learning suggests that learning is fundamentally prospective and goal-oriented, as need states prime expectancies for desired outcomes, and as those expectancies then organize working memory to shape successful and well-organized action plans.

As mammals evolved complex memory, their motivated self-regulation spanned longer intervals of time. As we consider the unique features of human self-regulation that shape each child's emerging personality within the progressive neurodevelopmental process, the evidence shows that the frontal lobe provides not just a memory buffer, but fundamental capacities for planning and executive control of behavior. Both clinical observations and psychological testing have shown that frontal lobe damage is most apparent when new learning is required under unstructured conditions, and when the person must therefore organize and evaluate a novel strategy for acting effectively (Luria, Pribram, & Homskaya, 1964; Wang, 1987). Throughout this process, learning results not from a passive response, but from the active integration of information that is relevant to goal-oriented—or threat-oriented—expectancies. By studying learning from the perspective of action regulation, we will see how the active expectancies examined in the learning studies of Chapter 1 are organized within the complex mechanisms of memory consolidation examined in Chapter 2.

We will find that the controls on working memory have inherent motive biases. The dorsal and ventral frontal networks are differentially specialized to organize unique motive controls from limbic and subcortical circuits to regulate consolidation, and thus neural differentiation. The dorsal and ventral networks of the cortex appear to have evolved from dual limbic systems, such that they now represent dual strategies of vertical integration, each with its own unique cybernetic properties. The dorsal corticolimbic pathways are organized around the cybernetics of feedforward control, operating as an expectant mode of working memory. The ventral corticolimbic pathways are organized around the cybernetics of feedback control, operating as a reactive mode of working memory and action regulation (Tucker & Luu, 2007).

In neurophysiological terms, we propose that what is typically described as *executive control* can be seen to reflect the frontal lobe's capacity for integrating not only posterior hemispheric cortex (Goldman-Rakic & Schwartz, 1982), but vertically integrated subcortical systems (basal ganglia, thalamic, limbic, and brainstem) (Goldman-Rakic, 1988). The subcortical systems, manifested through their connectivity to the frontal lobe, provide the necessary configuration of primitive controls on arousal, attention, and memory consolidation that are necessary to organize an ongoing conceptual model of values and goals. This integration can be called *representation of the regulatory functions.* It achieves representations of goals and plans (concepts) that simultaneously engage the integral neurophysiological control systems (motive controls). This idea can be traced to Nauta's reasoning for why the frontal lobe would have such well developed connectivity with the dorsal and ventral divisions of the limbic system (W. J. Nauta, 1972).

Although essential for the most abstract and complex human cognition, the frontal lobe's role in working memory emerges from mechanisms of motor control. Executive self-regulation can therefore be understood through a behavioral analysis, as it emerges from the concrete requirements for *action regulation*. Action regulation requires close support from perceptual controls, such that the posterior brain, and the parietal lobe in particular, are integral to motor control. Because it always operates from a motivational base, and yet is effected through cognitive expectancy, action regulation is achieved in large part through representation of the regulatory function, concepts of motives and their goals. The concepts that organize action include linked components of visceral evaluation, integrating ongoing control from limbic visceromotor and viscerosensory representations, operating at the dual dorsal and ventral bases of the hemisphere's cortical networks.

To emphasize the psychological significance of the specialization of dorsal and ventral corticolimbic divisions of the cerebral hemispheres, we will begin by considering the executive functions. We then turn to more elementary features of action regulation, in continuous motor control, which show the primitive yet fundamental cybernetic biases of the dorsal and ventral divisions that then provide insight into more complex psychological capacities.

## Executive Control of Working Memory

From clinical studies of the psychological deficits of patients with brain lesions at the end of the 19th century, John Hughlings Jackson (1931) concluded that the organ of mind is a great sensory motor machine. In his evolutionary-developmental analysis of the hierarchic organization of the brain's multiple levels, Jackson emphasized that even higher mental capacities must be emergent from a brain whose tissues are specialized for sensing and acting in response to a changing environment.

## The Working of Memory

The complex deficits in self-regulation seen with frontal lesions can be seen to imply that the normal human frontal lobe integrates cognitive and motivational processes of multiple brain regions to support action planning. This capacity for planning draws on the frontal lobe contributions to the regulation of what is often described as *working memory* (Baddeley, 1986; Zola-Morgan & Squire, 1993). Yet the clinical neuropsychological evidence suggests this is not simply a naked memory buffer, to support psychological functions in some unspecified generic way. Rather, the frontal lobe's capacity for working memory is emergent from action planning, such that it can be best understood as an elaboration of expectant motor control algorithms (Luu & Tucker, 2003a). Furthermore, the developing action plan is invariably guided by motivational control, such that it reflects the specific motive biases emergent from the limbic bases of frontal networks (Luu & Tucker, 2003b).

The notion that the frontal lobe contributes to working memory became well known through research with apes and monkeys. Frontal lesions impaired the animal's ability to shift the side of responding after a delay interval, in the *delayed alternation* testing paradigm (Konow & Pribram, 1970). Recordings of single unit neuronal activity during the delay interval led to the interpretation that the frontal lobe supports maintenance of the action plan (Fuster, 1985, 1989). Anatomical studies of frontal lobe connectivity showed patterns, such as the interdigitated representation of projections from both cerebral hemispheres in each frontal lobe (Goldman-Rakic & Schwartz, 1982), that could provide executive control of memory capacity in support of the action plan (Goldman-Rakic, 1987).

The concept of a frontal lobe working memory buffer, and even the favored anatomical designation of *dorsolateral prefrontal cortex* (DLPFC), have been adopted from the monkey literature by researchers studying human cognitive neuroscience. Although DLPFC is a common anatomical territory in the classical and recent literature, we interpret the anatomical connectivity (Pandya & Yeterian, 1990, 1996; Yeterian & Pandya, 1994) to suggest that the major functional division is between dorsomedial and ventrolateral frontal regions (Figure 3-1).

Although working memory is a primary concept of frontal lobe function, a number of studies have suggested that controlling distractibility, rather than simple working memory, may be the central component of frontal lobe

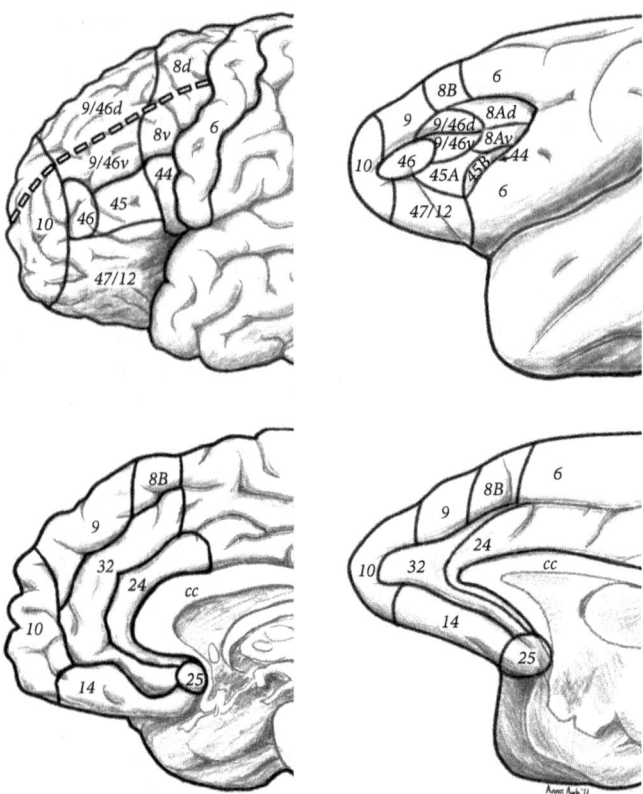

Figure 3-1. Frontal cortex in human (left) and monkey (right) with numbered Brodmann Areas. Much of the lateral surface reflects the ventral division (ventrolateral frontal lobe; below the dotted line), whereas much of the medial surface reflects the dorsal division (dorsomedial frontal lobe), with the exception of orbital frontal cortex (area 14) and possibly area 25 (which appears to be continuous with the anterior cingulate cortex but receives extensive ventral limbic input). We have extended the dorsal/ventral division through the frontal pole (area 10).

contributions to behavior, in monkeys and humans (Chao & Knight, 1995; Milner, 1984; Pribram, 1967). Consistent with these observations, a recent functional magnetic resonance imaging (fMRI) study in humans found that the experimental manipulation of distracting information during a retention interval enhanced hemodynamic activity in frontal cortex (DLPFC) compared to posterior cortex (Postle, 2005).

Thus, although the concept of working memory continues to be influential in explaining frontal lobe contributions, there are many findings that point to capacities having to do with motivation, controlling attention, or inhibiting responses rather than short-term retention per se. Current theoretical models have therefore attempted to reconsider frontal cognitive capacities, such as in relation to more elementary processes in sensory, motor, and representational control (Postle, 2006), or in relation to representations of goals that then guide developing action plans (Miller & Cohen, 2001).

Memory capacity is certainly critical to the complex human capacities for executive control of cognition, extending memory for past events to reach into the planning process. Ingvar has described this extension as *memory of the future* (Ingvar, 1985). In addition, the emphasis on control, inhibition, and attention in addition to memory in the frontal lobe literature shows that it may be the *working* component of working memory that must be understood to capture the frontal lobe's contribution. This conclusion from experimental studies is consistent with the clinical literature on the consequences of frontal lobe damage.

## *Cybernetics of Action Regulation*

The complex nature of frontal lobe functions has long been problematic for clinical neurology. Patients with frontal lesions often show normal intelligence test scores, including normal performance on simple memory tests, and they appear normal in conversation. At the same time, they prove incompetent in coping with jobs, families, and life circumstances (Lezak, 1983). In his paper "The Riddle of Frontal Lobe Function in Man," Teuber (1964) noted that frontal lobe lesions are not only subtle in the face of superficial clinical inspection, but produce a wide variety of deficits when they are tested in depth, including in visual search, in the perception of reversible figures, in a personal (body) orientation test, and in estimating line orientations when the body is tilted. To account for these deficits, Teuber provided a cybernetic explanation of expectant action regulation. He proposed that when actions that are projectional, and thus not a reactive response to an external stimulus, the frontal lobe produces a *corollary discharge*, informing the sensory regions of the brain of the impending effects of the developing action plan (Teuber, 1964).

At about the same time, considering the perseverative errors exhibited by frontal lobe patients on rule-learning and rule-following tasks, Luria and Homskaya proposed that many frontal lobe deficits reflect problems in the control of action (Luria & Homskaya, 1970). Specifically, they proposed that normal actions start with a *prestarting synthesis* of plans and intentions. In the course of action development (presumably through cortical communication similar to a corollary discharge), ongoing actions that are inconsistent with the prestarting synthesis are inhibited. To achieve this effect, there must be a comparator process that matches the outcome of the actions with the plan, although the nature of this process was not addressed by Luria and Homskaya directly.

Integrating the notions of corollary discharge and prestarting synthesis within an explicit anatomical model of the frontal lobe's pattern of afferent and efferent connections, Nauta (1972) noted that the frontal lobe is in a unique position to integrate sensory and visceral information. He speculated that the loss of this integration with a frontal lesion may result in the loss of the ability to anticipate the motivationally significant consequences of an action. Drawing on Teuber's corollary discharge concept, Nauta proposed that in addition to the preparation of sensory regions, a frontal lobe corollary discharge may engage anticipatory interoceptive responses or *somatic set-points* that serve then to keep action on course. Nauta suggested that without these set-points action plans may be evanescent, resulting in action sequences that are easily distracted by environmental demands. This proposal was developed further with the *somatic-marker hypothesis* (Damasio, 1996).

Whereas many authors have used the term *somatic* to refer to bodily processes generally, it may be useful to recognize that many important interoceptive responses are related more to visceral functions, and to maintain the classical anatomical distinction between the visceral and somatic divisions of the central nervous system (Yakovlev, 1948).

The early accounts of deficits in self-regulation with frontal lobe lesions emphasize the importance of expectancy in guiding actions in normal behavior. Without anticipatory control, behavior may become bizarrely perseverative or stimulus-bound (Stamm, 1987). Famous examples were given by L'hermitte (1986) of *utilization behavior*, in which patients with massive frontal lobe damage respond in a compulsory manner to environmental inputs. Given a syringe, for example, the patient is fully prepared to inject the doctor. Importantly, these patients are typically unaware that their obligatory behavior is peculiar. L'hermitte proposed that the utilization actions are implicit in the stimulus context, and that frontal lobe lesions prevent the inhibition of these salient motor plans as a result of more reflective, extended interpretation of the context.

## Limbic Roots of Action

The typical assumption about human intelligence is that it involves elaboration of some abstract intellectual quality, rather than the functions of limbic cortex, which are concerned at the most basic level with internal visceral regulation. As we have seen, however, from the degraded human behavior following frontal lesions, the self-regulatory functions of the frontal lobe require sensitive processes of motivational control. Nauta's emphasis on limbic set points as evaluative guides for behavioral planning may provide an instructive explanation (W. J. H. Nauta, 1971). In its recruitment of limbic motive control to guide the action plan, the frontal lobe draws on dual routes from limbic networks, a dorsal route through the cingulate cortex and a ventral route through the insula and uncinate fasciculus (W. J. H. Nauta, 1964). The elaboration of spindle cells at the limbic base of each of these routes may suggest that humans have evolved complex systems for motivating action regulation. Although for psychologists and cognitive neuroscientists motor control is often considered a trivial component of human frontal lobe function, we propose that even a cursory consideration of the nature of motor control provides important insights into the differential cybernetics of self-regulation provided by the dual, dorsal and ventral, frontolimbic pathways.

## Projectional and Reactive Vectors of Action

A considerable body of evidence, organized in a compelling theoretical formulation by Goldberg (Goldberg, 1985), suggests that there are two premotor systems: one centered on the supplementary motor area (SMA) of the mediodorsal frontal lobe and the other on the arcuate premotor area (APA) of the ventrolateral frontal lobe. Each premotor system can be defined in relation to its internal anatomical connectivity and its connections with subcortical structures (Barbas & Pandya, 1984). The SMA and mediodorsal pathway controls action through the development of action programs that are based on an internal model of the environment and *projected* outward, based on expectations inherent to the predictive model. In contrast, the APA and ventrolateral system organizes actions in a feedback manner, being *reactive* to the demands of the environment as represented through sensory data organized in the ventrolateral frontal lobe.

In addition to the SMA, motor areas of the anterior cingulate region exhibit properties of a forward-projecting mode of motor control (Dum & Strick, 1993). That is, very few neurons in the rostral cingulate motor areas are responsive only to stimulus cues or to the actual movement (Hoshi, Sawamura, & Tanji, 2005). Rather, neurons in this region are modulated by multiple control phases (e.g., cue, instructions, movement, feedback, etc.) of the task, suggesting that the cingulate motor area may be involved with the holistic, projectional representation of the action process.

More recent studies of the APA (also referred to as ventral premotor cortex or PMv) have shown that this region is fundamental to the control of actions by sensory input. Murata et al. (1997) showed that in an object grasping task, neurons in the APA were responsive to both the presentation of an object and the grasping of the object. Importantly, these neurons were responsive to the object even when no grasping was required. Functionally, Murata et al. interpreted the findings to demonstrate that the activity of these neurons represents object information in terms of motor actions. Examinations of neuronal activity in the APA and M1 show that cells in the APA showed preference for the type of object to be grasped (prior to the actual movement) and that this activity occurred earlier than activity recorded in M1 (Umilta et al., 2007). This preference was maintained until the actual grasp of the object.

In contrast, activity in M1 appears to vary according to different stages of the grasp, implying that M1 is involved in the execution of particular, sequenced movements of the grasp. The specificity of APA neurons to visuomotor representation is further exemplified by contrasting their responses to those of neurons recorded in more dorsal aspects of the premotor region. Yamagata et al. (2009) showed that APA neurons are involved in the coding of visuomotor responses when a visual stimulus directly specified the target action (e.g., simply reaching for a target when it is presented). In contrast, neurons in the dorsal regions of the premotor cortex responded to more symbolic cues (e.g., reach for left target when yellow square is presented), apparently requiring more internal regulation of the action program.

### Dorsal and Ventral Specializations for the Direction of Consolidation

An interesting theoretical possibility is that the projectional and reactive vectors of action regulation may emerge from the directional cybernetics of memory consolidation. We saw in Chapter 2 that the processing of consolidation must cross the network boundaries in each sensory and motor pathway, and that the direction or vector of this processing may take two forms. One proceeds through limbifugal connections, from visceral limbic networks toward somatic sensorimotor networks, engaging pyramidal projections from inferior (5) cortical layers to supragranular (2, 3) layers. In the organismic perspective on memory consolidation, this is the feedforward vector, allowing the motivationally charged memories and expectancies at the hemispheric core to shape the organization of representations across the linked networks of the hemisphere. Is this form of internetwork feedforward processing the same as the feedforward control describing action regulation in the dorsal division of the cortex? If so, does this imply that dorsal, archicortical networks have become specialized for one direction of consolidation, the limbifugal direction from hemispheric visceral core toward the somatic shell?

Similarly, connections in the opposite, limbipetal direction of consolidation, conveying the data from primary sensory and motor cortex to association and then limbic cortices, proceed from the supragranular layers of the originating network to granular layer 4 of the target. From the organism's perspective in action regulation, this may be the feedback direction of consolidation, providing feedback from the realities of ongoing motor articulation in M1 and S1 to constrain the developing action plan.

The cytoarchitectonic specializations of dorsal and ventral frontal networks are consistent with this interpretation. A distinguishing feature of the APA of the ventrolateral frontal lobe is the existence of an incipient granular layer (layer IV or 4), which is essentially absent in the mediodorsal premotor areas, such as SMA (Barbas & Pandya, 1986; Vorobiev et al., 1998). As noted by Barbas and Pandya (1986), the APA shows cytoarchitectonics that are more similar to sensory cortex than to motor cortex. With regard to connectivity, the APA is widely connected to many regions of the frontal cortex, including parainsular gustatory regions, whereas these connections are lacking for both the dorsolateral premotor areas as well as the SMA. With regard to thalamic connectivity, the APA receives projections from the ventral anterior pars magnocellularis nucleus, whereas the SMA receives projections from the ventral lateral pars oralis nucleus (McFarland & Haber, 2002).

The extensive sensory control over the ventrolateral frontal lobe includes not only insular but amygdalar integration of sensory data (Barbas et al., 2011). Jones (2007b) emphasizes the importance of the *triangular circuit* that integrates amygdalar projections to both the ventrolateral frontal lobe and the mediodorsal (MD) nucleus of the thalamus, with the third leg of the triangle provided by the MD projections to the ventrolateral frontal cortex (Barbas et al., 2011; Jones, 2007b). This triangular circuit may be particularly important in supporting not only ventral limbic consolidation generally, but resonant memory control of sensory input to guide actions specifically. This guidance would then integrate the amygdalar-MD thalamus memory circuit (Mishkin, 1982) to support *item* or *object* memory consolidation (Aggleton & Brown, 1999; Aggleton & Mishkin, 1986). In this way, there is anticipatory control over the feedback regulation of action, so that the ventral frontal networks are inherently prepared for future events even when the control process involves constraining actions in relation to sensory targets.

Just as the importance of sensory feedback for ventrolateral frontal networks is indicated by the presence of layer 4, the lack of sensory inputs to the mediodorsal frontal cortex (with the exception of certain auditory projections) is indicated by the absence of a layer 4 (Barbas & Pandya, 1984). In fact, cortical layer 3 is also minimally developed in this dorsal division of the frontal lobe, implying that the local cortical processing that marks the activity of sensory areas is missing in the dorsal frontal lobe (Shipp, 2005). Consistent with the functional interpretation of within-pathway projections in consolidation in Chapter 2, Shipp points out that the corticocortical projections between adjacent regions in the frontal lobe (in what we describe as the limbifugal or limbic-to-motor direction) proceed from lower cortical layers (5 and 6) to superior layers (2 and 3), a pattern that in sensory cortex is described as the "feedback" pathway. From the organism's perspective, this may be the feedforward direction of control, both in action as well as in perception. In the frontal lobe, the output of the local cortical columns (layers 5 and 6) is conveyed to the superficial, modulatory layers of the adjacent (toward motor) cortical network, apparently to condition the representation being formed there. Furthermore, this limbifugal, inferior-to-superior laminar projection pattern obtains for both ventrolateral (with a layer 4) and mediodorsal (without layer 4) premotor cortices; the difference is that the ventral frontal lobe is apparently guided by sensory input to layer 4 and the dorsal frontal lobe is not (Shipp, 2005; Hoshi et al., 2005). With the motor program organized in the mediodorsal pathway without the possibility of sensory guidance, the guidance must be provided by the original motive impulse, through feedforward control from the limbic networks of the anterior cingulate cortex.

Thus the several features of the differing cytoarchitectural patterns of the mediodorsal and ventrolateral divisions of the frontal lobe can be seen to be congruent with the projectional (internally guided) and reactive (externally guided) modes, respectively, as proposed by Goldberg (1985a). These specialized cybernetics of action regulation appear to reflect specializations of dorsal and ventral frontal networks for differing information biases, with a limbic dominance in the limbifugal direction of feedforward control in dorsal pathway and a somatic dominance in the limbipetal direction of feedback control in the ventral pathway. Of course, consolidation requires both directions of information flow; the specialization for one direction may imply that the other direction plays a subordinate, supportive role in that pathway.

As noted in a recent analysis by Haggard (2008), internally generated projectional control of actions (feedforward control from the organism's perspective) is the necessary mode of control when requirements are ill-posed, as in a novel or uncertain environment. For humans with considerable reflective capacity, a sense of agency or *will* may be the corollary of this internal mode of control in an uncertain environment. In the last section of this chapter, we review evidence on seizures (*absence spells*) that impair voluntary actions through disrupting frontopolar input to the rostral thalamic reticular nucleus and limbic thalamic nuclei. This circuitry is consistent with the notion of a sense of agency or intentionality associated with projectional control of action in the mediodorsal frontal lobe.

In contrast, when action is externally specified, determined by information already present in the stimulus, actions may be executed by an object-oriented feedback-control action system, in which criteria for adjusting actions are provided by the sensory targets. For Haggard, voluntary actions evolve through stages that are highly consistent with Goldberg's microgenetic analysis of action articulation: from motivations for action, to task and action selection, to a feedback check on the predictive model, and then to final execution. Modern single-unit studies with monkeys as well as noninvasive neuroimaging studies with humans have provided findings that are essentially consistent with Goldberg's proposal for dual modes of action regulation (Haggard, 2008; Passingham, Bengtsson, & Lau, 2010).

## *Motive Cybernetics and the Executive Functions*

Understanding the specific patterns of motor control in frontal networks thus provides insight into both representation, such as the presence of sensory data held in frontal networks, and control, setting the bias toward projectional or reactive modes of action regulation. Goldberg's analysis was important in bringing these specific neural mechanisms of frontal cortex to an integrative model of motor control. This analysis was explicitly developmental or microgenetic, framed in a way that is instructive for understanding the vertical integration of multiple cortical and subcortical systems within the progressive organization of the action plan. The action begins with global arousal of widespread limbic and cortical areas, setting the context for the action plan. The axial and postural muscles provide a platform for the movement (Yakovlev, 1948). Activity in the anterior cingulate as well as basal ganglia circuits provides converging inputs to the SMA to organize the pattern to be actualized in the primary motor cortex. In parallel with this progression, sensory representations are organized within ventrolateral frontal networks to set criteria for adjusting the action process through representations in APA that are transmitted to motor cortex. In parallel with these developing processes, cerebellar inputs to both the APA and motor cortex establish a spatiotemporal pattern for action coordination (Goldberg, 1985).

Goldberg's account of neural mechanisms of motor control requires a cybernetic analysis for understanding how ventrolateral and mediodorsal premotor pathways provide differing and complementary control biases on the microgenetic process. This cybernetic aspect of the model may be important for relating the specifics of motor control to the general problems of motivational control and executive self-regulation seen with human frontal lesions. Motivational control from the limbic base may be essential for both ventral and dorsal frontal pathways. In the dorsal frontal networks, with the dominance of the limbifugal direction of consolidation, the motive control emanates from the dorsal limbic base, perhaps consistent with the visceromotor functions of that base (Neafsey, 1990; Neafsey, Terreberry, Hurley, Ruit, & Frysztak, 1993). In the ventral frontal networks, with the dominance of the limbipetal direction of consolidation, the motive control may operate to constrain and align the sensory feedback for action guidance with the ventral limbic base, perhaps consistent with the viscerosensory function of the insula and ventral limbic networks (Neafsey, 1990; Neafsey et al., 1993). The motive base shapes the cybernetic process in both cases, but with different implications for the consolidation of the action plan.

"Cybernetics," derived from the Greek term for steersman, is a theoretical approach that emerged in the post–World War II era to integrate insights gained from physiology, mathematics, engineering, psychology, and anthropology (Heims, 1991). A basic tenet is that self-regulation results from purposeful behavior that is steered or adjusted in accordance with negative feedback signals (see Rosenblueth, Weiner, & Bigelow, 1943). As pointed out earlier, Teuber's (1964) corollary discharge proposal for frontal lobe function was a highly influential cybernetic model, instructive for Nauta's (1971) conceptualization of limbic set-points guiding the frontal action plan. Teuber's proposal drew explicitly on the cybernetic concept of purposeful anticipation (Heims, 1991). In turn, because Rosenblueth had previously collaborated with Walter Cannon (Heims, 1991), the pivotal connectionist reasoning introduced by

Rosenblueth and associates (1943) was likely influenced by Cannon's physiological yet explicitly cybernetic concept of *homeostasis*.

Homeostasis implies a goal, in the form of a balanced physiological state. *Allostasis* is a more recent conceptual extension of homeostasis wherein an organism actively anticipates challenges to internal parameters that can vary by a wider range (e.g., blood pressure) than those regulated by homeostatic processes. Allostasis emphasizes that an animal anticipates challenges to the system and adjusts both physiological and behavioral systems to meet the challenges (McEwen, 2000; Schulkin, McEwen, & Gold, 1994; Winn, 1995). Allostasis brings the concept of expectant control to the task of visceral regulation, through reasoning similar to the model of expectant control of learning in Chapter 1. Drawing on Nauta's (1970) notion of limbic set-points, it may be useful to formulate allostatic set-points that anticipate future states, and that thereby provide continuous direction to the action plan (Luu & Tucker, 2004).

In a similar sense, the allostatic process can be seen to extend the process of viscerosomatic consolidation to guide memory integration, and thus cognition, prospectively (Tucker & Luu, 2007). To be adaptive, memory and cognition must be motivated, and they must also provide prospective cybernetics, to allow behavior to anticipate events effectively. The cybernetics of dorsal and ventral frontal networks are different, and in many ways opponent, in the control of cognition as well as action. The dorsal frontal regions support a projectional, impulsive form of control, linking visceromotor urges with the feedforward control of ongoing cognition. In contrast, the ventral frontal regions support a reactive, constrained form of cybernetic process, with greater integration of sensory criteria, both visceral and somatic, for feedback control of the cognitive process.

## The Receptive Brain and Action

Cognition can thus be understood to draw on the fundamental mechanisms of action regulation, integrating both motive and motor cybernetics into the process of consolidating cognitive representations. Action regulation also draws on perceptual regulation, with different roles of dorsal and ventral perceptual networks of the posterior brain. Whereas the object sensory capacities of the ventral perceptual streams are critical to the ventral frontal feedback control, there are also integral influences of dorsal perceptual streams in representing the context for action. Understanding the roles of the posterior brain in action regulation may provide insight into the organization of cognition generally, particularly because it explains important aspects of the representational capacities of dorsal and ventral divisions of the cortex in concrete terms of motor control. From such concrete cybernetic explanations, we can easily understand the implications for more complex cognitive processes.

In their influential analysis, Ungerleider and Mishkin (1982) proposed that vision has two cortical pathways that start in primary visual cortex and diverge in the prestriate cortex (see Figure 3-2). A ventral pathway proceeds from the primary visual cortex toward prestriate and inferior temporal cortex. A dorsal pathway, also starting in the primary visual cortex, includes regions of the middle and superior temporal areas, superior temporal sulcal area, and inferior parietal region. Ungerleider and Mishkin proposed that

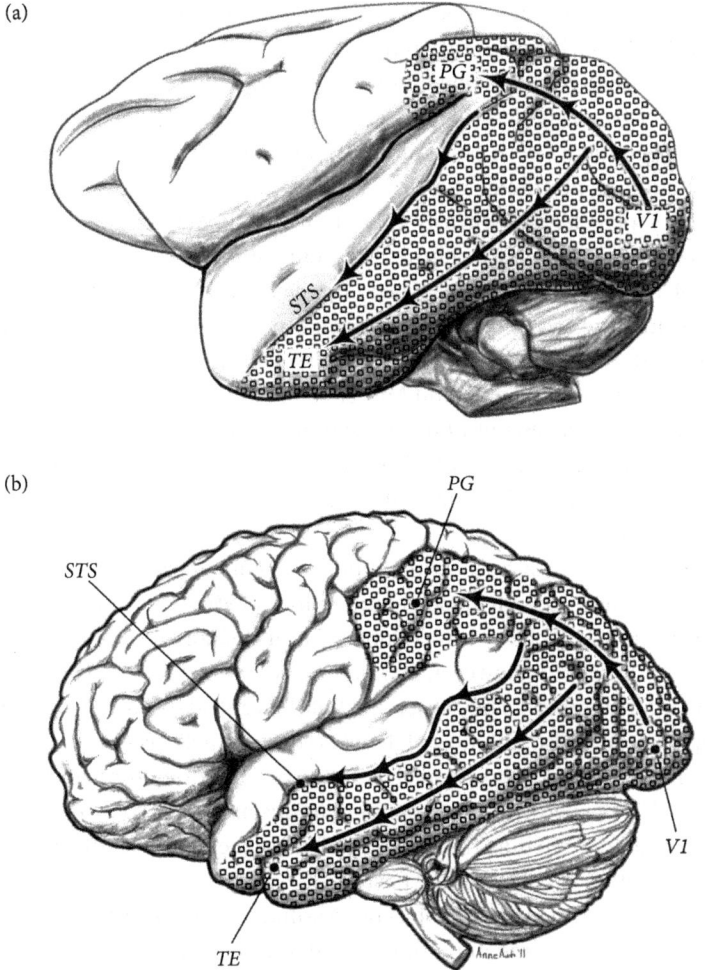

Figure 3-2. The ventral pathway from primary visual area (V1) proceeds down the superior temporal sulcus (STS) and inferior temporal area toward area TE, as well as toward parietal area PG in monkey (top). In the human, a similar ventral visual projection is found (labeled here with the monkey conventions). A considerable region of the human parietal lobe appears to be dedicated to object processing functions of the ventral pathway (see Figure 3-3).

the ventral visual stream is involved in the processing of *object* information for recognition and the dorsal stream for the processing of *spatial relations*. Considerable evidence and theory agrees that the function of the ventral system is in processing object information (Creem & Proffitt, 2001; Rizzolatti & Matelli, 2003). However, characterizing the function of the dorsal system has proved controversial.

### Frames for Action

A prominent alternative proposal to the spatial function of the dorsal stream was put forth by Goodale and Milner (1992). Their proposal comes from the observation that parietal lobe damage involves deficits in the motor domain that cannot be easily described as visuospatial deficits (at least not literally). Specifically, patients with parietal lobe damage display deficits in grasping of objects, such as matching the size of the distance between grasping fingers and size of objects. In contrast, patients with damage to the ventral stream of visual processing show deficits in recognizing the size, shape, and orientation of objects, described as *visual form agnosia*. Importantly, patients with visual form agnosia can accurately reach out and pick up the objects, unlike patients with parietal lobe damage. This dissociation of dorsal and ventral posterior functions led Goodale and Milner to conclude that the dorsal stream is involved in the processing and representation of *action-relevant* information from visual data. In agreement with Ungerleider and Mishkin (1982), Goodale and Milner concluded that the ventral stream participates in making sense of visual objects.

In reviewing these and similar findings, Jeannerod (1994; Jeannerod & Jacob, 2005) agreed with Goodale and Milner (1992) that the ventral system is involved in visual object perception and that the dorsal system is involved in visuomotor functions. However, Jeannerod and Jacob elaborated on this distinction in important ways. They observed that the ventral trend, in addition to perceiving objects, is also uniquely able to perceive the actions of conspecifics. The dorsal trend, rather than just solely emphasizing visuomotor transformations, is involved in the perception of spatial relations as well as contextual representations of action (such as the function of an object or the context in which a particular action occurred in the past).

Thus, in addition to visual object processing, the ventral processing pathway is capable of making meaning out of visual input, a function that Jeannerod and Jacob (2005) described as *semantic* processing. The dorsal system, in contrast, carries out roles of visuomotor representation and visuomotor transformation, described by Jeannerod and Jacob as *pragmatic* processing. Whereas the ventral pathway interprets the actions of conspecifics (an allocentric orientation), the dorsal pathway uses egocentric spatial information as the basis for the generation of actions.

In line with the notions of egocentric versus allocentric orientations, Jeannerod and Jacob delineated a vector of operation for each system. The ventral system's vector is from the world toward internal representation, wherein the goal is accurate and enduring representations (objects) of external visual stimuli. That is, the vector of operation in the ventral stream is to confirm representations of external events. In contrast, the dorsal system's vector of operation is intentional or projectional. An example used by Jeannerod and Jacob is reaching to grasp an apple. The goal is the representation of the apple in the hand. This representation is not the actual state of the world but rather how the world is expected to be (the goal).

Thus concepts of the function of the dorsal pathway have evolved from purely visuospatial toward more complex forms of visuomotor process. The dorsal system has been further divided to account for different aspects of visuomotor functions (Creem & Proffit, 2001; Jeannerod & Jacob, 2005; Rizzolatti & Matelli, 2003), based both on anatomical and functional evidence. Anatomically, the parietal lobe is divided into superior and inferior lobes by the intraparietal sulcus (see Figure 3-3). In primates, the superior parietal lobe has very limited visual inputs, originating mainly from area V6, which also projects to dorsolateral and mediodorsal premotor areas. In contrast, the inferior parietal lobe receives input from area V5/MT, the classic dorsal trend of Ungerleider and Mishkin (1982), and the temporal lobe (Rizzolatti & Matelli, 2003).

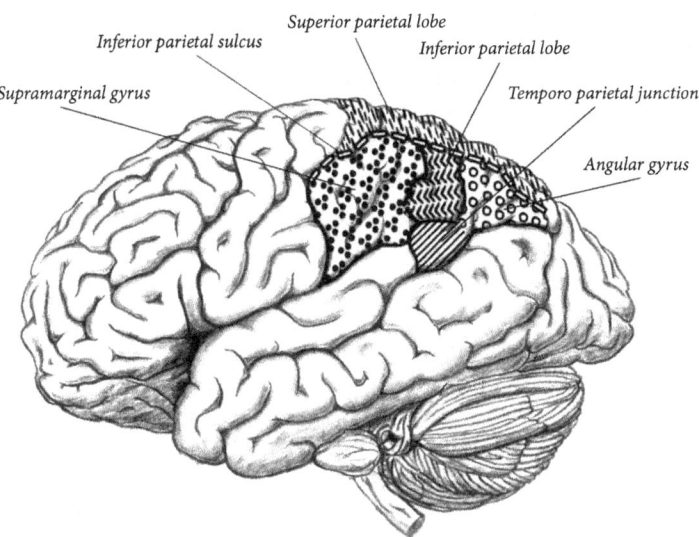

Figure 3-3. The dorsal division of the parietal lobe (above the dotted line) extends to the medial surface, whereas the ventral division extends to the temporal lobe, and the temporoparietal junction.

Functionally, lesions to the superior parietal lobe result in *optic ataxia*, wherein movements toward targets, particularly the periphery of the visual field, are poorly coordinated (Pisella et al., 2009). Patients with optic ataxia do not accurately reach for targets, have difficulty orienting their reach with the axis of the object, with a grip size that is not matched to the object. Interestingly, when these patients are presented with familiar objects, their reaching and grasping performance is greatly improved.

In contrast, lesions to the inferior parietal lobe result in *ideational* or *ideomotor* forms of apraxia. In ideational apraxia, patients have difficulty in forming movements based on objects (such as how to use a tool), even though they can imitate the use of the object when shown. In contrast, patients with ideomotor apraxia know for what and how to use an object, but they cannot execute the actions, as readily seen when they are required to pantomime an action. When presented with the actual object, they can execute the correct action. De Renzi, Liotti, & Nichelli et al. (1987) noted that ideational apraxia often involves lesions to the posterior aspects of the inferior parietal lobe, including the temporoparietal junction. Lesions that produce ideomotor apraxia, on the other hand, are often located in more anterior regions of the inferior parietal lobe, within the supramarginal gyrus.

## *Frames for Meaning*

It is apparent that the ventral to dorsal dimension of perceptual function in the posterior brain involves a complex set of differentiated cognitive capacities. Visual information is processed to delineate specific objects within the ventral-most aspect of the lateral hemisphere. These objects comprise semantic representations, in that the object has meaning separate from the context. Meaning is not entirely "objective" however, because it must engage a basis in motivational significance.

Considering functional specialization more dorsally, toward the superior temporal and inferior parietal lobes, information about objects in the world becomes integrated with representations of actions as well, with the result that actions also constitute the basis of object meaning (Rizzolatti & Matelli, 2003; Jeannerod & Jacob, 2005). This explains why lesions of these more dorsal networks produce ideational and ideomotor ataxias. Further dorsally, toward the superior parietal lobe, there is little or no representation of object information. Action, primarily of the reach and grasp type, is based predominantly on visuospatial information derived from the periphery. This distinction of spatial control of action can be extended to the medial parietal lobe (area 7), where lesions that encroach into this region also produce optic ataxia (Cavanna & Trimble, 2006).

The motivational significance for the operations of the dorsal and ventral contributions to perception and action can be found to be supported by the limbic bases of cortical networks, linked to posterior perceptual as well as to anterior motor regions. Vogt et al. examined the connectivity of ventral as well as dorsal regions of the posterior cingulate cortex (PCC) that integrate perceptual information with the visceral functions. It is the visceral functions that provide evaluation for motivational significance (Vogt, Vogt, & Laureys, 2006). The ventral limbic networks, including piriform cortex, provide input to the retrosplenial region and ventral posterior cingulate cortex. Links to limbic bases of the frontal lobe are seen for both ventral and dorsal divisions of the PCC. Extensions of the monkey tract tracing studies to humans were made through voxel-wise correlations of PET metabolic activity across persons. Voxels in the dorsal PCC showed metabolic correlations with widespread regions of dorsal parietal and frontal cortex. Voxels in the ventral PCC showed correlations with not only ventral limbic areas but also with both medial orbital frontal cortex and subgenual anterior cingulate cortex (ACC). This connectivity is consistent with the interpretation that important functions of the ACC involve not only connectivity of the dorsal ACC with mediodorsal frontal lobe, but also ventral limbic input to the subgenual (ventral) ACC in functions such as error-monitoring and discrepancy evaluation (Luu, Tucker, Derryberry, Reed, & Poulsen, 2003). Considering the role of PCC in integrative memory as well as visceral evaluation, Vogt et al. propose that the PCC is important to ongoing self-evaluation within the behavioral context (Vogt et al., 2006).

The representation of self in context is an interesting theme in functional analyses of the cortex, and there have been suggestions of differential perspectives provided by the dorsal and ventral divisions. From an inherently egocentric reference frame in the dorsal pathway there may be a shift toward an allocentric frame in the ventral pathway (Creem & Proffitt, 2001). The allocentric, externally oriented reference frame permits unambiguous, redundant, and enduring representation of external objects (Jeannerod & Jacob, 2005). More dorsally, within the inferior parietal lobe, where action also imbues meaning to object representation, both allocentric and egocentric reference frames are employed. This allows for actions to be object relevant as well as egocentrically guided. More dorsally, an egocentric reference is used for visuomotor transformation, requiring only limited visual inputs. Representations in the dorsal system may thus be less concerned with certainty or invariance of the external world and more concerned with the state of the world, particularly in relation to valued goals, as this state is affected by personal action. From these functional representations of the self in relation to perception and action, the vectors of operation described by Jeannerod and Jacob (2005) may be understood more fully: from the mind to the world for the dorsal system and from the world to the mind or the ventral system.

## *Complementarity of Pragmatic and Semantic Modes*

These vectors of information flow in the posterior brain are thus congruent with the projectional (toward the world) and reactive (from the world) modes of controlling action in the dorsal and ventral frontal lobes. The complementarity of the two divisions of posterior cortical networks to those of the frontal networks thus allows the percepts of sensory integration to mirror the concepts of action regulation. Consistent with the divisions between dorsal and ventral regions of limbic cortex observed by Vogt (2005), the anatomical connections are segregated for the dorsal and ventral networks linking posterior and anterior regions of the neocortex. Rizzolatti and Matelli (2003) noted that cortex of the inferior parietal lobe, particularly that in the intraparietal sulcus, has strong connections with the ventral premotor regions of the frontal lobe. In a complimentary fashion, the superior parietal lobe has connections with the dorsal premotor regions as well as the supplementary motor area (Dum & Strick, 2005). The segregation of the dorsal and ventral divisions of neocortex appears fairly complete. For example, there are no known projections from the inferior parietal lobes to the dorsal premotor area or to the supplementary motor area.

Through these linked anterior-posterior network alignments, both sensation and action are elaborated through the dual and opponent modes of dorsal and ventral corticolimbic cybernetics. These modes of action regulation may underlie the organization of the frontal executive function through elaboration of dual routes to limbic networks (W. J. H. Nauta, 1964). Through its boundaries in both visceral and somatic functions, cognition can thus be understood as emergent from the sensorimotor substrate, rather than resident in specialized and arbitrary cognitive modules. The dorsal and ventral contributions are in many ways opponent. They organize opposite modes of control on elementary action regulation, and opposite perspectives on the flow of information with the external context.

The opponent control modes of the dorsal and ventral divisions of the cortex appear to be aligned with opposite directions of information flow in the process of consolidation. The egocentric or self-to-world direction of cognition in the dorsal division of the posterior brain described by Jeannerod may arise through a literal direction of information flow in the process of consolidation, manifesting a dominance of limbic influence, mediated through the limbifugal (5 to 2-3) direction of interregional laminar connections. This may be congruent with a visceromotor dominance in the dorsal networks. In perception as well as action the motivation for the representation of the context is the egocentric organization of goal-oriented action.

In this way the cognitive representation formed in the distributed representations of the cortex may not be separable from the motive cybernetics that generate it. The goal opportunities are represented as integral components of the context for action. Jeannerod's description of cognition in the dorsal division as *pragmatic* fits with this integral, visceromotor, motive direction.

In contrast, Jeannerod describes the cognition of the ventral stream as *semantic*. This may fit well with the allocentric perspective of the ventral networks, organizing cognitive representations in relation to the invariant, object realities of the world. Such a perspective could emerge from a limbipetal bias in the information flow in consolidation, modifying cognition to fit the world, mediated by the specific projections between networks, from 2-3 to the granular layer 4.

In linguistics, "semantics" refers to the invariant, objective meanings of words within the culture, which each individual must adhere to. In contrast, pragmatics refers to the more opportunistic and adaptive use of language by the individual. Each of these modes of language may reflect a more generic mode of meaning. Each may be understood as having a different relation to somatic and limbic constraints, in the varying degrees shaped by the dominant direction of consolidation.

Dominated by the limbifugal direction of influence, likely through recurrent interactions, the consolidation in dorsal networks begins with the highly distributed representations at the limbic core, such that a more holistic and widespread pattern of connectivity then supports the cognitive representation of the context for action in the dorsal networks. In contrast, with greater influence by the limbipetal direction, the consolidation in ventral networks may be more differentiated in relation to the discrete elements of the sensory mirror of the world, with inhibitory specification of sensory objects, either directly in the posterior networks or linked to the perceptual objects through the projections to granular layer 4 of the ventrolateral frontal networks to provide feedback guidance of the action plan.

To the extent that the general process of cognition emerges from the brain's networks of action regulation, it may be possible to clarify specific qualities of cognitive structure and content in relation to the connections and processing in cortical networks. These qualities may include the breadth of cognitive representation as well as the balance of egocentric motives versus fidelity to the constraints of perceptual reality. The networks of the cortex are constrained by the anatomical structure and physiological data at the dual boundaries, marked by the visceral limbic function at the core of the hemisphere and the somatic networks of primary sensory and motor cortices mapping the interface with the world. Cognition, and the memory consolidation that underlies it, must arise through the continuing and dynamic arbitration of these constraints. This arbitration takes projectional and reactive forms, emphasizing the advantages of the limbifugal and limbipetal directions of viscerosomatic consolidation, respectively.

This general organization of the cortex is more or less consistent across mammals, such that the process of action regulation in related species such as primates is highly informative for action regulation and cognition in humans. However, there are interesting clues to unique human capacities of attentional monitoring and volitional control to be gained from examining the unique association networks of the human frontal pole. These frontal polar networks are more fully elaborated in humans than in other primates (Semendeferi,

Armstrong, Schleicher, Zilles, & Van Hoesen, 2001), possibly suggesting insights into the hierarchic properties of executive control that may become complex and powerful in the course of human learning. Remarkably, the functions of the human frontal pole appear possible only through coordination with specialized circuitry of the thalamus, creating a kind of dynamic frontothalamic resonance that may support the highly evolved capacities of human conscious self-regulation.

## Parallel Architectures for Attention and Intention

We have seen at the beginning of this chapter that careful study of the clinical observations on human frontal lesions has emphasized the importance of self-regulation to the executive functions. Self-regulation requires a balance of motivational control with socially sensitive inhibitory control in the ongoing behavioral process. In understanding the unique human abilities in self-regulation, it is interesting to examine the expansion of the human frontal lobe, specifically the frontal pole (BA 10), to understand how this region interacts with multiple cortical and subcortical systems in regulating the behavioral process.

It is often assumed that the human frontal lobe has expanded in recent evolution compared with other cortical regions. However, direct anatomical comparisons with extant monkeys and apes have suggested that only the frontal pole (BA 10) has expanded in humans, and it has expanded considerably (Semendeferi et al., 2001). In this section, we consider evidence that the capacities of the human frontal pole are closely linked to thalamic circuits, specifically those of the limbic thalamus. Whereas certain lateral frontal networks engage the thalamic mechanisms of the searchlight of attention, the networks at the frontal pole appear to engage more generic controls on the intentional monitoring of actions. Together, attentional differentiation and intentional monitoring form a foundation for the conscious control of action regulation.

Studies in the macaque have shown that each frontal lobe includes connections with both hemispheres, arranged in interdigitated columns that would allow local frontal networks to span bilateral perceptual fields, and perhaps to exert integrated control over both hemispheres (Goldman-Rakic & Schwartz, 1982). In studying the expansion of BA 10 in hominoids, Semendeferi et al. point out that in gibbons the only analogous structure to BA 10 is the orbital frontal cortex (Semendeferi et al., 2001). This may imply that the orbital region, with its integration of limbic and autonomic viscerosensory input (O'Doherty, Kringelbach, Rolls, Hornak, & Andrews, 2001; Rolls, 2000), provides the foundation for the elaboration of the frontal pole in hominoid evolution. On the other hand, connectional studies in macaque have suggested that BA 10 includes connections with mediodorsal as well as orbital (ventrolateral) frontal regions

(Barbas & Pandya, 1989), indicating that the dorsal division of the frontal lobe also became elaborated at the frontal pole with the expansion of frontal cortex in macaques.

In considering the relative importance of the frontal pole in human evolution, it is interesting to note that the endocasts of the *homo floresiansis* hominins (hobbits) show an elaborated frontal pole compared to that of the *homo erectus* (apparently the hobbit's closest relative, see Figure 3-4). This elaboration includes vertical parasagittal convolutions at the medial frontal pole, separated from lateral frontal "swellings" by large sulci (Falk et al., 2005). In light of the evidence of frontopolar control of intentionality reviewed below, it is interesting to speculate on the cognitive capacities, including intentionality and self-awareness, allowed by such frontopolar expansions in these small creatures.

## At the Human Frontal Pole

The expansion of BA 10 in humans is considerable in quantitative terms, even in relation to total brain volume. Semendeferi et al. plotted BA 10 volumes against total brain volume for each of the hominoids in their small sample: the BA 10 volume for humans was twice that expected from the hominoid regression line assuming a linear increase with evolutionary complexity (Semendeferi et al., 2001).

In recent years, cognitive neuroscience studies have attributed a number of sophisticated functions to the human frontopolar cortex, including processing self-relevant information, multitasking, prospective memory, and interpreting others' perspectives (mentalizing) (Gilbert et al., 2006). In organizing their meta-analysis of neuroimaging studies, Gilbert and associates considered evidence of functional differentiation within BA 10; this differentiation in turn raised questions of the interaction of BA 10 with adjacent frontal networks.

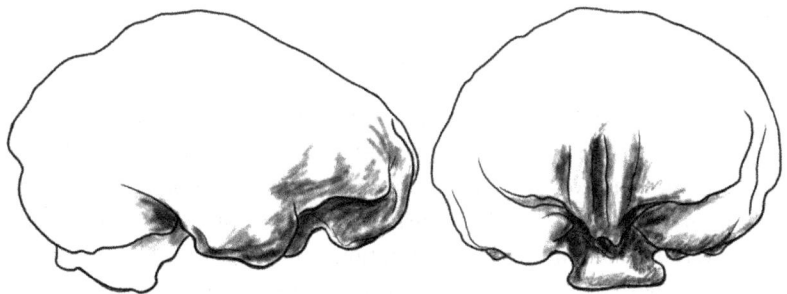

Figure 3-4. Enlargement and extension of orbital frontal cortex into frontal pole in *homo floresiansis*. The midline orbital sulci appear to extend superiorly in the frontal pole bilaterally. After Falk et al. (2005).

Tasks requiring episodic memory appeared to engage more lateral regions of BA 10, whereas tasks requiring coordinating of multitask demands engaged anterior medial as well as lateral regions of this area. Both emotional processing and mentalizing required more medial areas of BA 10, judging from the functional imaging activations, particularly when mentalizing required emotional judgments (Gilbert et al., 2006). Gilbert and associates considered a hierarchic organization in the medial frontal lobe. Emotional evaluation engages the anterior cingulate cortex, whereas more complex reflections on emotions, including the awareness of one's own emotions and estimates of those of others, engage more rostral networks within medial BA 10.

To appreciate the role of the human frontal pole within the distributed networks of the frontal lobe, or within the cortex generally, the hierarchic relationships considered by Gilbert, et al. may be extended to the general network architecture such as illustrated in Figures 2-7 to 2-9 of Chapter 2. BA 10 is a heteromodal network, interposed between limbic regions (the ACC in the mediodorsal direction and the orbital and insular cortices ventrally) and the premotor areas. Its internal functional differentiation appears to reflect its position within these multiple frontal networks, apparently integrating between the emotional and interpersonal functions represented in the medial areas and the memory and cognitive task requirements represented in more lateral frontal cortex. If we take the Floresian frontal pole as an intermediate form, BA10 in humans would include considerable influence from orbital frontal cortex, including the regulation of the reticular nucleus of the thalamus, and thus thalamocortical relations generally (Tucker, Brown, Luu, & Holmes, 2007; Tucker & Holmes, 2010).

Thus there appear to be both ventral limbic (ventrolateral frontal) and dorsal limbic (mediodorsal frontal) networks engaging respective divisions of the frontal pole. Although these cortical links are obviously important to executive self-regulation, the human frontal pole appears to have unique capacities for recruiting functions of subcortical control systems, particularly those of the thalamus.

## *Concepts of Thalamic Regulation*

The network representations instantiated at the frontal pole appear to integrate multiple neural control systems, including brainstem, limbic, and thalamic systems, within the goal-directed action plan. The action plan can be seen to integrate *goal concepts*, engaging integral neurophysiological controls as well as sensory, motor, and motive representations, thereby assembling representations of the regulatory functions. The orbital frontal lobe of mammals has a particularly important role in organizing thalamic control of the cortex generally. The human frontal pole seems to have elaborated this capacity for thalamic control in important ways.

As reviewed above, the thalamic recruiting responses involving the intralaminar nuclei of the thalamus may be an important candidate for the regenerative, self-organizing neurophysiological activity that is necessary for learning and memory consolidation. Although widespread cortical regions are interconnected with the intralaminar as well as more specific thalamic projections, the frontal lobe has long been known to play a unique role in regulating recruiting responses. When Morrison and Dempsey searched for cortical sites in the cat brain where electrical stimulation caused thalamic recruiting responses, these sites were limited to the orbital frontal cortex (Morrison & Dempsey, 1942, 1943).

In addition to the intralaminar thalamic nuclei, the thalamic reticular nucleus is important to regenerative neurophysiological responses, such as sleep spindles and epileptic seizures (Steriade & Amzica, 2003). The thalamic reticular nucleus, and its gating of sensory input, has also been found to be controlled through projections from the orbital frontal lobe in cats (Yingling & Skinner, 1976, 1977).

Although the human frontal lobe is elaborated in important ways over that of the cat, seizure responses have been observed in human patients that may parallel thalamocortical recruiting responses (Steriade, 2003, 2004; Steriade & Amzica, 2003) and that include discharges that engage regions of the frontal cortex specifically (Holmes, Brown, & Tucker, 2004). *Absence seizures*, formerly termed petit mal epilepsy, involve a brief loss of consciousness, at which time large spike-wave discharges appear with maximal amplitude over frontal regions (Gloor, 1978; Rodin, 1999). Mapping these discharges in a series of absence patients with dense array (256-channel) EEG showed that the slow waves were maximal over broad frontal and temporal areas, but that the positive spikes of the spike-wave complex were focal over the pole of the frontal cortex (BA10) (Tucker et al., 2007). Interestingly, these spikes typically showed a rapid propagation from BA 10 inferiorly toward orbital frontal cortex (Tucker et al., 2007).

One implication of the frontal involvement in the spikes of absence discharges may be that the brief loss of consciousness and voluntary control during the absence spell may be due to the disruption of frontal lobe control over the thalamus, including the reticular nucleus and its inhibitory control over thalamocortical projections. The classical evidence on frontal stimulation of recruiting responses suggests there may be engagement of the intralaminar nuclei as well. Animal models of spike-wave discharges have emphasized that the reticular nucleus and thalamocortical neurons comprise a circuit with the timing properties that could generate the 3/s cycles of these discharges (Futatsugi & Riviello, 1998). At the same time, cortical activity remains critical to the onset and maintenance of the seizure (Steriade, 2003), implying that the frontal pole engages the thalamus to create a complex corticothalamocortical loop.

In considering the functional significance of frontothalamic control, it may be important to note that whereas limbic seizures create a retrograde amnesia that impairs ongoing memory and orientation to context, absence seizures do not. The absence patient is able to continue a conversation after the spell, with good orientation to the conversation and the context. The implication may be that the frontothalamic circuitry engaged by the absence seizure does not impair ongoing memory for context (as limbic seizures do). Rather, this circuitry of the absence spell appears to control a specific component of cerebral self-regulation that is critical to voluntary control and conscious intentionality (Tucker et al., 2007). It is interesting to recognize that a role in intentional control of behavior for this frontothalamic circuitry, engaging the rostral thalamic reticular nucleus, anterior nuclei of the thalamus, and cingulate gyrus, would be consistent with the egocentric mode of dorsal networks in action regulation pointed out by Jeannerod and Jacob (2005).

## Differentiated Attention Versus Integrative Intention

There may be insights into the neural mechanisms of cognition and learning if we could differentiate between expectancies formed through limbic representations of the ongoing, implicit context in contrast to expectancies formed through frontothalamic control of a more explicit, conscious control of cognition. The fact that nonhuman primates do not exhibit absence seizures (Niedermeyer, 2000) may suggest that the human frontothalamic circuitry may be particularly excitable, particularly in the early developmental years when absence spells first appear. The fact that absence seizures disrupt intentional behavior while leaving ongoing memory intact suggests that the excitability of this frontothalamic circuit may be relevant to unique human capacities in conscious intentionality (Tucker et al., 2007; Tucker and Holmes, 2010).

Although human frontal lobe may be unusual, it appears to elaborate basic features of the primate brain. Recent studies of frontothalamic connectivity in the macaque may be interpreted to suggest that the selective attention to specific objects and sensory modalities is supported by lateral frontal networks, connecting preferentially to the sensory nuclei of the thalamus, whereas more general aspects of the arousal state and motive engagement are supported by specific connections from the frontal pole to the rostral thalamic reticular nucleus and then the limbic nuclei of the anterior thalamus. Zikopoulos and Barbas (2006) examined projections from frontal cortex to the thalamic reticular nucleus (TRN) and observed that the frontal pole (BA10) projects specifically to the rostral TRN. The rostral TRN then projects to the anterior (limbic) nuclei of the thalamus, including the anterior medial, anterior dorsal, and anterior ventral nuclei, which are in turn interconnected with the precuneus, anterior cingulate, and posterior cingulate regions of the limbic lobe, respectively (Zikopoulos & Barbas, 2006). The projection of BA 9 to the anterior dorsal nucleus of the thalamus, which projects to the anterior cingulate cortex, is consistent with a dorsolateral frontal engagement of anterior cingulate control in support of the motor initiative underlying intentional behavior.

In contrast, other frontal areas (BA13, 46) project to more caudal and lateral areas of the TRN, and these then connect to more specific sensory and motor nuclei of the thalamus, where they interdigitate with projections from sensory cortex. Whereas projections to the thalamus from sensory cortices are exclusively from layer 6, the projections from frontal areas include those from layer 5 as well (Zikopoulos & Barbas, 2006), suggesting that there may be unique regulatory influences on the TRN and thalamus (including the thalamic matrix projections) applied by the frontal lobe that are not found for other cortical areas.

Given the inhibitory control applied by the TRN to the thalamocortical projections, the pattern of projection from lateral frontal networks to specific sensorimotor nuclei would be consistent with ventrolateral frontal control over the searchlight of selective attention (Crick, 1984; Crick & Koch, 1992). In contrast, considering the specificity of absence seizures for engaging the frontal pole in the human patients, it may be that the more dorsal frontopolar control over the rostral TRN and limbic thalamic nuclei may be more important to the general state of arousal and voluntary intention, rather than to selecting specific sensory modalities or their contents (Tucker et al., 2007). To the extent that human consciousness is differentiated by attention and integrated by intention, we may find that the differing forms of seizure disorders produce an instructive fractionation of otherwise opaque components of the frontal executive function (Tucker & Holmes, 2010).

## Complementary Opposition in the Dual Limbic Systems

In this chapter, we began with a consideration of executive self-regulation, and adopted Nauta's view that understanding the motivational control from limbic networks is critical to understanding the frontal lobe's contribution to human cognition (W. J. H. Nauta, 1971). Furthermore, the dual routes from frontal to limbic regions imply that dorsal and ventral limbic networks comprise different substrates of executive control (W. J. H. Nauta, 1964). By examining the evidence on motor control, we found that the duality of frontolimbic anatomy is associated with a duality of control processes, such that the projectional and reactive cybernetic vectors apply differing and complementary modes of self-regulation to the ongoing organization of behavior. From this emphasis on behavior, and specifically action regulation, we reasoned that the dual cybernetic modes can explain important features of the contributions of the dorsal and ventral divisions of the cortex to human cognition. These contributions include the differential modes of organizing perceptual information in the posterior brain, which are closely complementary to the differential modes of action regulation in the frontal brain.

With unique circuitry engaging limbic and diencephalic structures as well as the basal ganglia, the dorsal and ventral networks could be described as dual limbic systems. Then again, given the integral role of these limbic networks in the organization of consolidation and function in their associated neocortex, it becomes difficult to separate limbic from neocortical networks on functional grounds. What does seem clear is that the dorsal and ventral systems are not only complementary, but opposed. The projectional, feedforward control mode of the dorsal frontal lobe is opposite to the restrictive, feedback control from the ventral frontal lobe. These are balanced in opposition, in a way that is complementary for the overall task of action regulation. Neither of these control modes is adequate without the other, and each seems to have evolved to elaborate its unique cybernetic properties because its operation is balanced by its complement.

The dorsal-ventral complementary opposition seems to extend from control to representation. The configural representation of the context for action in the dorsal parietal lobe seems to require an opposite structure of information representation than the delineation of perceptual objects, and differentiation of each object from its embedding surround, in the ventral perceptual streams. Similarly, the extension of action regulation from simple motor control to cognitive representation in frontal networks appears to organize a fluid and holistic concept of the goal-oriented impulse to action in mediodorsal frontal networks that is inherently opposed to the more differentiated, sequenced, and criterion-based reasoning in the ventrolateral frontal networks. For both perception and action, the direction of consolidation seems to reflect the dominance of visceral or somatic constraints. In dorsal networks, the elaboration of the impulse seems to reflect dominance of the visceral function, as motives frame actions directly. In ventral networks, the constraints of sensory models of the environment cause an opposite, somatic-dominant and limbipetal, direction of influence in the consolidation process.

For both posterior and anterior regions, the dorsal and ventral networks are able to organize different patterns of representation, largely because they are segregated from each other. Yet their points of contact are critical: in limbic and heteromodal networks and in the two regions where their unique cytoarchitectonics are mixed, including the temporal-parietal-occipital junction in the posterior brain and the lateral surface of the frontal lobe anteriorly (Eidelberg & Galaburda, 1984). Relying on network constraints at these points of contact, forms of cognitive structure must be organized across the hemisphere in ways that are both differentiated and hierarchically integrated (Werner, 1957).

The dorsal-ventral visceral motive (visceromotor vs. viscerosensory) complementarity appears to operate in an opponent mode, whereas the somatic (anterior motor vs. posterior sensory) balance may reflect a more congruent complementarity. The dorsal posterior cortex's organization of the configural context for action is inherently congruent with the holistic, impulsive action plan of the anterior premotor networks. The ventral posterior cortex's organization of the significant objects of perception is congruent with the ventral frontal specification of the criterial objects for the feedback control of action. The segregation of the somatic sensory and motor networks means that there is duality in both dorsal and ventral somatic domains. The mirroring of perceived action patterns in frontal networks illustrates the duality of somatic representations: perceiving another's actions involves covertly synthesizing those actions. In language, the differing constraints of expressive and receptive regions of the ventral left hemisphere do seem balanced with differing if not opposite constraints (Tucker, Luu, & Poulsen, 2009). Thus in cognition and sensorimotor function generally, the somatic input-output specialization seems to be organized around a congruent duality or complementarity, in which the representational forms (holistic configuration dorsally and differentiated objects ventrally) are similar between anterior or posterior networks. In contrast, the dorsal-ventral complementarity is more clearly opponent, both in representation (holistic vs. differentiated) and control (feedforward vs. feedback).

## *Integral Mechanisms of Primitive Cognition*

In human and even primate research the cognitive processes are often fairly complex. In contrast, the cognition examined in most rat learning studies is more primitive, including the hedonic and aversive/discrepant expectancies that can be inferred from the learning results examined in Chapter 1. However, these species differences may reflect differing levels of complexity in the manifestation of similar underlying mechanisms of learning and cognition.

The evaluation of stimuli in relation to hedonic expectancies in learning experiments is a primitive form of cognition that may reflect the learning/expectancy bias of the dorsal division of the mammalian brain. This form of cognition appears to support the slow, asymptotic learning process described by Gabriel and associates (Gabriel et al., 1996; Gabriel, Vogt, Kubota, Poremba, & Kang, 1991). Lesions of dorsal cingulothalamic networks in rabbits (including the posterior cingulate cortex, hippocampus, and anterior ventral nucleus of the thalamus) were specific in impairing the context-updating capacity, described by Gabriel et al. as slow learning because it supported a gradual acquisition of the behavior. The more focused cognition in anticipation of aversive (threat) events or need-driven (hunger) states is specifically impaired by lesions of the anterior cingulate cortex, mediodorsal thalamus, and amygdala, reflecting the ventral limbic input to the anterior cingulate. This ventral limbic input may be important to supporting the fast learning process, in which the context model is inadequate and sustained activity in working memory is required for organizing a new response to the discrepant predictions.

The elementary mechanisms of rabbit learning described by Gabriel et al. may thus reflect the fundamental limbic-diencephalic circuits underlying the dorsal and ventral systems of cognition and action regulation characterized in primate and human studies. Although the memory and cognition of rabbits may be primitive, it appears based on the same duality of corticolimbic mechanisms of representation and control that determines human cognition. And in both primitive and more complex forms, memory and cognition appear to be organized in their dorsal and ventral extents through opponent complementarity.

The cybernetics required for holding and maintaining the hedonic context model of the environment within dorsal networks—and for updating this model to (generally congruent) new input—appear to have evolved from the impulsive mode of action regulation associated with the visceromotor function of the dorsal brain. Visceromotor functions act directly on their visceral targets. Similarly, the projection of motive goals from the limbic base outward to more differentiated neocortical networks may provide the expectant priming of the dorsal context model.

This is an opposite cybernetic bias from the constraint and feedback mode of control that operates in ventral networks. This mode seems closely linked to viscerosensory functions of the ventral limbic networks, where the control is not a direct action (as with visceromotor) but an indirect process, as the visceral response sets up the criterion for evaluating ongoing actions. The ventral limbic networks appear to have evolved to apply feedback guidance from somatic sensory as well as visceral sensory data, such that the elaborated layer 4 (frontal granular cortex) becomes integral to the ventrolateral frontal contribution to reactive control of the motor plan. This mode of feedback control seems well suited for disrupting the context model, for focusing attention to accommodate internal representations to fit discrepant and therefore potentially threatening sensory events, and for sustaining working memory in order to organize a new cognitive model of self-in-context (Jeannerod, 1994; Tucker & Luu, 2007; Vogt et al., 2006).

## Self-Regulation Through Activation and Arousal

Opponent complementarity thus seems to arise from the working of fundamentally different mechanisms of regulating neural activity in order to modify the connectivity in cerebral networks. We think the theory of primitive cybernetics of action regulation developed in this chapter can be integrated with the theory of primitive cybernetics from Chapter 1 that described spatiotemporal tradeoffs in controlling neural activity. Specific strategies of action regulation may emerge from primitive control systems that modulate neural activity over time: the habituation and redundancy biases of tonic activation and phasic arousal.

The cybernetics of memory consolidation in ventral corticolimbic networks appear to achieve unique capacities of *object specification* in perception. Perhaps through similar mechanisms, these cybernetics also achieve unique capacities in the feedback control of action, where discrete perceptual targets are linked with successively more accurate constraints of the action plan. We propose that both these properties emerge through corticolimbic networks that have evolved to self-regulate through the primitive cybernetics of a redundancy bias. The ventral limbic frontotemporal networks include the triangular circuit comprising the amygdala input to the mediodorsal nucleus of the thalamus, the mediodorsal thalamus projections to the frontal lobe, and the direct amygdala-frontal projections. These ventral limbic circuits are closely related to the subgenual anterior cingulate and not only orbital but also ventrolateral frontal cortex in humans. They also appear to include the projections from the basal ganglia, through pallidal and thalamic routes to mediate sequencing and inhibitory specification of actions in frontal networks (Gurney, Prescott, & Redgrave, 2001; Helmich et al., 2005). The maintenance of ongoing neural activity over time through the subcortical circuits underlying ventral limbic control, including the brainstem and mesencephalic dopamine and acetylcholine projection systems, appears to generate a focused, restricted cell assembly, and thus a tight pattern of spatial representation in cortical networks. With the integral spatiotemporal trade-off of neural activity control through the redundancy bias, active neural assemblies are maintained in time through restriction in representational space. This focus and restriction may be well suited to the differentiation of objects through inhibitory specification, to feedback control of action sequences, to a narrow attentional focus under threat or strong need state, and to the sustained maintenance of working memory to support fast learning of new behavioral patterns under conditions of discrepancy and uncertainty.

An opposite, and complementary, cybernetic bias appears to be supported by the habituation bias of the dorsal networks. The primitive learning mechanisms in the animal studies for gradual learning in support of the context model were centered on the posterior cingulate cortex, with subcortical projections of the classical Papez circuit, including the anterior ventral nucleus of the thalamus and the hypothalamus (Gabriel et al., 1983). In primates and humans, this dorsal limbic circuitry engages the parietal lobe, dorsal regions of the anterior cingulate, and mediodorsal frontal lobe (Pandya, Seltzer, & Barbas, 1988; Vogt et al., 2006). As neural activity in dorsal networks is modulated by the habituation bias, apparently through the influence of noradrenergic and serotonergic neuromodulators, the activity in a given cell assembly is transient. The limbic motives are allowed to dominate the consolidation process, through impulsive, projectional feedforward control. As a result of the inherent spatiotemporal trade-off of the habituation bias, over a given interval of time the rapidly dissipating and reconfiguring neuronal assemblies of the consolidation process allow the effective working memory capacity to engage a wide range of new information from ongoing perception and exploratory behavior. The result is a holistic and expansive context model, well suited to the hedonic expectancy and configural control of action regulation in the dorsal cortical networks under conditions of competence and success.

## Summary

We have seen in this chapter how complex abilities in self-regulation, often described as executive functions, can be understood to emerge through the human elaboration of more elementary capacities for motivational control of action. The capacity for working memory is integral to complex plans, and to complex choices in self-control. The frontal lobe's contribution to working memory is closely linked to action regulation, and to the motive base of action in the limbic networks. As recognized by Nauta, there are dual limbic bases of the frontal lobe, one dorsal (anterior cingulate) and one ventral (orbital, insular, anterior temporal).

The dorsal and ventral divisions of the frontal lobe operate in tandem with their dorsal and ventral counterparts in the posterior brain. Linked in a kind of opponent reciprocity, the frontal-temporoparietal networks of each division provide coherent organization to action regulation. The frontal projectional control of dorsal networks appears well integrated with the parietal configural control, grounded as it is in an egocentric reference frame. The frontal reactive control of ventral networks appears well integrated with the temporal lobe object representations that help set the targets for criterial control of action. The shift of attentional perspective toward the allocentric frame within ventral networks may be a literal basis for objectivity.

Understanding the mechanisms of control and representation in action regulation may provide a basis for understanding the control of cognition generally. The motive biases from limbic networks may suggest how cognition emerges from visceral control. The feedforward mode of control, for both frontal and posterior networks, may be the natural cybernetic vector emerging from the visceromotor function of dorsal limbic networks. The feedback mode of control, operating in action with differentiated guidance from perception, may be integral to motives organized around viscerosensory criteria.

We speculate that the dorsal and ventral modes of cognition and action regulation emerge from primitive mechanisms, redundancy and habituation, for regulating neural activity in time. Within neurodevelopmental theory, these are the same mechanisms that regulate neural activity to organize the connectivity of the brain in fetal development. By understanding neurodevelopmental continuity, we may appreciate that the dual modes of action regulation progressively articulate holistic and differentiated cognitive capacities through the course of lifespan development.

## Looking Ahead

Two general questions arise from this conclusion, and we will examine them in the following chapters. First, can we build on this model of motivated action regulation to formulate principles of neuropsychological theory that explain major domains of psychological function? In Chapter 4 we will assert that this is possible. We will provide rough theoretical illustrations in the context of the current and historical literatures in psychology and cognitive neuroscience to support that assertion.

Second, how can we understand the underlying anatomical and neurophysiological mechanisms of dorsal and ventral specialization in the mammalian brain? Chapter 5 will enumerate the anatomical and physiological features that may explain important concrete properties of the differing neural substrates of dorsal and ventral divisions of the cortex.

Once these features are characterized, we will argue that we can gain important insight into how the dorsal and ventral divisions of the cortex and limbic system organize cognition through an evolutionary-developmental analysis of their origins. Chapter 6 will look to the process embryological development itself to theorize on the role of dorsal and ventral divisions in the emergence of mammalian memory. The developmental process is ontogenesis, both for growing the brain and for using it. As a result, understanding ontogenesis at its embryonic roots may have interesting implications for understanding its course over the life span.

Chapter 7 will then draw general conclusions, for neuropsychological and psychological theory, of the evolutionary-developmental analysis of cognition as a neurodevelopmental process.

# 4
# Opponent Complementarity in Psychological Function

The elementary organization of action thus involves dual and opponent control processes. One is impulsive and projectional. The other is constrained and reactive. These processes are perhaps most apparent in the operation of the dorsal and ventral divisions of the frontal lobe, but they also figure importantly in the operation of dorsal and ventral posterior networks for sensation and perception.

The congruence between functional modes of operation across frontal and posterior cortical networks appears to be achieved through integrated cybernetic modes, of feedforward dorsally and feedback ventrally, applied by limbic networks to organize both input and output functions of the cortex. The functional specialization of somatic sensory (posterior neocortex) and somatic motor (anterior neocortex) networks of the cortex is thus balanced by visceral sensory (ventral limbic) and visceral motor (dorsal limbic) motivational and memory systems. Considered in relation to the primitive cognition of animal learning, the projectional mode is well suited to support the hedonic expectant cognition underlying approach learning, whereas the feedback mode is well suited to support the focused learning and vigilance required when events are discrepant with the ongoing context model (Tucker & Luu, 2007). An important theoretical question is why these differential cybernetics emerged in the dorsal and ventral divisions of the mammalian hemisphere. One explanation is that the cybernetics are integral to visceral controls. The projectional cybernetics are a unique extension of visceromotor control processes, which emerge directly and impulsively, in the limbifugal direction of consolidation originating at the limbic base. In contrast, feedback cybernetics seem to emerge from viscerosensory functions, orienting to the limbipetal direction of consolidation starting at the sensory boundary, providing criteria for actions that must be sustained and organized over time (Luu & Tucker, 2003; Tucker & Luu, 2007).

As reviewed in the last section of Chapter 3, the dorsal and ventral cybernetic modes appear integral not only to elementary sensorimotor functions; they are also integral to the conscious monitoring and elaboration of self-regulation in the specialized networks of the human frontal pole. The dorsal projectional mode motivates goal-oriented intention. The ventral reactive mode motivates differentiated, critical attention. There are important implications of these cybernetic modes for social orientations (Jeannerod & Jacob, 2005). Consistent with their differential roles in intentional versus attentional control (Tucker, Brown, Luu, & Holmes, 2007), the dorsal projectional mode is egocentric; the ventral reactive mode is inherently responsive to actions of others (Jeannerod & Jacob, 2005).

In this chapter, we propose that the dual modes of self-regulation can be understood as general psychological controls. They evolved as complementary neural mechanisms for regulating the activity-dependent differentiation of neural architecture in embryogenesis. They were then conserved in mammalian evolution as the opponent, yet balanced cybernetic bases for the cognitive processes underlying learning and behavioral organization throughout the life span. As constructs for human psychological theory, we suggest new terms for these neurophysiological mechanisms of self-regulation. We describe the dorsal projectional mode as the *impetus*, Latin for impulse. The ventral reactive mode is the *artus*, Latin for constraint.

We consider how these elemental, opponent and complementary, neural control mechanisms can be understood to shape broad domains of human psychological function, from elementary learning to abstract cognition to complex challenges of social relations. We begin with concepts of self-regulation in the cognitive neuroscience literature, drawing on the unique evidence from modern neuroimaging technologies. Although most of the neuroimaging studies have been framed in relatively narrow, descriptive terms, there is clear overlap with the more general questions of neuropsychological theory that may be explained in important ways by the dual modes of self-regulation.

## Executive Control of the Cognitive Process

The development of cognitive neuroscience over the last several decades has integrated systematic experimental studies of cognition with powerful methods of neuroimaging, particularly functional magnetic resonance imaging (fMRI). We think that many of the findings in this literature that pertain to self-regulation can be interpreted within the action regulation framework of Chapter 3. There are important theoretical advantages in recognizing that cognition is not comprised of isolated mental skills, but rather emerges from neural networks that evolved for motivational control of sensation and action, negotiating between both visceral and somatic constraints.

An important recent development in the neuroimaging literature has been insight into the functional networks of the cortex, revealed by patterns of correlation in fMRI activations in multiple brain areas in resting states and task demands (Dosenbach et al., 2007; Fox et al., 2005). Unexpectedly, one of the clearest network patterns was observed not during cognitive tasks, but in a *resting state* between tasks, leading to the term *default mode* network (Raichle & Gusnard, 2005). Although it may seem reasonable that limbic regions, such as the PCC and ACC, would be active in an internally focused resting state, in contrast to an externally focused state supporting task performance, other evidence shows limbic regions play critical roles in controlling cognition, particularly when *executive control* is required (Corbetta, Patel, & Shulman, 2008; Dosenbach et al., 2007; Fair et al., 2007; Fox, Corbetta, Snyder, Vincent, & Raichle, 2006). Executive control, also known as cognitive control, refers to processes that guide attention, working memory, stimulus selection, response preparation, and response initiation. Executive control was a key theme in classical neuropsychological studies of frontal lobe damage, as reviewed in Chapter 3, and it remains a central issue in the modern cognitive neuroscience literature.

In 1986 Norman and Shallice proposed an influential cognitive model of multiple subsystems that regulate behavior. The first is a contention scheduling system that regulates behaviors via selections of "schemas." A schema is an internal model, based on past experiences, of the external world. A schema is selected by the contention scheduling system via local inhibition of competing schemas. Once selected, actions are automated because they are part of the schema. When there is conflict between a selected schema and environmental demands, a supervisory system was proposed to intervene so that a more appropriate schema can be selected. This system, Norman and Shallice argued, has access to the overall representation of the environment and goals of the person. Moreover, they define situations in which this system should be engaged. These include: planning and decision making, error correction, response novelty, danger, and overcoming of habitual responses. Essentially, these situations require learning what to do because actions (practiced or automated) that were adaptive are no longer relevant.

The Norman and Shallice model did not address the neural substrates of its subsystems and processes. With widespread availability of noninvasive neuroimaging technologies (such as fMRI), this model was employed by scientists to study control mechanisms of the human brain. In this section, we review the extant neuroimaging literature and show that the results from studies of executive control processes are consistent with complimentary opponent processes of the dorsal and ventral divisions. We reframe the neuroimaging findings within an explicit neurocybernetic model that emphasizes executive control as an extension of the process of action regulation.

## The Cognitive Neuroscience of Executive Control

In recent meta-analytic reviews, Schneider and colleagues (Chein & Schneider, 2005; Cole & Schneider, 2007) found that tasks that require executive control consistently engage the ACC, PreSMA, dorsolateral prefrontal cortex (DLPFC), dorsal premotor cortex (dPMC), and posterior parietal cortex. These regions are in the dorsal division of cortex. Schneider and colleagues also found that executive control tasks often engage regions in the ventral division of the cortex, including the inferior frontal junction (IFJ; roughly similar to the arcuate premotor area or APA from Chapter 3), as well as anterior insular cortex (AIC). Taken together, these regions form a cognitive control network. Each region appears to contribute a unique functional subcomponent of cognitive control.

A popular model of the ACC/PreSMA in cognitive control proposes that this region is involved in the evaluation of action outcomes generally and response conflict in particular (Botvinick, Cohen, & Carter, 2004). The outcome of the evaluation is communicated to other components of the cognitive control network when control is required. A similar proposal was put forth by Ridderinkhof, Ullsperger, Crone, and Nieuwenhuis (2004), who proposed that the ACC/PreSMA monitors action in relation to anticipated rewards. The process being monitored could include negative feedback, conflicts, or decision uncertainty. On the other hand, it is noted that the ACC/PreSMA is involved in aspects of cognitive control other than monitoring, such as preparation for task performance (Dosenbach et al., 2006; Coles & Schneider, 2007). Similarly, others have noted that the ACC/PreSMA appears to be involved in a more general process of integrating reward values with actions (Rushworth, Walton, Kennerley, & Bannerman, 2004).

Brass, Derrfuss, Forstmann, and von Cramon (2005) reviewed evidence that implicate the IFJ including Brodmann area 6 (ventral premotor) in task representation. For example, in a task such as naming the color of a word (the Stroop task) it is argued that the IFJ is involved in representing the stimulus-response rules. The involvement of the AIC in cognitive control has only been recently recognized, and this region appears to share a close functional relation with the IFJ (Dosenbach et al., 2006). This proposal of task representation in the ventral premotor cortex is consistent with the findings that arcuate premotor cortex (APA) of the ventral frontal lobe is involved in object-response representations (Umilta, Brochier, Spinks, & Lemon, 2007; Yamagata, Nakayama, Tanji, & Hoshi, 2009).

As reviewed in Chapter 3, it was originally believed that the function of the DLPFC is to support working memory and the maintenance of information to guide the action plan. Single-unit studies showed that cells within the DLPFC exhibit sustained activity during the stimulus-response interval (Fuster, 1989; Goldman-Rakic, 1987). The tonic firing of DLPFC neurons during the delay interval, coupled with effects of DLPFC lesions in impairing behavior when a delay is imposed between stimulus and response, were consistent with the concept of working memory. However, as noted in early lesion studies, the performance impairment is not observed if during the delay interval the animal is not distracted by events in the environment. For example, after darkening the experimental room to eliminate visual distractions, the animal performs normally (Pribram, 1987).

Rather than focusing on working memory, recent models emphasize that the DLPFC is involved in the maintenance of task-related goals, using the goal representations to bias processing pathways, including those that lead to action (Miller & Cohen, 2001). There is now good evidence that the DLPFC is involved the inhibition of signals that may compete with the goals of an individual, as represented in the DLPFC (Postle, 2005). Thus rather than retention of information, in a working memory buffer sense, the DLPFC prevents interference through inhibition. Through inhibitory control, the rostral DLPFC selects actions that are consistent with task-related goals. Koechlin, Ody, and Kouneiher (2003) proposed that the rostral DLPFC uses episodic information (on past and ongoing events related to the task) to select appropriate task-response set associations, where these associations are represented in the caudal DLPFC. In turn, the caudal DLPFC selects actions linked specifically to the present stimulus, where these actions are represented in premotor regions. The linked, hierarchic relations between these adjacent networks of the dorsal frontal lobe, consistent with the general corticolimbic architecture in Figures 2-6 through 2-9, appear to be effected in large part through inhibitory mechanisms.

## *Attention and Cognitive Control*

The distinction between attention and cognitive control is often blurred. Indeed, *executive attention* is sometimes used to describe applications of cognitive control, including learning (Posner & Rothbart, 1998). Executive attention refers to those processes that exert exogenous control over the allocation of attention, such that brain regions involved in processing the relevant information are affected (Posner & Dehaene, 1994). Posner and Dehaene review evidence from cellular recordings that attention suppresses the activity of neurons in primary sensory processing regions that are responsive to unattended stimuli. In secondary, unimodal association areas, attention both enhances activity in cells involved in processing of the relevant stimulus and inhibit those cells not involved. In regions further downstream, involved in task performance rather than sensory processing, attention appears necessary for activating the relevant cognition or action.

In addition to these several *sites* where the effects of attention are observed, a number of findings suggest that major *sources* of executive attention are the ACC and basal ganglia (Posner & Dehaene, 1994). In a hierarchical model of cognition, executive attention would originate from a component at the top of the hierarchy, and it would then be applied to sites of operation at lower levels. Whereas it may make sense with such a model to attribute executive control to the frontal lobe generally, it is curious that the evidence points to the limbic base of the dorsal frontal lobe, the ACC, as well as the subcortical basal ganglia, for this higher level control.

In animal research, cognitive control is also referred to as *associative attention*, the process by which an animal learns which stimulus in the environment predicts the delivery of reward or punishment (Gabriel, Burhans, Talk, & Scalf, 2002). As reviewed in Chapter 1, neurons within the ACC show differential firing to the conditioned stimulus (CS+ and CS-) events during the learning process. When a CS is made less salient (for example, when it is presented for only 200 ms), ACC cells may compensate for the decreased stimulus saliency by increasing their activity.

Gabriel et al. (2002) noted that the conditions under which associative attention is required in animal studies are similar if not identical to the conditions under which executive attention is invoked in human studies. However, there is less need in the animal studies to consider attention as an exogenous process, emerging from a higher center to control a lower one. Rather, in the animal studies of action regulation, associative attention is a consequence of the cybernetics of learning within the motor control pathway itself, with a critical role played by the motive control within the limbic base (ACC). We suggest that the critical role of the ACC in the recent cognitive neuroscience literature should be interpreted in light of our general knowledge of action regulation. The cingulate cortex is limbic cortex, providing motive direction for the operations of consolidation within corticolimbic networks. Its role in executive function is well documented, and arises because executive tasks such as planning actions, making decisions, and managing conflict all require integration of the motive basis for behavior and cognition.

As noted, reviews of the current literature on cognitive control also identify the involvement of additional brain structures, including regions of ventral cortex such as the IFJ and AIC. How are these structures involved in attentional control? Based on a meta-analytic review, Corbetta et al. (2008) proposed a dual model of attention that emphasizes goal-directed (expectancy driven) versus stimulus-driven attentional control. Goal-directed attentional control involves the superior parietal cortex and the dorsolateral aspects of the frontal lobe (near the frontal eye fields). Goal-directed attention is concerned with orienting to relevant external stimuli based on internally represented goals. This system is believed to provide top-down control of attention through its integral ability to establish expectations. Complementary to this system is a ventral frontal-parietal attention system that responds to behaviorally relevant objects in the external environment and is particularly active when expectations are violated. The temporal-parietal junction, the ventral frontal cortex, the IFJ, and AIC are the cortical components of this system.

The model of attention proposed by Corbetta et al. (2008) recognizes that attentional control may be integral to the basic mechanisms of learning and memory. It thereby avoids the homunculus problem inherent to the concept of executive control, described most clearly by Posner (Posner, 1978). The homunculus is the internal executor that takes control when confronted with uncertainty. The framework of Corbetta et al. (2008) is consistent in many respects with the model of dorsal and ventral contributions to action regulation in Chapter 3.

In the context of frontopolar control over thalamic mechanisms, we suggested that the dorsal frontoparietal network is important to *intentional* control whereas the ventral frontoparietal network is involved in *attentional* control. Conceptually, intention and attention could be seen to emphasize internality and externality, respectively, as vectors of control. Understanding these vectors of orientation, shifting between egocentric intentions versus attentional differentiation of external information, may be important for characterizing the differing modes of cognitive control exerted by the impetus and artus.

## *Internal World Models and Action Control: Creative Hypotheses*

In the dual model of attentional control (Corbetta et al., 2008) the attentional systems are responsible for guiding of appropriate actions based on internal plans and reconciling them with emerging external demands. This model has important parallels to Goldberg's model of feedback and feedforward of action control. In this section, we build on these theories with a focus on the role of the medial frontal cortex. This region of the frontal lobe is part of a larger mediodorsal system that extends to posterior brain structures, including the posterior cingulate cortex (PCC) and hippocampus. We argue that the elemental role of medial frontal cortex in action regulation also underlies expectant cognition, in the process often described as hypothesis formation.

The function of the frontal components of the dorsal system is to provide *active* guidance of action, in support of the dorsal system's feedforward function. We propose that the generative result of the mediodorsal system is a *hypothesis*, an expectation for what might be. The hypothesis can be seen as a kind of context, generated through hedonic expectancy from the internal model, rather than the external environment. The hypothesis allows the individual to determine appropriate actions under conditions of uncertainty (Luu, Tucker, & Stripling, 2007).

There are several aspects of this proposal. First, we suggest that integral to the generation of an action plan is an active, though not necessarily explicit, formation of a hypothesis that serves as a goal for how the action plays out in the world. Most proposals regarding monitoring functions of the ACC rely on the notion of cognitive expectancies, but few discuss how they are developed (e.g., Ridderinkhof et al., 2004). Second, we can understand now that these expectancies are not object expectancies but rather contextual expectancies, derived from the feedforward cybernetic control in the dorsal division of the hemisphere. Third, this process of hypothesis generation is integral to learning, and in fact can be inferred to exist from the learning evidence reviewed in Chapter 1.

Feedforward control is a projectional mode of control in which expectancies are formed based on internally generated models of the external world that embody positive expectancies of action outcomes. These expectancies are the bases by which outcomes are compared, such as in the successive negative contrast effect (Papini, 2003). Although the elemental hypothesis is goal-oriented, this is not a goal orientation set by external demands, as often assumed by top-down cognitive models. Rather, it is an orientation to goals that emerge from internal desires and urges.

### Studies of Medial Frontal Cortex Function

Thus we propose that the elemental expectancies that guide animal learning are also engaged in the feedforward control of human cognition, through generation of hypotheses with primary representations in mediodorsal networks of the frontal lobe. To examine this proposal in relation to the current cognitive neuroscience literature, we sampled four classes of studies: learning, probability tracking, social economic games, and internal-external control of actions (see Table 4–1). Common to all of these studies is that subjects were presented with ill-posed problems, a condition under which feedforward control is most likely engaged (Haggard, 2008; Ridderinkhof et al., 2004). We predicted that under these conditions there would be activation of mediodorsal frontal regions, specifically the ACC, SMA, and preSMA. Following are brief descriptions of each group of studies.

> Learning Studies: In these studies, subjects are presented with a task that requires them to learn correct stimulus-response mappings. Subjects are often explicitly informed of the requirement to learn, or the stimulus-response mappings are consistent enough such that learning can be achieved. Some tasks can be probabilistic, without explicit instructions, such as the Iowa Gambling task. However, these tasks have criteria for choice that can be learned, such as with decks that are associated with gains over the long run (Christakou, Brammer, Giampietro, & Rubia, 2009).
> 
> Probability Tracking Studies: These studies are similar to learning studies in that subjects have to make a decision, but there are no consistent mappings that could be learned. Rather, decisions must be based on probabilities for the outcome. Often, the decisions are framed within a gambling framework that involves gains or losses, implying that a choice is correct by the statistical pattern of the outcomes.

Social Economic Game Studies: These are similar to the gambling studies in that subjects have to make decisions that will either earn or lose money. A major difference is that the uncertainty lies in the reciprocation that is made by another person. The prototypical paradigm is referred to as the "Trust Game" (e.g., Tomlin et al., 2006). The game involves interactions between an "investor" and a "trustee." At the start of the game, the investor is endowed with a fixed amount of money. In the first round, the investor has to choose how much money is to be transferred to the trustee. The amount is multiplied by a factor (e.g., 3) and then sent to the trustee. The investor and trustee are aware of the multiplication factor. The trustee then decides how much to send back to the investor. This interaction is carried out for multiple rounds.

Internal Versus External Generation of Actions: In these studies, contrasts are made between conditions under which subjects decide for themselves what to do and conditions in which they are instructed by stimulus parameters. For simple actions, generation of responses may be specified by stimulus parameters (such as making a movement toward a specified target) or the actions may be self-generated (such as making a movement toward a self-chosen target). For more complicated actions, subjects can decide on the tasks that they perform. For example, in contrast to directions to perform a particular task (e.g., make a semantic judgment), subjects can decide what task to execute (e.g., make a semantic or word syllable judgment).

Because the question is brain activity associated with the generation of internal models (i.e., hypotheses) that guide action, we focused on results showing the hemodynamic response between the time of stimulus presentation and the execution of the response. We do not include data analyzed in response to performance feedback (i.e., outcome stimulus). Of the regions active in these studies, we list those that are along the medial frontal cortex. For those studies that report results in the Montreal Neurologic Institute (MNI) coordinates, we translated the MNI coordinates to Talairach coordinates using GingerALE v2.0 (Eickhoff et al., 2009).

Figure 4-1 shows the activation coordinates from each class of studies. The medial prefrontal cortex can be seen to be active in all these studies. In particular, activity is clustered around the caudal aspects of the ACC and superior medial frontal gyrus (MFG), which includes BA 8, 9, pre-SMA, and SMA (see Rushworth et al., 2004). The regions that show substantial overlap for all tasks are the caudal ACC, pre-SMA, and caudal aspect of BA 8 (demarcated by the red box in Figure 4-1). There are several notable findings from these studies that we highlight here.

To our knowledge, Elliot and Dolan (1998) were the first to demonstrate that the dorsal ACC is involved in hypothesis generation. These authors presented

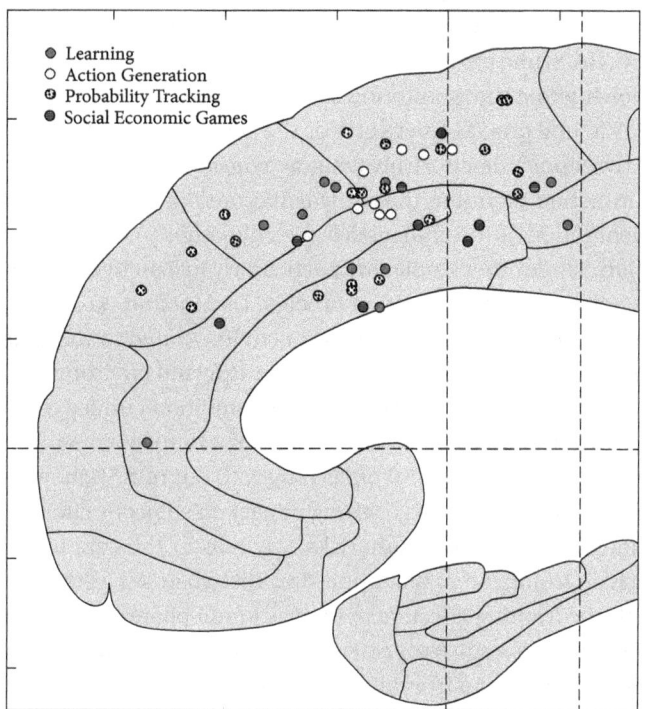

Figure 4-1. Summary of centers of activations in dorsomedial cortex in studies involving various forms of the executive control of cognition.

subjects with two checkerboard patterns on every trial and asked participants to generate a hypothesis that would guide their selection of the correct target to choose from the pair. They were asked to generate a hypothesis for every pair and update it on each trial based on feedback. As a control task, subjects were presented with a pair of identical checkerboard pattern and were told to respond according to a cue that guides their response. Elliot and Dolan found that hypothesis generation was associated with activity in the dorsal ACC. One could argue that this dorsal ACC activation occurred because subjects were required to internally generate actions, rather than responding to external directions. However, these researchers found that the ACC was active during hypothesis generation regardless of whether subjects had to make an actual response. Thus, it is not the motor act alone that activates the ACC. Rather, it appears as if the generation of the hypothesis is itself part of the action plan, and this cognitive process engages the ACC.

In a more recent study, Tachibana and colleagues (2009) employed a task that involved a detection phase, wherein subjects are presented with a series of stimulus response mappings that would show which of two rules could be used to determine a correct response in subsequent trials. After this detection phase, subjects were then presented with trials that would allow them to test their hypotheses about the stimulus-response mapping rule. Activity centered on the ACC, BA 8, and pre-SMA regions was found during the hypothesis testing phase, implying that the monitoring and evaluation of the hypothesis occurs in the same ACC and pre-SMA networks as the generation itself.

These two studies demonstrate what we consider to be a central component involved in action planning during learning and probabilistic decision making: the generation of internal models (i.e., hypotheses) about future events to guide action. Three other studies are particularly instructive for illustrating the cognitive nature of this generation process. De Martino, Kumaran, Seymour, and Dolan (2006) examined the neural activity related to the framing effect. The framing effect describes the reliance on external environment context to frame the basis of decision, particularly for conditions under which information needed for the decision is incomplete. For example, when faced with the option of losing $30 out of $50 or keeping $20 out of $50, how the choice is framed (as a loss or a gain) has a strong impact on subjects' choices. In the loss frame, subjects tend to take more risks, whereas in the gain frame they tend to avoid risk. DeMartino et al., found that the when subjects chose in accordance with the framing effect, activity was found bilaterally in the amygdala. This would be consistent with external sensitivity in action regulation, relying on ventral limbic and frontal networks for feedback guidance of action. In contrast, when subjects made choices that ran counter to the framing effect, showing greater internal control of the decision, activity was observed in the ACC. In an apparently similar effect, Venkatraman et al. (2009) showed that the regions around the ACC and medial frontal gyrus (MFG) were active when subjects made choices that were inconsistent with their preferred decision strategies.

In a trust game, Delgado, Frank, & Phelps (2005) provided subjects with information about the moral quality (praiseworthy, suspect, neutral) of their partner. Participants tend to share with praiseworthy partners and withhold sharing with suspect partners. When subjects went against this bias, activity within the ACC was observed.

The common theme illustrated in these three findings is the distinction between actions that are biased by external information and actions that overcome this bias through the exertion of internal control. Separation from the external context and exertion of internal control involved activity in the ACC and MFG. Action control under conditions of salient external constraints requires an internal model of the world that is used to guide selection of appropriate behavior. The internal model can be automatically generated based on previous experience, or it can be deliberately and effortfully formulated. Once the internal model is generated, the corresponding actions are simply projected into the world.

Table 4–1.

| Study Type | First Author | Year | Task |
|---|---|---|---|
| Learning | Blair | 2006 | Differential Reward/Punishment |
| | Christakou | 2009 | Iowa Gambling Task |
| | Elliot | 1998 | Hypothesis-Testing |
| | Hampton | 2006 | Probabilistic Reversal Learning |
| | Tachibana | 2009 | Logical Principles |
| | Volz | 2004 | Rule Learning |
| Probability Tracking | Behrens | 2007 | Probability Tracking |
| | DeMartino | 2006 | Financial Decision Making Task |
| | Hsu | 2005 | Probability Tracking |
| | Rustichini | 2005 | Lottery Decisions |
| | Venkatramen | 2009 | Mixed Gamble Task |
| | Zysset | 2006 | Multi-attribute Decision Making |
| Social Economic Games | Delgado | 2005 | Trust Game |
| | King-Casas | 2005 | Trust Game |
| | Kuo | 2009 | Coordination Game |
| | Kruger | 2007 | Trust Game |
| | Tomlin | 2006 | Trust Game |
| Internal Versus External Generation of Action | Bengtsson | 2009 | Task Selection |
| | Cunnington | 2002 | Button Pressing |
| | Forstmann | 2008 | Task Selection |
| | Forstmann | 2005 | Task Switching |
| | Hunter | 2003 | Button Pressing |
| | Lau | 2004 | Action Selection |
| | Nachev | 2005 | Saccade Choice/Conflict Task |

## Actions Constrained by Reality: Critical Thinking

In complementary opposition to the system for generation of an internal model (hypothesis) for action is the system for action control by constraints from the external world. This is the form of cybernetic control specified by Goldberg (1985) as feedback control. Because Goldberg focused exclusively on motor control, he emphasized the role of the arcuate premotor area (APA) of the ventral frontal lobe in this process. Clearly, the APA has strong inputs from the sensory regions, particularly the visual cortex, and it contains neurons that are responsive to sensory inputs as well as motor execution (e.g.,

Murata et al., 1997). In describing feed-forward control, we emphasized the role of not only the medial premotor areas but also the limbic cortex, namely the ACC. The equivalent limbic regions for feedback action control in the ventrolateral frontal cortex are the orbitofrontal cortex, AIC, and amygdala. These regions have a role in learning, and in cognition under uncertainty, that is diametrically opposed to that of the feedforward system. Elaborated in human cognition, this form of motivated cognition is essential to critical thought.

Feedback control is a reactive form of control wherein actions are specified by the external environment, particularly with object-related information. Inherent to feedback control is the detection of negative outcomes. In new learning situations, such as when expectancies are violated, the role of the feedback system is to orient attention to the relevant stimulus (Corbetta et al., 2008). Reorienting attention to environmental stimuli is required for the processing of environmental stimuli that must be associated with new responses. The function of the feedback system during learning is to associate a given stimulus with correct behavior that restores the previous level of reinforcement, that is, is consistent with expectations.

The feedback function of ventral frontal networks may extend from motor control to cognitive control, providing ongoing constraint of cognition in relation to external criteria. This form of control may be integral to the capacity for critical thinking (Tucker, 2008; Tucker & Luu, 2007).

Just as dorsal mechanisms of feedforward control and hypothesis generation can be seen as integral to elementary cognition required in animal learning, the ventral mechanisms of feedback control and critical constraint may provide complementary opposition to control the learning process. Studies by Timothy Bussey and his colleagues (e.g., Bussey et al., 2001; 2002) show that, in rodents as well as primates, damage to the ventral and orbitofrontal cortex impairs rapid learning of stimulus-response associations. These results are generally similar to those obtained in rabbits by Gabriel et al. (2002). Because of its role in the retrieval of stimulus-response association (Bussey et al., 2001), the ventrolateral prefrontal cortex continues to be important even after the early stages of learning. Removal of the inferior temporal lobe, which provides visual input to the ventrolateral and orbitofrontal cortex, clearly demonstrates the dependency of the ventrolateral frontal cortex on sensory stimuli for feedback guidance of learning (Bussey et al., 2002). In a similar effect in the cognitive domain in humans, the left ventrolateral prefrontal cortex is engaged early in learning and remains active even after learning is established (Chein & Schneider, 2005).

## *Interaction Between Feedforward and Feedback in Learning*

Although it is useful to separate unique motive biases, they must be coordinated in normal learning and cognition. In the case of learning to discriminate between two stimuli, one of which predicts delivery of punishment, it is clear that amygdala is critical for rapid learning (Poremba & Gabriel, 1997). The amygdala's role seems closely linked to the aversive expectancy in this learning task. The feedforward system (the medial prefrontal cortex) would provide only a supporting role for aversion learning, in that the CS+ generates a positive expectancy for relief from punishment (Gray, 1990). As reviewed in Chapter 1, approach learning does not appear to be directly dependent on the ventral feedback system; discriminant firing patterns to CS+ and CS- do not develop with learning under purely approach conditions (Smith, Freeman, Nicholson, & Gabriel, 2002). The deficits in approach learning that are seen with amygdala lesions seem not to involve the CS+ (the approach per se), but rather the CS- (the stimulus to avoid; Smith et al., 2002, Bussey et al., 2001; 2002).

The importance of positive expectancies and feedforward control in approach learning is revealed in a study by Bussey, Muir, Everitt, and Robbins (1996). Rats were first trained to press a lever for a delivery of a sucrose solution. Simultaneous with the delivery of the solution, a light was illuminated. After this pretraining period, animals were trained to discriminate between fast and slow flashing lights, with one type assigned to be the CS+ and the other to be CS-. Surprisingly, after the ACC was lesioned, the animals quickly learned to respond appropriately to the CS+ and CS-. This is in stark contrast to the repeated findings that ACC lesions impair the acquisition of discriminant learning.

Key to understanding these results is the effect of pretraining. During the pretraining stage, the animals associated the light with the reward. During the training stage, they were required to differentiate between the light-flash rates. The improved learning rate after ACC lesions is due to the fact that animals with this lesion made *fewer* commission errors to the CS-. An interpretation of this apparent paradoxical finding is that ACC lesions removed the approach (feedforward) bias acquired during the pretraining stage, thereby allowing the animals to learn strictly through stimulus-response associations (Bussey et al., 1996). The lack of a feedforward bias would facilitate rapid learning to avoid responses to CS-, a negative event to which the feedback control is tuned.

Under approach learning conditions that do not have these pretraining properties, ACC lesions retard learning (Bussey, Everitt, & Robbins, 1997; Smith et al., 2002). Without pretraining, lesions to the ACC remove the ability to develop positive expectancies against which the CS- can be compared, producing indiscriminate approach behavior to both CS+ and CS-.

Thus, it appears that the dorsal limbic base of the feedforward system provides positive expectancies to serve as the background against which negative outcomes, represented particularly by the feedback system, are detected. In animals and humans, the cingulate motor area (CMA; between the ACC and motor cortex) has been shown to be involved in behavioral adjustments when ongoing actions produce a reduction in expected reward. Shima and Tanji (1998) showed, in monkeys, that the CMA contains neurons that are involved in the selection of alternative responses to restore the original reward level when the original action is associated with reward reduction. Deactivation of this region produces perseveration, in that the animal fails to change behavior in the presence of reward reduction. In humans, Williams et al. (2004) reported that there are cells in the dorsal ACC that were sensitive to actions that produce a reduction in expected reward. These cells are not merely responsive to cues that signal behavior change, nor are they simply responsive to detection of reward discrepancies. Williams et al. also showed that increasing the reward level beyond expectations does not significantly increase the activity of dorsal ACC neurons. From a control perspective, better-than-expected outcomes should not modify actions; the animal simply has to continue to perform the same behavior.

In humans, dense-array EEG recordings have shown that signals arising from the dorsal ACC are reliably elicited when subjects make an erroneous response (Dehaene et al., 1994; Luu, Flaisch, & Tucker, 2000). Known as the error-related negativity (ERN, Gehring et al., 1993) or error negativity (Falkenstein, Hohnsbein, Hoormann, & Blanke, 1991), the onset of this signal is in close proximity to the execution of the erroneous response (~50–100 ms post response). This signal is distinct from signals elicited by error feedback, such as those provided when subjects are not aware of the accuracy of their response (Luu et al., 2003; Luu et al., 2009). Functional MRI studies confirm that feedback signals following errors elicit responses from the rostral ACC, as opposed to the dorsal ACC (Lauren, Ngan, Bates, Kiehl, & Liddle, 2003; Menon, Adleman, White, Glover, & Reiss, 2001). Just as in animal studies that have shown activity in the dorsal ACC to be preferentially sensitive to loss, the amplitude of the ERN appears to be predominantly affected by negative outcomes and negative affect (Luu et al., 2000; Gehring, Himle, & Nisenson, 2000; Johannes, Wieringa, Nager, Dengler, & Münte, 2001). The interaction between feedforward (dorsal limbic) and feedback (ventral limbic) motive biases is critical to learning, and the evidence in both animals and humans suggests that this interaction is represented strongly within the ACC, drawing on the considerable ventral limbic input to the subgenual ACC.

Thus the interaction between dorsal and ventral corticolimbic systems is critical to the learning process. Nonetheless, these two systems provide differing, and in many ways opponent, biases on learning and elementary cognition, so that understanding their coordinated function requires first understanding their differences. We think many of the findings in current research on the executive control of cognition, including both EEG and fMRI methods, may be interpreted within the framework of action regulation reviewed in Chapter 3. The feedforward, projectional control from dorsal cortical regions may be integral to the expectant control of cognition, through the psychological mechanism of the impetus. In the elementary cognition of animals and people, this is a mechanism of hedonic expectancy. In the more complex cognition of humans, this mechanism supports the generation of hypotheses. In contrast, the feedback, reactive control from ventral cortical regions may be integral to the constraint of cognition as well as action regulation, through the psychological mechanism of the artus. In the primitive cognition of animals and people, this is a mechanism of constraint by attention to the facts, and threats, of reality. In the more complex cognition of humans, this mechanism supports the analytic control of cognition, through the process of critical thinking.

## Dimensions of Psychological Self-Regulation

We argued on the basis of the animal learning literature in Chapter 1 that the cybernetic properties of feedforward and feedback control emerge from elementary motive biases in learning. In humans, the specific motive biases of the impetus and the artus also may be associated with specific motivational and emotional states, and with the primitive cognitive qualities of expectancy and discrepancy recognition. Self-regulation by the impetus may emerge as an integral feature of greater elation. Self-regulation by the artus may be an intrinsic result of greater anxiety. Several lines of psychological research point to dimensions of emotion that we think can be interpreted in relation to these dual dimensions of elation and anxiety. Because affective states seem to apply intrinsic biases to intention and attention, it may be possible to build a theoretical model for how the dimensional structure of affect and personality is simultaneously a dimensional structure of cognition and intelligence.

### *Affect and Arousal*

The proposal of activation and arousal theory (Tucker & Williamson, 1984) is that there is no physiological arousal separate from affect. Whereas psychological theory has often separated orthogonal dimensions of emotional valence (good/bad) and physiological arousal (low/high), this may be a logical fiction that fails to reflect the way that human moods vary in nature. Within activation and arousal theory, the depression-elation dimension of phasic arousal describes both affect and neural arousal. Similarly, the calm-anxious (anxious/hostile) dimension of tonic activation also describes both affect and neural arousal.

Research on the dimensional structure of emotion with a psychometric (self-report) methodology has produced results consistent with this framework. In the psychometric methodology, the measures are self-report of endorsement of psychological states or traits, described either by single words ("lively") or in statements ("I have a good imagination"). The dimensional structure of these characteristics is inferred through statistical correlation among the endorsements. Once the overall pattern of correlation is established (with N items this is an N by N correlation matrix), techniques of matrix algebra are used to find patterns of correlations that go together. These patterns of correlations are called *factors*. When the factor analysis is organized appropriately, a few factors can describe the underlying dimensions that explain most of the correlations among the endorsement items. An important technical fact is that the number of factors is a matter of interpretation, so that the same data can be described by many or few factors. In some cases, a set of many factors can be refactored to construct a few *higher order* factors. Another technical fact, perhaps best explained through the examples below, is that the factors or dimensions can be *rotated* such that the fit of a few dimensions to the correlational patterns of the data is as simple as possible. This goal of simplicity is described in the factor analysis literature as *simple structure*.

In selecting the rotation that describes the higher order dimensional structure of emotion and personality ratings, most investigators have settled on two factors, *arousal* (varying from calm to excited) and *valence* (varying from good to bad). This seems to be a logical arrangement for factors, with qualitatively different properties (arousal and valence) on the two dimensions. Thus an emotion like anger would have a high score or loading on arousal and a high loading on the bad end of the valence dimension. However, in his research on psychological ratings of arousal, Thayer (Thayer, 1978, 1989) observed that the ratings were better described by factors rotated 45 degrees from the arousal/valence dimensions, producing two forms of arousal, one positively valenced and the other negatively valenced. Thayer described the positive arousal dimension as varying from the experience of *tiredness* to that of *energy and vigor*. This seems congruent with the depression-elation dimension of phasic arousal. Thayer described the negative arousal dimension as varying from the experience of *calm* to that of *anxiety* and *tension*. This is congruent with the calm-anxiety dimension of tonic activation (Tucker & Williamson, 1984). Thayer's psychological analysis of the experience of arousal provided the important suggestion that these dimensions are experienced subjectively (we might say affectively), and that they provide important and continual signals to guide coping efforts. In many decisions during the day, the felt level of energy-vigor guides a person's choices for coping efforts in relation to the perceived level of energy resources (Thayer, 1989). These are, of course, psychological rather than metabolic energy resources. Similarly, the experience of anxiety provides a subjective signal that may color perception of a conscious challenge or threat, or when unattached to a specific object, may guide the effort to identify the threat that engenders it.

At the same time as Thayer was studying the psychological experience of forms of arousal, Watson and Tellegen applied a similar dimensional analysis to both emotion and personality (Watson & Tellegen, 1985). In describing current ratings of emotion (affect) state, the factors most descriptive of variation across people were Positive Affect and Negative Affect. In describing self-ratings of emotion and personality traits, the corresponding dimensions were Positive Emotionality and Negative Emotionality (Tellegen, 1985; Watson, Clark, & Tellegen, 1988; Watson & Tellegen, 1985). Although the emphasis on affect or emotion led to different items and a somewhat different self-rating task than in Thayer's studies, the differences may not be that meaningful (Thayer, 1989). Not only are the psychometric results quite similar, but, more importantly, the distinction between affect and arousal may not be one that people make in subjective experience.

## The Simple Structure of Personality

The higher order factors of Positive Emotionality (PE) and Negative Emotionality (NE) are descriptions of personality as well as characteristic emotional responses (Tellegen, 1988; Watson et al., 1988), and relate to the conventional personality factors of extraversion (PE) and neuroticism (NE) (Tellegen, 1993). Tellegen pointed out that evaluative trait words were dropped by Allport in his personality trait studies (Allport, 1924), and that this convention was maintained by Norman and other psychometric researchers who found the "Big Five" dimensions of personality. The effect of this choice was to minimize the two evaluative dimensions of PE and NE that Tellegen argues are integral to personality.

The Big Five dimensions are *openness, conscientiousness, extraversion, agreeableness*, and *neuroticism* and have been widely replicated in many studies (L. R. Goldberg & Rosolack, 1994; McCrae & Costa, 1985; Saucier & Goldberg, 1998). Recently, however, cross-cultural studies have suggested that the Big Five may be more typical of English and American samples, rather than people generally (Saucier, 2009; Thalmeyer & Saucier, 2011). The factors that are proving to be consistent cross-culturally are described by Thalmeyer and Saucier as *dynamism* and *social propriety*. On the surface, these factors appear similar to the traditional extraversion and constraint dimensions. They would be more complex than captured by a simple alignment with good and bad arousal dimensions, such as elation and anxiety, or PE and NE, or energy-vigor and tension. However, there are more specific cybernetics, beyond the evaluative dimensions, associated with the differential modulation of dorsal and ventral corticolimbic systems by habituation (elation) and redundancy (anxiety) from Chapter 3. From the perspective of action regulation, dynamism may be an appropriate concept for the energetic, fluid, and impulsive mode of the impetus, whereas social propriety may reflect the greater constraint, vigilance, and allocentric attentional perspective associated with the artus.

## The Emotional Basis of Cognitive Evaluation

Another important line of evidence on the dimensionality of human emotion has come from studies of how people evaluate emotional stimuli. Lang, Bradley, Cuthbert, and their associates have conducted a series of experiments on the cerebral and autonomic processes involved in perceiving and rating affective pictures (Bradley, Codispoti, Cuthbert, & Lang, 2001; Bradley, Codispoti, Sabatinelli, & Lang, 2001; Cuthbert, Bradley, & Lang, 1996; Lang & Bradley, 2009; Lang, Bradley, & Cuthbert, 1998a, 1998b). The International Affective Picture System (IAPS) includes photographs of a range of emotionally pleasant (for example, nudes), unpleasant (mutilated bodies), and neutral (scenery) photographs. The ratings of these stimuli on arousal and pleasantness dimensions have been examined in many studies. A consistent and theoretically significant observation is that, when plotted on arousal and valence (pleasantness) axes, the ratings clustered in a "boomerang" pattern, with one lobe of the cluster extending to the quadrant defined by high pleasantness and high arousal and another lobe extending to the quadrant defined by high unpleasantness and high arousal (Figure 4-2). This

Figure 4-2. Dimensional structure of ratings of emotional pictures. A logical organization of the dimensions includes Pleasure (valence) and Arousal. However, the clustering of ratings (in what's called the "boomerang" pattern) suggests two dimensions of arousal/affect, one pleasant arousal (Positive Affect or elation) and the other unpleasant arousal (Negative Affect or anxiety). After Lang (1995).

pattern would be consistent with the interpretation that the variance of emotional responses (as shown by picture ratings) is aligned not with arousal and valence dimensions, but with two valenced arousal dimensions, one pleasant (such as elation = energy-vigor) and the other unpleasant (such as anxiety = tension).

An important difference between the affective picture ratings and the emotion or trait self-ratings of psychometric research is that the picture ratings were evaluations of the properties of the pictures. These properties are largely objective, to the extent that people agree which pictures are pleasant, arousing, and so forth. Nonetheless, it may be significant that the same dimensional structure, of valenced arousal, emerges in ratings of external objects as in ratings of emotional states and personal traits. The implication may be that judgments of the emotional qualities of things involve an implicit affective response. We judge the value of something by how it makes us feel. If so, then it would be natural that judgments would have the same dimensional structure as affective arousal.

Within action regulation theory, this dimensional structure of evaluation is more complex than just valenced arousal. Each form of arousal includes an inherent neuromodulatory influence, habituation or redundancy, that structures the cognitive appraisal at the same time as it motivates it. The impetus is a hedonic expectancy, an expansive, hopeful hypothesis, and objects congruent with it are not just good, but confirming of the self. In contrast, the artus provides feedback constraint, with an inherent critical appraisal, so that objects engaging the artus are not just criticized, but differentiated from the self.

## *Self-Regulation in Psychological Development*

Thus from the perspective of action regulation, two basic constructs in conventional psychological theory—arousal and affect—are fundamentally misleading in the way they are typically used. Arousal is not just more or less activity. Rather, it involves controls on activity—cybernetic modulations—that appear to be integral to the mechanisms of neural arousal. Similarly, affect is not just the subjective experience of emotion, but the subjective experience of a motivational process that, as it is engaged, regulates action and cognition in specific ways. Furthermore, it is also wrong to assume that affect and arousal are separable. At least in the major vectors that regulate waking function, they appear to be manifestations of the same underlying mechanisms.

The study of psychological dimensionality has provided another finding that extends the framework of normal emotion described above. This is particularly relevant to a neurodevelopmental theory, revealing the patterns of symptoms in the assessment of child psychopathology. Achenbach applied factor analysis to determine the typical covariance among symptoms assessed in children with psychological disorders (Achenbach, 1982). He found these could be grouped in two clusters, one at either end of an *internalizing-externalizing* dimension.

Externalizing disorders involve problems of impulse control, hyperactivity, and conduct disorder. They are "externalizing" in the sense that the child's problems are manifested in problematic behavior in the social environment. Internalizing disorders involve anxiety, depression, and the prodromal obsessive and compulsive disorders of childhood. They are "internalizing" in the sense that the psychological problem is to some extent contained within the child's personal experience, rather than acted out in the social context.

In the developmental approach to psychopathology, these childhood patterns are important in signaling the impairment of the developmental process of self-regulation that may lead to adult personality disorders (Cicchetti & Tucker, 1994; Tucker, 1989). They may also be relevant to the more severe disorders of manic-depression (bipolar disorder) and schizophrenia, although those disorders take such a catastrophic, qualitative course in the transition to adulthood that the relevance to normal personality organization is less clear. The personality disorders, for example in the American Psychiatric Association *Diagnostic and Statistical Manual* (DSM), show continuity with the childhood forms of psychopathology, and can be grouped into internalizing (anxiety, depression, obsessive-compulsive, and paranoid disorders) and externalizing (histrionic, impulsive, narcissistic, and psychopathic personalities) patterns.

When the characteristic emotions are considered, the internalizing-externalizing dimension can be seen to be consistent with dimensional models of emotion and personality. Positive Emotionality and dynamism are normal characteristics that become exaggerated in externalizing. Negative Emotionality and social propriety are the normal characteristics that become exaggerated in the internalizing disorders. Importantly, the context of socialization is always critical in the child's self-regulation, whether adaptive or pathological, framing the background for the operations of the impetus and the artus in childhood. Human self-regulation then negotiates the impulse for hedonic gratification and the constraint required for social propriety.

Research on temperament has considered how individual differences in affective self-regulation may be relevant to the development of skills in attention and cognition (Derryberry & Rothbart, 1987, 1988, 1997). Both affective responses (Derryberry & Rothbart, 1988) and attentional orienting (Derryberry & Reed, 1994, 1998) have been related to the classical temperament (personality) dimensions of extraversion and neuroticism (H. J. Eysenck, 1967; M. W. Eysenck, 1976). Neural systems analyses of temperament and motivation have emphasized both dorsal and ventral networks in self-regulation. These analyses led directly to the formulation of the dorsal and ventral frontolimbic cybernetics presented here as the impetus and the artus (Derryberry & Tucker, 1990, 1992, 1994; Luu, Tucker, & Derryberry, 1998; Tucker & Derryberry, 1992; Tucker, Derryberry, & Luu, 2000). Although the notion of self-regulation is often used in the developmental literature in the folk psychology sense of social appropriateness (Eisenberg, Spinrad, & Eggum, 2010; Eisenberg et al., 2009), it should be apparent that, in a cybernetic analysis, self-regulation involves impulsive behavior as much as it does constraint. Furthermore, the evidence on child psychopathology indicates that excessive constraint, leading to internalizing disorders, can be as maladaptive as the impulsiveness of externalizing disorders.

A basic question for understanding psychological development is how a successful child draws on increasing cognitive capacities at each stage of development to organize abstract and complex forms of self-regulation from the primitive cybernetics of impulsiveness and constraint. An important theoretical model by Derryberry and Rothbart (Derryberry & Rothbart, 1997) emphasized that in addition to the reactive or primitive modes of self-regulation—impulsiveness and anxiety—children develop the capacity for *effortful control*, in which cognitive resources are engaged voluntarily in the organization of complex and deliberate coping efforts. This approach to self-regulation has been influential on both theoretical and empirical work on self-regulation in the developmental psychology literature (Derryberry & Tucker, 2006; Eisenberg et al., 2010; Eisenberg et al., 2009). The traditional view of developing cognitive appraisal skills has been reformulated in recent theoretical models to include more dynamic models of motive-cognitive interactions that are framed in terms of specific corticolimbic networks (Lewis, 2005). Could it be that each of the primitive motive biases—impulsiveness and anxiety—support a more complex form of effortful self-control? In the next two sections, we develop a neuropsychological framework for dual modes of controlling cognition—intentional and attentional—that may provide dual modes of effortful control.

## Structure and Process in Neuropsychological Theory

Self-regulation has also been a central concept in psychoanalytic models of child development, with clear implications for the internal versus external locus of control in experience and cognition. In *object relations* theory, the self is organized from object relations, relations with the significant interpersonal figures in the child's life (Bowlby, 1997; Mahler, 1968; Spitz, 1965). A fundamental object relation is *attachment*, in which the child forms a close and syncretic or undifferentiated bond with the mother or caregiver (Bowlby, 1997). The challenge of individuation from this bond is an essential step in the development of autonomy (Mahler, 1968). Because it is formed from these early primitive object relations, the self of the developing child builds on the social orientations of infancy and childhood as templates that shape both social interactions and cognitive appraisal of the world (Kohut, 1978). An important theoretical question is how the primitive motive controls of impulsiveness and anxiety, or more developed forms of the impetus and the artus, frame the developmental basis for object relations.

Modern psychiatry has for many years distanced itself from psychoanalysis. The goal seems to have been to become scientifically acceptable by adopting a descriptive, rather than theoretical, approach to psychopathology. The DSM clearly aspires to a descriptive, atheoretical, approach. Yet, although it is clearly atheoretical, the level of psychological description is fairly crude. The DSM is limited to symptoms that a physician may observe in a medical context, rather than the psychological processes that would explain the patient's life problems. Consistent with the medical, disease, model of mental illness, there is little recognition of the continuity of personality disorders with normal personality.

In contrast, psychological descriptions of personality disorders in the psychoanalytic tradition clearly recognized the continuity with normal personality. Furthermore, some of these accounts provided detailed descriptions of psychological mechanisms that could explain how exaggerated self-regulation in personality development could lead to chronic psychopathology (Shapiro, 1965). From psychological testing with intellectual as well as personality measures, psychologists working in this tradition observed patterns of cognitive organization that appeared consistently in people with certain motivational patterns.

Often derived from performance on cognitive and IQ tasks, these patterns of personality were often described as cognitive styles. In a classic summary, Shapiro (1965) described them as *neurotic styles*. Impulsive and hysteric (now called histrionic) personalities were found to show impressionistic and holistic cognition, with the most important issue for adjustment being their poor skills in analytic thought (Shapiro, 1965). Self-regulation in these persons appears loose, impulsive, and would be described in popular culture as a lack of self-regulation. Yet from a cybernetic perspective, impulsivity and emotional reactivity reflect an active form of self-regulation under motive demands, the native mode of the impetus. In contrast, obsessive-compulsive and paranoid personalities show exaggerated self-control, in cognition and behavior, to the point of pathology. Again, all cognition is controlled. These disorders may be understood as imbalanced control, an excess of the artus.

## Control Process and Cognitive Structure

We think that action regulation theory, as outlined in Chapter 3, provides a way to understand the intrinsic motive structuring of cognition, as seen in the unbalanced cognitive styles of the personality disorders. The qualitative features of the intrinsic cybernetics of the motive arousal mechanisms, including both the feedforward control of the habituation bias and the feedback control of redundancy bias, create structural modes of representation that become characteristic of the personalities who are dominated by one mode or the other. We can consider these intrinsic influences of control process on cognitive structure within the framework of mammalian frontolimbic mechanisms of action regulation.

In the literature on posterior brain contributions to action regulation reviewed in Chapter 3, the unique forms of memory and cognition in the dorsal and ventral *posterior* networks have been found to be complementary to the forms of motor control in the dorsal and ventral *anterior* networks. The projectional, feedforward mode of control of the dorsal motor process is congruent with the spatial, configural representations of the superior parietal networks, drawing on the holistic apprehension of the context provided by peripheral vision. There is a complementarity of process and structure in this integration, such that the rapid, fluid, and vectoral action motive of the dorsal pathway (control) is framed within the holistic and global apprehension of the current environmental context (representational structure). These are the balanced, synchronized somatic input/output channels of the impetus. The configural representation of the context for action is egocentric, consistent with the integral motive direction of memory from the limbic base: the context representation does not just appear spontaneously, but is continually shaped by the perception of personal affordances.

Thus the cybernetic process of projectional, vectoral, and intentional control of action in the dorsal pathway is closely aligned with the holistic and configural cognitive structure of the dorsal division of the brain. It seems likely that these organizational mechanisms of behavior extend beyond more elementary perception-action sequences to organize more abstract processes and structures of cognition. Holistic cognitive skills may emerge from the cybernetics of the impetus, as these are engaged in states of elation and hopeful expectancy. Consistent with this proposal, the creativity of manic persons has been documented by the high incidence of cyclothymia and bipolar disorder in creative writers (Andreasen & Canter, 1974; Andreasen & Powers, 1975) The creativity of mania, as measured by a remote associates test, is significantly decreased by administration of lithium (Shaw, Mann, Stokes, & Manevitz, 1986).

A similar alignment may be seen between the constrained and focused cybernetic process of the ventral frontal networks and the differentiated, analytic structure of cognition supported by the object perception of the posterior ventral networks. These complementary constraints on perception and action (somatic input and output) define the artus. The dynamic cognitive process, of feedback control of actions in relation to discrete perceptual targets, requires a specific cognitive structure, of analytic differentiation of the informational array into discrete yet meaningful objects. These elemental structure-process relations of action regulation may be extended to more complex mechanisms of cognition. The process of sequential cognition in relation to discrete references, such as in expressive grammatical language, must be tightly integrated with the analytic structure of perceptual and mnemonic representations that are differentiated as discrete semantic objects, such as words. This is an integral mode of control in normal cognition, yet when exaggerated may lead to the overly constricted and fragmented cognition and behavior of obsessive-compulsive disorders (Tucker & Derryberry, 1992).

## Spatiotemporal Complementarity in the Neurocybernetics of Activation and Arousal

Is this alignment of process and structure a result of simple functional convenience? Such that one form of representational structure just happens to be suited to the corresponding form of control process? Or is there a more mechanistic explanation, wherein the same neural operations of the posterior brain that support a certain form of representational structure are uniquely suited to generate—and be regulated by—the parallel form of control process in the frontal lobe?

A mechanistic theory of neural arousal that proposes an integral structural influence of elementary neural activity controls on cognition is the Tucker and Williamson (1984) model, of tonic activation and phasic arousal, that we introduced in Chapter 1. Based on the restriction of the range of behavior observed in animals given dopaminergic agonists, with higher doses leading to repetitive and stereotyped actions, Tucker and Williamson theorized that a *tonic activation system* in the brain produces a sensitization or *redundancy bias* on working memory, restricting the range of information in the current store and thereby increasing the sustained representation of that restricted information.

Because of the spatiotemporal complementarity inherent to primitive controls on neural activity, the redundancy bias of tonic activation leads to an integral cognitive structure. The focused attention from the redundancy bias limits the scope of active cell assemblies, and thereby necessitates a differentiated, analytic cognitive structure. Although Tucker and Williamson theorized that this mechanism could explain the analytic cognition of the left hemisphere, it could also explain the properties of the ventral divisions of both cerebral hemispheres. The object focus and differentiated cognitive structure of the ventral perceptual networks are aligned with the constrained, inhibitory, feedback control from the ventral frontal networks, and both may reflect the inherent restriction of scope applied by the redundancy bias (Tucker & Luu, 2006, 2007). As we will see in the more detailed analysis of neural mechanisms in Chapter 5, the control of both forebrain cholinergic and mesolimbic dopamine projections by ventral limbic pathways may be important to understanding this specific attentional mechanism. Constraint and differentiation in time lead to constraint and differentiation in representational space, through what appears to be a spatiotemporal trade-off in neuronal cell assemblies. This may occur because representations in corticothalamolimbic networks are not instantaneous, but are formed through the continuing operations of neural representation and control (and thus working memory) in time.

An interesting paradox arises in relating the notion of the tonic activation system to the ventral frontal control over action regulation. Whereas the redundancy bias fits well with the focused attention required for feedback control, and with the restriction of attention to motivationally salient objects, the tonic activation system was related to *motor readiness* (not perceptual arousal) in the Tucker and Williamson model. In contrast, in our theoretical analysis of action regulation the ventral frontal contribution is marked by the unique presence of *sensory* input to layer 4 of the ventral frontal networks (G. Goldberg, 1985; Tucker & Luu, 2007). The evolution of complex forms of working memory in mammals thus seems to have involved a paradoxical mix of the primitive cybernetics of motor control (the redundancy of dopamine and the basal ganglia) with a new capacity for sustaining the sensory model (in the ventrolateral frontal lobe), with its bias on the limbipetal consolidation of sensory constraints. The effect of this capacity is to provide sustained and focused feedback guidance of action. We will return to this paradox in Chapter 7, with additional insights from neurodevelopmental theory to be gained in Chapter 6.

An opponent and complementary control on cognition in the Tucker and Williamson model is the *habituation bias* emergent from the more phasic arousal supported by noradrenergic brainstem projection systems. When working memory is controlled by the habituation bias, representations are not sustained, but change rapidly. As a result of this form of process control, there is an inherent structural control, leading to a more expansive and holistic allocation of working memory. Although the habituation bias and its expansive attentional mode were applied to explain the holistic cognition of the right hemisphere in the Tucker and Williamson model, the specificity of noradrenergic projections for the dorsal networks of both hemispheres (Foote & Morrison, 1987; Morrison & Foote, 1986) suggests that the confluence of process and structure from the habituation bias may apply well to the alignment of the projectional control process of dorsal frontal networks with the holistic representational process of dorsal posterior networks (Tucker & Luu, 2006, 2007).

A similar paradox arises in the emphasis on the sensory versus motor bias if we align the phasic arousal system with the psychological construct of the impetus. In the action regulation formulation of the dorsal brain, the impetus is the impulse toward action, with a clear emphasis on motor control. The limbifugal direction of consolidation in the dorsal division is based in the viscero*motor* expression from limbic networks. Yet in the Tucker and Williamson model the phasic arousal system and its habituation bias are seen as the primitive basis for the control bias of *sensory* systems. We will see in the next chapters how this apparent paradox may also resolve with additional insight from neurodevelopmental theory. As with the effect of the redundancy bias, the operation of the habituation bias on the dorsal frontal lobe may have evolved to bring a primitive cybernetic mode (the habituation bias for phasic sensory arousal) to a new and unusual purpose, the fluid and projectional control of dynamic action in the motor system.

Thus the mammalian cortex has evolved with what may seem to be a paradoxical crossing of cybernetics, in which the habituation bias, apparently linked to primitive sensory control, becomes elaborated to support the projectional mode of motor control in the anterior brain. Similarly, the redundancy bias, linked to the motor sequencing and control of the basal ganglia, seems to have become integral to the constancy in working memory of the ventral frontal lobe that allows sensory guidance to be held in time to provide feedback control to the action plan. Although the controls on tonic activation (redundancy) and phasic arousal (habituation) are clearly primitive, we can see that mammalian cortical evolution has achieved new capacities in working memory through aligning new forms of complementary opposition within the cortical architecture. The habituation bias that evolved to allow sensory systems to adapt to, and ignore, constancy becomes redirected in a form of motor control (the impetus) that supports a fluid and impulsive form of action regulation. The redundancy bias that evolved to support constancy and continuity in motor systems becomes redirected to mediate a form of sensory constraint that provides a powerful and complex capacity in focused feedback control over action regulation.

### *Asymmetric Hemispheric Specialization*

An important theoretical question is whether the differing motivational and cognitive properties of dorsal and ventral corticolimbic systems—which are the key emphases of action regulation theory—could be related to the psychological specializations of the left and right hemispheres—which are the original emphases of the Tucker and Williamson theory of tonic activation and phasic arousal. If these models can be rationalized as mutually congruent, then we would have to consider a new view of hemispheric specialization, in which the right hemisphere elaborates the skills of the dorsal cortical division, and the left hemisphere elaborates the skills of the ventral cortical division. The psychological process of the impetus would then support the cognitive structure not only of the dorsal configural mode, but of the right hemisphere's specialization for holistic cognition. The psychological process of the artus would then support the cognitive structure not only of the ventral object mode, but of the left hemisphere's specialization for analytic cognition.

This framework for hemispheric specialization, which has apparently become exaggerated in recent human evolution, may be compatible with the differential alignment of the dorsal and ventral systems with limbic versus neocortical biases, respectively, in the direction of consolidation. Within the Goldberg and Costa model, the right hemisphere is specialized for intermodal integration, consistent with greater influence by the densely interconnected limbic base (Tucker,

1992; E. Goldberg & Costa, 1981). A right hemisphere elaboration of the dorsal networks would be consistent with this emphasis on a limbic bias in consolidation. Similarly, the Goldberg and Costa model proposed that the left hemisphere has become specialized for processing in unimodal sensory and primary motor networks. This would be compatible with a left hemisphere elaboration of the ventral pathways, with the corresponding emphasis on the limbipetal direction of consolidation, implementing the dominant influence of representations in primary cortices.

The initial differentiation of spatial memory in the dorsal visual pathway and object memory in the ventral visual pathway (Mishkin, 1982; Ungerleider & Mishkin, 1982) came at a time (the 1970s) when Pandya and his associates were demonstrating the anatomical segregation of the primate cortex into archicortical (dorsal) and paleocortical (ventral) moieties (Barbas & Pandya, 1984; Galaburda & Pandya, 1982; Pandya, & Seltzer, 1982; Pandya & Yeterian, 1984). The interesting question for neuropsychologists who studied anatomy carefully was how human hemispheric specialization, which also involves holistic versus analytic (object-based) cognition, could have elaborated on this more fundamental functional division between dorsal and ventral pathways.

Galaburda (Galaburda, 1984) took on this question with a creative formulation that integrated the cognitive, representational aspects of hemispheric specialization with the evidence of the right hemisphere's unique capabilities in understanding and expressing emotion. Beginning with an evolutionary analysis of the emergence of archicortical and paleocortical divisions of the brain, Galaburda (1984) pointed out that in reptiles and amphibians the archipallium (on the medial wall of the telencephalic vessicle) receives major input from the hypothalamus, whereas the lateral pallium does not. This suggested that the archipallium may be specialized for processing information about the organism's internal state. The lateral pallium (progenitor of the mammalian paleocortex) receives olfactory input, suggesting to Galaburda that it may be specialized for processing information about the external environment. Extending these specializations to the dorsal (archicortical) and ventral (paleocortical) divisions of the mammalian brain, there is a close and interesting parallel with the Jeannerod and Jacob (2005) model of dorsal cortex handling of internal—and ventral cortex handling of external—motive and cognitive perspectives.

The hypothalamic influence over the dorsal brain may be relevant to right hemisphere specialization for emotion. Galaburda proposed that the human right hemisphere has become specialized to elaborate the functions of the dorsal cortical networks. In support of this notion, he pointed out that a region of the dorsal parietal lobe with an elaborated pyramidal layer (a cytoarchitectonic feature of archicortical derivation) was found to be larger in the right hemisphere of the human brain (Eidelberg & Galaburda, 1984). In support of a corresponding left hemisphere specialization for the functions of the ventral brain, Galaburda (1984) reviewed anatomical evidence that BA 39 on the angular gyrus, a region of association cortex with granular (layer 4) cytoarchitectonics (characteristic of paleocortex), is larger in the human left hemisphere.

A neurodevelopmental analysis then provided important anatomical evidence for Galaburda's theory of functional specialization. In examining Fontes's photographs of human fetal brains, Galaburda (1984) discovered hemispheric asymmetries in the emergence of sulci in the dorsomedial wall and perisylvian fissure. The sulci on the dorsomedial wall are *slower* to develop in the right hemisphere, suggesting to Galaburda that the right hemisphere develops these networks not only more slowly but more extensively than does the left. Similarly, the perisylvian sulci develop more slowly in the left hemisphere, possibly consistent with eventual left hemisphere elaboration for these paleocortical networks in human development. In both cases, slower development would be consistent with greater elaboration in the neotenous (retarded) developmental course of human juveniles.

Integrating the dorsal/ventral differentiation of internal and external control with these observations on developmental asymmetries, Galaburda (1984) proposed that the right hemisphere's capacities in nonverbal emotional communication may stem from its elaboration of the dorsal networks, with their intrinsic hypothalamic connections, consistent with a specialization for internal control. In contrast, the left hemisphere's language skills may emerge in large part from its elaborated ventral cortical networks, reflecting a specialization for external information exchange.

Galaburda's theorizing, although not widely influential at the time, provided a clear illustration of the theoretical insights that could be gained by interpreting hemispheric function in light of more the fundamental anatomical divisions of corticolimbic networks. In following this line of reasoning, Liotti and Tucker (Liotti & Tucker, 1994) emphasized that both dorsal and ventral pathways may have unique motivational and emotional properties. Drawing on theoretical work considering psychological implications of lesion, imaging, and animal studies (Derryberry & Tucker, 1994; Tucker & Derryberry, 1992), Liotti and Tucker proposed that anxiety and negative affect closely modulate the processing of ventral limbic networks, and that both anxiety and ventral limbic control are particularly important to the function of the left hemisphere. Consistent with the earlier Tucker and Williamson (1984) model, the Liotti and Tucker formulation continued to emphasize the alignment of motivational process with specific biases on conceptual structure. Greater motivational influence on dorsal networks (elaborated within the right hemisphere) would lead to a more holistic representational scope. Greater engagement of ventral networks (elaborated within the left hemisphere) would lead to a more restricted, focused representational scope. As we will see next, these general concepts of dorsal and ventral corticolimbic (and hemispheric) networks can be seen as extensions of the primitive neuromodulator controls on neural activity.

## *Integral Affective Qualities of Intention and Attention*

Thus several lines of reasoning suggest that there may be an integral relation between the process of neural control and the structure of conceptualization. Formulated in some contexts, the motivational control biases have been related to hemispheric asymmetries, and in others to dorsal and ventral corticolimbic mechanisms. We propose that the corticolimbic analysis of action regulation, as provided in Chapter 3, may provide a systematic theoretical basis for understanding psychology of emotional orientations, motivational biases, and the cognitive operations of cortical and subcortical networks. The hemispheric asymmetries of dorsal and ventral networks in fetal development of the cortex, as pointed out by Galaburda, may be essential in extending the theoretical analysis from basic processes of sensory and motor consolidation to account for more complex neuropsychological functions.

The *impetus*, emergent from dorsal corticolimbic networks, may be integral not only to the impulsive form of motivational control associated with elation, but also to an expansive and holistic representational structure that has become integral to human right hemisphere specialization. In the primitive form, this is the egocentric impulsivity of reactive temperament (Derryberry & Rothbart, 1997). Yet it also may become elaborated through the integration of cognitive structure with motive impulse, leading to a form of effortful control that is closely aligned with *intentional* direction of behavior. Personal expectancy, with an integral positive hedonic quality, guides action. The observer of the child who is self-regulating by the impetus describes the behavior as externalizing, yet with this motive bias, cognition is drawn to internal, personally attractive and intentional, goals. With the affective modulation of a certain degree of elation, decisions are made with confidence.

The *artus*, operative within ventral corticolimbic networks, provides the opponent control, of constraint and restriction associated with anxiety. With the structural effect of constraint cybernetics, the artus creates a differentiated and analytic representational structure that has become a foundation of human left hemisphere specialization. In the elaboration of effortful control within this mode, the shift to an allocentric perspective is integral to the focus on external criteria. This allocentric shift provides a motive bias toward sensitivity to external events that is essential for the *attentional* control of cognition. For the child whose experience is dominated by this control mode, the external observer describes behavior as internalizing. Yet the inhibition and constraint of behavior are achieved through an allocentric focus, in which actions are linked to focused attention to the objects of the world.

## Visceral and Social Contexts of Human Self-Regulation

A neurodevelopmental analysis of cognition thus may address the psychological mechanisms of intentionality and attentional control that mediate between personal needs and the interpersonal context that shapes developmental self-regulation. We have argued that the neural architecture of memory consolidation points to an arbitration between visceral needs at the limbic core and the somatic requirements for contact with the external context at the sensory and motor networks of the neocortical shell (Tucker, 2001, 2007; Tucker & Luu, 2006). In mammalian evolution, the capacity for memory emerged in parallel with the capacity for social attachment. Fundamentally, the opportunities for neurodevelopmental learning were provided by an extended juvenile period under parental care.

The general process for developmental self-regulation, through the viscerosomatic mechanisms of memory consolidation, is a process of negotiating concepts that mediate between internal visceral needs and the complex demands of organizing a self in an interpersonal context. Although both these boundaries, visceral and social, provide essential constraints, cognition is the primary arena for human self-regulation. Extending the mammalian trend, humans do not interface the world through simple stimulus-response reflexes operating in immediate negotiations between the world and the internal biological milieu. Rather, we operate through cognition. Through cognition, past experience is consolidated into integrated values, beliefs, and attitudes, in order to anticipate future events through motivated (worrisome, hopeful, playful) expectancies.

In this process, motive self-regulation is first and foremost a process of regulating cognition. Even in complex and abstract acts of intelligence, the frontolimbic mechanisms of hypothesis formation, for example, must self-regulate the formation of hedonic expectancies through engaging adequate motive control, such as the phasic arousal that is elaborated and experienced subjectively as hopeful optimism. We have seen that critical reasoning, in another example, requires the engagement of focused cognition of ventrolateral frontolimbic networks, supporting the reference to external criteria. This form of cognition requires the motive direction of the ventral networks by degrees of anxiety and hostility, the integral affects of reason.

Of course, in the effort to demonstrate scientific objectivity, cognitive psychology, like cognitive science and cognitive neuroscience, has proceeded under the assumption that cognition is inherently objective, and can be controlled fully without recourse to affect and motivation. But within the neurodevelopmental approach, as we examine the neural mechanisms of elementary processes of action regulation, movement requires motivation. The brain's cognitive capacities have evolved only because they serve adaptive ends. Motivational control is essential for forming representations, for structuring them within working memory, and for relating them one to another.

This is not to say that there are not levels of representation and control, where certain (limbic) levels are strongly affective and motivationally charged (and concerned more with control), whereas other (neocortical association) levels are less charged (and concerned more with representation). As we saw in Chapter 2, classical neuropsychological observations on the differential effects of neocortical versus limbic lesions support this distinction of levels of motivational charge (Monrad-Krohn, 1924). A frontal neocortical lesion impairs voluntary control of facial movements contralateral to the lesion, but leaves intact the spontaneous emotional expression, emergent primarily from limbic regulation (Jürgens & Ploog, 1970). Nonetheless, even though each level of the cortex (limbic, heteromodal association, unimodal association, primary sensory and motor) engages differing degrees of limbic charge, the effective control of cognition requires engaging the limbic base. The same limbic mechanisms that consolidate memory do so through the process of motive engagement, operating through the motive charge provided by the visceral limbic functions (Luu & Tucker, 2003; Tucker & Luu, 2006).

Interpreting the anatomical architecture literally, the complementary boundary to visceral self-regulation of the hemisphere is not an abstract cognitive representation in the neocortex, but the somatic function, linked to the primary sensory and motor cortices (Luu & Tucker, 2003; Tucker & Luu, 2006). Although it has been traditional in neuropsychology to consider the cortex in an opponent balance with subcortical and limbic influences, exerting inhibitory control over them (Monrad-Krohn, 1924), the modern anatomical evidence on network organization implies that the cortex must integrate four levels, within which the two (heteromodal and unimodal) association areas must form representations that link the visceral milieu with the somatic surface, thereby integrating present values and historical knowledge for ongoing self-regulation in the sensorimotor flux.

Because the context for both values and knowledge is social, the enduring concepts for cognitive self-regulation are those that reflect the residuals of significant social interactions. Understanding the dorsal-ventral differentiation within this developmental framework may require appreciating the social significance of the specialization of dorsal limbic networks for visceromotor control, versus ventral limbic networks for viscerosensory control (Neafsey, 1990; Neafsey, Terreberry, Hurley, Ruit, & Frysztak, 1993). Elaborated as cognitive controls, the motivational biases emergent from dorsal and ventral limbic systems may be as fundamental to the organization of cognition as the differentiation of somatic interface systems between anterior motor and posterior perceptual systems (Tucker & Luu, 2006, 2007). Although we can only outline the implications within the scope of the present treatment, these implications are complex, at first paradoxical, and important for understanding the locus of control in the social context.

## Paradoxes of Internal and External Control in the Social Context

The projectional control of action, with its integral hedonic expectancy, appears to arise from the visceromotor function of the dorsal limbic networks. We have argued that complex psychological processes of intentionality and volition may emerge from the primitive motive basis of the impetus. Consistent with the motive control of action regulation, the intentionality of dorsal networks is typically egocentric, reflecting the individual's internal motivation for action (Jeannerod, 1994). The basic description of the impetus and the artus within the child's developing socialization would then have an interesting psychodynamic quality, pitting the egocentric urges of the impetus against the social constraints represented by, and mediated by, the artus.

Yet, as we have seen for the paradoxes in the cybernetics of sensory and motor systems, mammalian (and human) complexity seems to have elaborated control modes for new purposes. In studies of how children learn words, for example, Baldwin and her associates have observed that the toddler learns that a word has meaning through attending to the mother's intention when she speaks (Baldwin, 1991, 1993; Baldwin & Markman, 1989). Hearing the word associated with an object does not convey meaning. The word is not an arbitrary associate of the reference, but is rather a signal of the mother's communicative intent.

Gaining language thus requires the capacity for sharing intentionality. The child's internal representation of the mother's intentionality would appear to emerge from the mirroring functions (Rizzolatti, Fadiga, Fogassi, & Gallese, 1999) of the child's own mechanisms for representing intentions. These mirroring functions are then applied to the challenge of understanding the mother's intentions. Through his logical analysis of the cognitive requirements for interpreting speech, the philosopher Paul Grice (Grice, 1957) concluded that forming a model of the speaker's intentions is an essential first step. This process of modeling the other's communicative intent remains important in modern biological analyses of language (Givon, 2009).

Sharing intentionality is a complex capacity in the cognitive representations that are motivated by social attachment. This capacity is another paradoxical effect in the psychological locus of control emergent from the differential cybernetics of the impetus and the artus. As with the basic analysis of action regulation, the dorsal networks organize action on the basis of internal control, as urges arise directly through visceromotor control mechanisms of the dorsal limbic networks (Tucker & Luu, 2007). This is fundamentally an egocentric mode of control, organizing cognition in the limbifugal direction, projecting actions from their motive base toward the world. However, perhaps paradoxically, interpersonal attachment appears to align the representation and control properties of the dorsal networks with the hedonic expectancy of the attachment relation, such that the immediacy of affective response and projectional cybernetics support a primitive, direct synchronization of feelings and actions within the relational bond (Mahler, 1968). This appears to be an essential foundation for communication, and without it the child remains autistic (Mahler, 1968).

In a parallel yet opposite sense, the artus arising from ventral corticolimbic networks is fundamentally oriented to external control (Jeannerod, 1994). The sensory objects of the environment are maintained within ventral frontal networks as criteria for action monitoring. These motor and cognitive cybernetics are also closely aligned with the limbic motive base, where the insula and anterior temporal networks regulate viscerosensory regulation (Tucker & Luu, 2007). However, even as there is a dominance of the sensory representations on the consolidation process (a limbipetal direction of control), the focused differentiation of objects that is created by ventral limbic (viscerosensory) control seems to lead to a separation of self from context, seen in extreme form in autism, and in more typical form in the vigilance of anxiety (Tucker, 1989; Tucker & Williamson, 1984). The motive bias of viscerosensory control, anxiety, focuses attention on external criteria but the integral allocentric perspective differentiates the self from the social context. In the extreme form this is autism, but in the adaptive form it is individuation (Mahler, 1968).

The self-regulation of social adaptation is achieved over time through the self-regulation of cognition. When the interpersonal context is congruent with the hedonic context model, then feedforward control by dorsal networks is adequate for synchronizing internal and external information in a kind of intersubjective whole. But when there is a discrepant prediction, as when the child's intentions diverge from the parent's, then anxiety and ventral limbic engagement leads not only to a cognitive representation of the external control, but a paradoxical separation of the self from the represented external context.

### Losing the Locus of Control

The autistic child has a specific deficit in sharing interpersonal intentionality, a capacity known as intersubjectivity or "theory of mind" (Carlson & Moses, 2001; den Ouden, Frith, Frith, & Blakemore, 2005; Grossberg, 2000; Moses, 2001). The autistic child is not only isolated from intersubjective sharing, but is at risk for poor language development. The perceptual fascinations and behavioral stereotypies of autism suggest exaggeration of the perceptual constraints of the ventral frontal lobe and basal ganglia. The core features, of cognitive autism and aversion to interpersonal contact, themselves suggest fundamental deficits in the hedonic social arousal and attachment that in normal children are closely aligned with the capacities for intersubjectivity and sharing of intentions that are mediated by dorsal frontal regions.

The external control delusions of schizophrenia may suggest a different but related form of neurodevelopmental impairment of the impetus within mediodorsal frontal networks. Without the sense of personal agency associated with the intentional control of action, the schizophrenic fills in the blanks, as it were, and experiences control of thoughts and actions as arising from an external source (den Ouden et al., 2005; Farrer et al., 2004; Farrer & Frith, 2002). Although the neurodevelopmental dysfunction in schizophrenia is more complex than a simple brain lesion, it is similar in form to the delusions of external control with an extensive mediodorsal frontal lesion. In the *alien hand sign* resulting from a mediodorsal frontal lesion, the patient fails to perceive ownership of the action. With motor control restricted to the intact ventral frontal networks, the action appears to arise externally, not from the self (DellaSala, Marchetti, & Spinnler, 1991; G. Goldberg, Mayer, & Toglia, 1981). Through a neurodevelopmental disorder not yet fully understood, the schizophrenic person loses the sense of agency associated with the volitional control of actions, leading to a loss of the intentional coherence in cognition, an autistic disintegration of the self, and, as a result, delusions of external control.

### *Motive Process, Locus of Control, and Cognitive Structure*

The integral alignment of affective self-regulation, locus of control, and cognitive structure can be seen not only in the severe neurodevelopmental disorders of autism and schizophrenia, but in normal personality and its less severe pathological exaggerations. We considered the integral influences of motive state (elation and anxiety) on cognition in the section above on neural mechanisms of normal personality. There also may be predictable shifts in the locus of control created by motivational mechanisms in normal persons. Although normal human self-regulation is organized through a kind of tensile balance of the impetus and the artus, children may develop a reliance on one mode or the other as a function of complex patterns of genetic bias interacting with the exigencies of environmental challenges presented by their specific family contexts.

The motive mechanisms regulating neural activity in development seem to entail an inherent spatiotemporal complementarity, thereby shaping the structure of cognitive representation. This shaping occurs during specific emotional states, and yet it may produce the personality structures that endure over a lifetime. The motive mechanisms are also affective mechanisms, of tonic activation (anxiety) and phasic arousal (elation). Each of these applies an inherent bias on the structure of cognitive representation.

The dynamic modulation of cognition by motive activation and arousal are best seen in the exaggerated cognition associated with strong mood states (Thayer, 1989). The phasic arousal dimension, experienced as Positive Affect or elation, expands the scope of working memory, and thereby facilitates holistic concepts (Tucker & Williamson, 1984). In strong states of elation, such as clinical mania, this broad scope and inherent optimistic affective bias lead to grandiosity, which is not only broad in scope (Shaw et al., 1986), but also reflects the inherent egocentrism of dorsal corticolimbic processing (Jeannerod & Jacob, 2005; Tucker & Luu, 2007). In contrast, the same individual when depressed shows a deflation not only of affective arousal, but of the felt importance of the self.

A similar shaping of cognitive structure by motive state can be seen in the exaggerated states of anxiety and hostility. In this case, the redundancy bias on working memory leads to a pathological focusing of attention and cognitive structure. Here the chronic effects of unbalanced neurodevelopmental self-regulation through chronic Negative Affect are apparent in the cognitive styles of certain personality disorders, including paranoid and obsessive-compulsive disorders (Shapiro, 1965). The persons with these disorders show not just isolated symptoms, but integrated patterns of cognitive distortion and personality disorder. They become obsessed with detail, and develop cognitive skills, including analytic reasoning, that reflect long experience with the discipline of focused attention. The rigidity of cognitive function, directly traceable to the redundancy bias, becomes an essential theme of the person's character (Shapiro, 1981). Although pathological in many ways, the obsessive rigidity can be an effective self-regulatory device, allowing the person great tenacity in the persistent application of the restricted coping effort (Shapiro, 1965).

An opposite and contrasting self-regulatory pattern is seen in the histrionic and impulsive personality disorders, in which the defining deficit is a lack of self-organization, caused by inadequate influence of the artus. Without the focusing of attention and working memory of the ventral limbic influence, these persons show not only the lack of impulse control and poor anticipation of negative consequences (such as in the ineffective anxiety of the psychopath) but structural cognitive deficits in analytic reasoning (Shapiro, 1965). The apparent denial of problems and Pollyannaish optimism of the histrionic personality seem to reflect a chronic self-regulation through the impetus, with its holistic, impressionistic cognition, limbifugal infusion of affect in the cognitive process, and positive hedonic expectancy. The problem, of course, is not just the self-regulation by the impetus, but the fact that self-regulation in these persons is unbalanced by the countervailing anxiety of the artus.

Psychological process (motivational control) is thus integral to cognitive structure (the organizational pattern of representations). As we have seen, the motivational bias also has inherent implications for the person's locus of control, producing disorders of internalizing, related to excess anxiety and depression, and externalizing, related to impulsiveness and conduct disorder (Achenbach, 1982; Cicchetti & Tucker, 1994). In adult personality, and in its pathological variants, the personalities displaying opposite poles in the locus of control were known as *introversion* and *extraversion* in classical models, similar to the dimensions of social propriety and dynamism in modern psychometrics (Thalmeyer & Saucier, in press). These personality types and their characteristic self-world orientations can be seen as emerging from the ventral and dorsal motive biases, respectively. Personality is thereby formed by the confluence of affective orientation, structure of cognition, and personal locus of control.

Consistent with strong self-regulation by the artus, the introvert is sensitive to external constraints, is typically anxious and careful, and is pessimistic about the likelihood of hedonic success. This *attentional* mode is concerned with external constraints, with the vector of cognition being from the world to the self (Jeannerod, 1994). At the same time, however, the paradoxical effect of this attentional bias is to provide the introvert with autonomy from the interpersonal context. The well-developed capacity for constraint over his own impulses allows the introvert a kind of deliberate control over behavior that may be integral to effective autonomy.

In contrast, by self-regulating primarily through the impetus, and its limbifugal projection, the extravert acts freely, *intentionally*, on the world, readily actualizing personal impulses within a context model that expects hedonic success. As we described above for the paradoxical locus of control of the impetus, although the vector of cognition is from the self to the world (Jeannerod, 1994), the extravert's self-regulation through the habituation bias leads to a close fusion of experience and working memory with the flow of changes in the environment. It is only in the short range that the extravert's cognition is exclusively expressive, self-to-world. Over time, the integral cybernetics of this control mode cause the extravert's cognition to be externally directed, captured by the salient events in the world.

Internality and externality thus have somewhat paradoxical relations to the modes of introversion and extraversion. What is important for each is the cybernetic mechanism that emerges from the elemental control of neural activity over time, through habituation or redundancy, together with the direction of consolidation that is thereby engaged. For the extravert, the loose, feedforward mode of control of the habituation bias causes behavior to be internally directed in the short term, because it is shaped by the internal impulse, and projected onto the world with minimal constraint. Yet in the longer term, this mode of cognition becomes externally directed. This is because, given the spatiotemporal complementarity of the habituation bias over time, cognition is soon captured by the flux of external events, and there is little historical continuity of internal direction.

For the introvert, cognition is shaped by the cybernetics of constraint. Behavior is externally directed in the short term, because attention is focused on external sensory events as the criteria for actions, and personal impulses are restricted. Yet, because of the same constraint, in the longer time frame behavior becomes internally directed, because the introvert is separated from experiential fusion with the flow of events, and is constrained from acting in impulsive response to those events.

Although personality disorders reflect chronic imbalance in motive self-regulation, they clearly illustrate the motive processes that shape both the locus of control and cognitive structure dynamically as they are elicited. The cognition of the histrionic person becomes highly impressionistic, undifferentiated, and Pollyannaish under emotional arousal (Shapiro, 1965). The psychopath's cognition becomes pathologically optimistic, and incapable of anticipating obvious threats when engaged by the expectancy of hedonic pleasures.

In contrast, the obsessive-compulsive person experiences anxiety, and the obsessions or compulsions are a direct result. Although clinicians often interpret the symptoms as ways of controlling the anxiety, and anxiety does become intense if symptoms are not allowed to express, the more fundamental causality is from the motive influence of anxiety and its redundancy bias on the action and cognition pattern generating mechanisms of the ventral limbic and basal ganglia circuitry. Anxiety structures the mind. Similarly, in paranoid personality disorders, anxiety and hostility produce the cognitive and attentional vigilance of this disorder in direct proportion to their motive levels. The world-to-self direction of consolidation becomes intense to the point of chronic paranoid vigilance. Of course, an increase in vigilance and suspicion are normal consequences of anxiety and hostility. What becomes disordered in the paranoid personality is the cognitive process that should allow relaxation of the suspicious state as the threat is not confirmed. In large part, it is the rigidity of cognition, ironically caused by the intense focus of the redundancy bias of anxiety and hostility, that leads to the vicious circle of chronic suspiciousness.

### *Drug Manipulations of Motive Controls*

Compelling scientific evidence of direct effects of modes of neurophysiological arousal on cognitive structure and the locus of control has come from studies of stimulant drug use. Consistent with the model that norepinephrine is associated with a habituation bias, whereas dopamine produces a redundancy bias (Tucker & Williamson, 1984), animal studies have shown that, whereas both norepinephrine and dopamine are released by amphetamine, the norepinephrine release is transient (habituating rapidly), whereas the dopamine release continues with repeated drug ingestion (consistent with a redundancy bias) (Kokkinidis & Anisman, 1980). These different time courses may explain the euphoria (phasic arousal) produced by the initial amphetamine uptake, which soon habituates to uncover sustained anxiety and vigilance (tonic activation) with chronic use. These are mechanistic effects of the control of neural activity, yet they are pervasive throughout the individual's psychological organization. Amphetamine and cocaine abusers often emphasize that it is not simple pleasure, but the feeling of personal importance and power that is highly addictive in the initial drug state euphoria.

As the elation wanes, the chronic anxiety of exaggerated dopaminergic modulation produces predictable distortions of cognition and behavior. Amphetamine given to normal persons in high doses soon produces a paranoia that makes them fear for their lives (Kokkinidis & Anisman, 1980). The person under chronic amphetamine intoxication shows fixation of stereotyped cognition and routinized, compulsive behavior, not unlike rats given continued high doses of dopamine agonists (Iversen, 1977). In the course of an amphetamine binge, for example, motorcycle gang members have been observed to dissemble and reassemble their motorcycles compulsively (Ellinwood, 1967). At the same time, the chronic vigilance of anxiety (unbalanced by the elation that soon habituates) distorts cognition and the locus of control predictably, leading to drug-induced alienation and paranoia (Ellinwood, 1967).

### Deficits of Social Self-Regulation With Dorsal and Ventral Brain Lesions

Evidence that the primary vectors of self-control can be related directly to differential influences of dorsal and ventral frontolimbic systems comes from studies of the effects of frontal lobe lesions on personality. Lesions of the ventral limbic and orbital frontal networks lead to what is classically described as the *disinhibition syndrome*. The textbook example is Phineas Gage, whose frontal lesion led to impulsivity and loss of social self-awareness (Damasio, Grabowski, Frank, Galaburda, & Damasio, 1994). A systematic survey of the effects of orbital frontal brain lesions (Blumer & Benson, 1975) concluded that the common psychological effect could be described as the *pseudopsychopathic syndrome*. The patients lost critical control of their impulses and their capacity to predict negative social consequences of their actions. These are capacities that in normal persons are provided by the critical constraint of the artus, emerging from intact ventral frontolimbic networks.

In contrast, lesions to the mediodorsal frontal lobe lead to the *pseudodepression syndrome* (Blumer & Benson, 1975). Similar to the loss of initiative in the related disorders of akinetic mutism and transcortical motor aphasia, the patient shows a paucity of behavior and blunted affect. The integral role of hedonic expectancy in the impetus is clearly illustrated in this disorder. From another clinical perspective, the integral role of the behavioral impulse in the hedonic affect state is illustrated by the *psychomotor retardation* of psychiatric depression. In this disorder there is no obvious brain damage, but the exaggerated affective depression impairs the essential motive control of the impetus. As a result, the person loses all behavioral initiative (Tucker & Luu, 2007). The primary defect in depression is a disruption of the mood system, yet there is a direct manifestation of the affective disruption in the cognitive and motor systems as well, showing the integral and specific motive basis of action regulation.

## Opponent Complementarity in the Organization of Experience

Thus the impetus and the artus provide opponent and complementary modes of control that must be balanced in the normal personality. Each of them creates cybernetic modes of self-regulation in the social context that are at the same time familiar to everyday experience, and yet somewhat paradoxical in the way that needs and impulses are balanced with the constraints of the social environment. Although we have only outlined these modes of self-regulation and their psychological implications briefly, it should be clear that each of them is associated with unique structural influences on cognitive representation, whether it is the holism of the extravert's characteristic elation and habituation bias or the differentiation of detail provided by the introvert's anxious, repetitive ruminations.

## Cognition Is the Neurodevelopmental Process

Thus a framework of motivated action regulation, such as we developed in Chapter 3, can be extended to more complex questions of psychological theory informed by neuropsychological evidence. There is an integral relation of the control process, whether projectional or reactive, to the psychological structure, whether holistic or differentiated. Furthermore, there is an integral motive bias to the self-regulatory mechanism that is key to regulating learning. The hedonic charge of the impetus—the integral hope of elation—provides the valued goal orientation of expectant learning. The aversive charge of the artus—the integral dread of anxiety—supports the critical constraint integral to feedback control of behavior in relation to external requirements. These elementary control systems are opponent. Each can be biased with its unique cybernetics because its influence is continually balanced by that of its opponent and complementary mechanism.

Although these motive control systems may be easiest to recognize in relation to the self-regulation of psychological performance, we assert that they also describe psychological, and neural, development. We hypothesize that there is at least a strong continuity, and very likely a complete identity, between the elementary neurodevelopmental mechanisms in embryonic development and the continuing control of mammalian activation, arousal, and learning throughout the postnatal life span. The motive modes of regulating mammalian neural development may also be the mechanisms of human psychological self-regulation.

### Self-Regulation of Neural Morphogenesis

The challenge for neurodevelopmental mechanisms in embryogenesis is to organize widespread cerebral networks coherently, drawing on elemental mechanisms of exuberant synaptogenesis and activity-dependent pruning. Important aspects of embryonic neural plasticity are regulated by brainstem neuromodulator projection systems (Trevarthen, 1985). A similar influence is applied by the neuromodulator systems, such as NE and ACh, on neural plasticity at critical periods of postnatal development (Bear & Singer, 1986). The neuromodulator influences on neural plasticity thus appear to continue in postnatal development, where they guide the neurodevelopmental process.

To keep the present theoretical analysis tractable within our limited knowledge, we have retained a simplistic model of neuromodulation, in which several specific influences are summarized under the two notions of redundancy and habituation. A more explicit theoretical analysis is clearly required. Each of the more complex patterns of neural architecture and function, including the limbic and thalamic mechanisms of controlling consolidation in the cortical architecture, can be understood as elaborating the elemental cybernetics of the neuromodulator systems.

The parsimonious interpretation is then that the regulation of neural plasticity continues, guided by essentially the same mechanisms, throughout development. Learning is neural morphogenesis continued. If so, then the mechanisms of memory consolidation can be understood as the mechanisms of extended morphogenesis, shaping the brain's architecture as the cumulative actualization of life experiences. Similarly, the motive control systems guiding consolidation and learning, what we have described as the opponent mechanisms of self-regulation, may be understood as neurodevelopmental mechanisms, shaping neural activity and thus the organization of memory throughout life.

The structural requirements for organizing corticolimbic networks through embryonic activity-dependent specification may be similar to the requirements for organizing memory consolidation in the process of action regulation. Engagement and synchronization of large-scale networks is essential to provide an effective global organization of embryonic neural systems. In embryogenesis, the core reactivity of diencephalic (hypothalamic and thalamic) circuitry is essential to recruit the brainstem neuromodulator systems to condition the plasticity of the widespread corticolimbic networks to which they project. A similar conditioning may be essential in postnatal learning, mediating the specific modes of reactivity in limbic networks that then guide specific modes of corticolimbic arbitration within the hemisphere. Although global synchronization is essential to provide the initial framework, the specification of network representations, largely through local cortical inhibitory networks, is also necessary to the synaptic pruning process in embryogenesis, and perhaps in cognition. The integral alignment of the control process, whether impulse or constraint, with the representational process, whether holistic or discrete, may be a general feature of neural mechanisms at multiple frames of the neurodevelopmental continuum. As we will see in the next two chapters, understanding these multiple frames requires an evolutionary-developmental analysis. In each frame, the organizational process shaping the network anatomy is invariably motivated.

## *Shaping Connections Through Expectant Intention*

As reviewed in Chapter 1, the modern literature on animal learning shows that the learning capacities of mammals imply elementary cognitive processes, including holding expectancies for valued goals, and evaluating the discrepancy of familiar predictions (Papini, 2003; Papini & Bitterman, 1990; Rescorla & Wagner, 1972). These cognitive processes of learning may be aligned with the dual modes of action regulation. To support the projectional or feedforward control of action in the mediodorsal frontal lobe, there may be a kind of projectional expectancy, in which the organism anticipates future goals. Consistent with the lack of sensory feedback in this motor pathway, the mediodorsal expectancy is vectoral and impulsive, actualizing the initial internal urge toward a desired goal, with minimal adjustment by ongoing feedback. Perhaps because of this control property, the projectional learning mechanism appears to be tied to approach behavior particularly (Tucker & Luu, 2007). Given the evidence from Gabriel's studies of cingulothalamic lesions (Gabriel et al., 1996; Gabriel, Vogt, Kubota, Poremba, & Kang, 1991), the projectional control from dorsal corticolimbic networks, consolidated by the hippocampus, can be described as an incremental, gradual learning mode, with the conceptual organization of the context for action achieved within the configural, spatial representations of these networks.

Although it may seem anthropomorphic when applied to simpler mammals, this dorsal learning mode could be described as *intentional*, in that it is organized in a vectoral, projectional fashion with an inherent, egocentric, goal orientation. Within the impetus, the representation of the goal-in-context is an early, syncretic guide to the feedforward control of the microgenesis of action. Characterized in terms of neural control theory, it may be that even rats have intentions, in the form of vectoral action motives engendered by elemental cognitive representations of rat goals. This may be the elemental form of expectant animal learning (Krechevsky, 1932). More fully elaborated in humans, this dorsal frontolimbic mechanism of intentionality may be an integral component of the conscious, volitional control of experience and behavior (Tucker et al., 2007). The elaboration of frontopolar networks in recent human evolution (Semendeferi, Armstrong, Schleicher, Zilles, & Van Hoesen, 2001) can be interpreted in relation to the unique anatomy of the frontal poles. The frontal regions include bilateral control of both hemispheres, not just the ipsilateral hemisphere (Goldman-Rakic & Schwartz, 1982). Furthermore, the frontal pole of each hemisphere exerts control over thalamocortical regulation through its projection to the thalamic reticular nucleus or TRN (Zikopoulos & Barbas, 2006). For the dorsal division of the frontopolar region, the projections to the rostral TRN proceed to the anterior nuclei of the thalamus, which regulate the cingulate cortex. In this way the dorsal frontal polar region can be seen as effecting the *representation of the regulatory function*, a form of cognitive process that directly engages subcortical controls in a way that directs corticolimbic processing hierarchically. This may be a way to understand the neural mechanisms of effortful control emerging from impulsivity. With a base in the elementary motive impetus from dorsal limbic networks, the dorsal frontal pole allows a hierarchic organization of intentionality to achieve a more cognitively mediated and flexible form of self-regulation.

Thus the primitive control of neural activity, such as with the habituation bias of phasic arousal, may provide a foundation for more complex forms of self-regulation in psychological development. We have seen that this motive bias supports holistic or expansive cognition, such as that of the configural representations of the dorsal division of cortex, and such as may be elaborated particularly by the right hemisphere. This manifestation in cognitive structure may also reflect neurodevelopmental structure. The primitive neural cybernetics may be integral to neurodevelopmental mechanisms for regulating activity in developing neural networks. Under the principle of spatiotemporal complementarity, the habituation bias attenuates ongoing activity within a current neuronal cell assembly. In a developing neural system, the effect is to minimize the dominance exerted by current activity on the connectivity of the system. Such an effect may be appropriate as an integral cybernetic mechanism of a feedforward approach motive, to sustain developmental activity only long enough for adaptive completion of the approach behavior, and to habituate rather than become fixated by pleasure. Under rapid habituation, working memory would be well suited to the novel exploratory behavior that is appropriate to successful, unstressful adaptation. As we saw in Chapter 1, over an interval of time the habituation bias expands the range of connections engaged by the developing network, allocating activity-dependent specification broadly. Intentional self-regulation may be a kind of effortful control that organizes the neurodevelopmental process through creative impulses, confident self-assertion, and optimistic expectancy.

## *Shaping Connections Through Reactive Attention*

In contrast with the dorsal approach mode, the learning process associated with the ventrolateral frontal lobe's control of action appears to be specialized for feedback guidance, with persistent and direct sensory input. This control mode, and the anxiety and redundancy bias it entails, may be particularly important to the rapid learning required when environmental events are discrepant with the animal's expectancies and avoidance is required (Gabriel et al., 1996; Gabriel et al., 1991). In fact, we may speculate that the focused, sustained memory and attentional control required for rapid adjustments to discrepant events is effected by the triangular resonant circuit linking the amygdala and MD thalamus with the ventrolateral frontal lobe (Barbas, Zikopoulos, & Timbie, 2011; Jones, 2007). This resonance and stability of control may be required for the frontal regulation of the ventrolateral networks of the frontal lobe (Zikopoulos & Barbas, 2006), linked to the caudal TRN and thalamic mechanisms of the searchlight of sensory attention (Crick & Koch, 1990). Although involved in most instances of learning and action regulation, the ventral limbic and ventrolateral frontal controls may be particularly important for focused, selective attention to discrepant and threatening environmental events (Tucker & Luu, 2007).

The spatiotemporal complementarity in the control of neural activity leads the redundancy bias to restrict the range of engaged connections, even as it sustains the activity that consolidates this restricted range. The neurodevelopmental effect may be one of rapid differentiation of the connections within the child's developing neural networks. In an environment dominated by threats, the continual engagement of the redundancy bias would lead not only to a focusing of attention reactively, but to the restriction of brain growth to neural networks that are immediately relevant to coping with threat. Because brain development is continuous, the adaptive scope of cognition and neurodevelopment provides continuing constraints on the possibilities for future development.

Thus with the principles of neurodevelopmental theory we assert that psychological self-regulation, including sophisticated qualities such as intentionality and selective attention, can be seen to be emergent from elemental controls on neural activity that shape the developmental process. Although human learning may be deliberate or volitional only under the best circumstances, it is important to understand that even the most sophisticated forms of human self-regulation, those under effortful, conscious self-direction, can be understood as building on elementary motive controls. The impetus engenders an integral intentionality, which under optimal conditions may be elaborated as a complex and even reflective capacity of volitional self-control. In complementary fashion, the artus is accomplished through a developed capacity for selective attention, which under optimal conditions provides a refined and differentiated capacity for inhibitory self-control. A neurodevelopmental theory of human learning may thus allow reasoning from primitive mechanisms to explain how the continuing differentiation of cerebral network architecture is achieved at least in part through conscious intentionality and attentional sophistication in the development of human psychological function.

## *Opponent Complementarity in Developing Language*

Language has long been the epitome of uniquely human cognition that cannot be reduced to biological mechanisms that we share with lesser creatures. Could the cognitive basis for language be understood in relation to elementary neurobiological mechanisms of memory control and action regulation? We recognize language as an essential tool of self-regulation, as the expression of thoughts and feelings in words provides the reflective capacity for delay, for extended interpretation, and thus for a more deliberate ownership of behavior. Could language itself be subject to self-regulation through elementary opponent mechanisms of motivational control?

In Brown's microgenetic theory of language, each speech act is developmental in each progression from motive to utterance, and in each progression from motivated expectancy to articulated comprehension (Brown, 1979, 1988). Interpreting the microgenetic process within the corticolimbic network architecture revealed by Pandya's studies, Tucker (Tucker, 2001, 2002) considered the interaction weaving the holistic representations in limbic cortex with the more differentiated representations in expressive or receptive regions in association cortex. These representations, or *concept components*, in association cortex are in turn interdigitated with the articulation of speech perceptions or actions in primary sensory and motor cortices. Abstract concepts are then multicomponent structures created through reentrant processing across the linked corticolimbic networks and their individual concept components (Tucker, 2007). The holistic and syncretic representations in limbic regions (semantic feelings) are thereby articulated in relation to the sensorimotor requirements for matching the evidence of the external world (words). A microgenetic account neatly describes the progressive specification across this linked corticolimbic hierarchy, particularly if we recognize that it must be recursive and reentrant. In speech production, the more general semantic representation develops into the more specific representational requirements for lexical and grammatical forms and their articulation in speech acts. Through reentrant feedback, the constraints required for effective grammatical organization provide a primitive yet formative logical structure for the cognitive process.

In speech comprehension, although the process is often considered to start in auditory cortex and end in association cortex, the developmental organization must follow the corticolimbic architecture, and it must fit the requirements of contextual cognitive interpretation. Semantic, contextual expectancies at the limbic level shape the more differentiated lexical forms in association cortex that in turn constrain the differentiation of linguistic forms from the continuous auditory stream in auditory cortex. In Roger Shepard's terms, perception is hallucination constrained by the sensory data (Shepard, 1984).

Just as in action regulation, the anterior expressive and posterior receptive functions are unique, but are linked in a dynamic reciprocity. With lesions of Broca's area in the left frontal lobe, language expression is impaired, but the evaluative constraints from posterior areas are fully operable. The result is that the person struggles to speak, but any residual speech production capacity is highly appropriate to the semantic context (Goodglass, 1993). In contrast, with lesions of Wernicke's area in the left temporoparietal region, comprehension is impaired, and so too is critical self-regulation. The intact expression networks of the left ventral frontal lobe are free to produce paraphasia and jargon (Goodglass, 1993).

Just as action regulation engages a recursive and reentrant process across the corticolimbic networks, so must language emerge as a network property of multiple representational levels. Within the corticolimbic pathways, action regulation links the general adaptive representations at the limbic base to the articulation of motor and somatosensory-kinesthetic patterns. So does language, with the additional requirement for ordering the sensorimotor patterns of speech in relation to the structured, symbolic and arbitrary, conventions of the culture. Chomsky (1965) argued for a *deep structure* of language, a generic innate form. A microgenetic analysis suggests there are multiple levels of structure, with the deepest level embedded in the syncretic limbic representation of motivational and semantic significance (Tucker, Frishkoff, & Luu, 2008). The intermediate (heteromodal and unimodal) association areas appear to include the well-known specialized object representation networks for language.

Importantly, these specialized networks for speech, in expression (Broca's) and in comprehension (Wernicke's), occupy *ventral* regions of the left hemisphere. The ventral corticolimbic skills in object perception and cognition are integral to parsing speech objects in linguistic discourse. The creation of an object requires separation of the informational elements of the object from the embedding frame—whether the frame is spatial, as in visual perception, or temporal, as in the flow of speech. The classical example of the fundamental speech object is the phoneme, which is automatically parsed by the native speaker/listener from the continuous speech stream. Speech is unlike other perception in the discreteness of its *categorical perception* (Lieberman, 1984). For native speakers, phonemes (for example, ba and pa) are perceptual objects with sharp categorical boundaries.

The critical cognitive operation for parsing speech objects, in semantic as well as perceptual space, may be described as *inhibitory specification*, in that the inhibition of the surround allows working memory to be focused on the object (Tucker, Luu, & Poulsen, 2009). The artus, the feedback control from the ventral hemisphere, as seen in regulating action in relation to discrete perceptual targets, may be essential to inhibitory specification in language, allowing the personal meaning represented in limbic networks to be articulated in relation to the lexical and syntactic structures of the linked linguistic networks in association and primary sensorimotor cortices. In line with the functional capacity for object memory, the left hemisphere specialization for language may be seen as a specific example of the left hemisphere specialization for the object cognition of the ventral corticolimbic networks more generally.

Language may thus require the focused attention from ventral limbic networks. This may include the regulation of dopaminergic and cholinergic controls integral to the redundancy bias. It may also include the specific circuitry for memory consolidation for the ventral limbic networks, including the triangular circuit from amygdala to MD thalamus, MD to frontal lobe, and amygdala (and associated insular and anterior temporal areas) to frontal lobe. The integral links of the amygdala and ventral limbic networks with the basal ganglia may be important to specific temporal sequencing capabilities. These capabilities are important to motor control generally (Redgrave, Prescott, & Gurney, 1999) and they may be integral to language specifically (Kotz & Schwartze, 2010). Selective sequential control, guided by ongoing semantic interpretation, is essential for complex linguistic reasoning, such as in syllogistic logic. It is also an integral capacity for the routine operation of grammatical constraints, such as demonstrated with remarkable felicity by any normal four-year-old.

Although the core cognitive capacities for speech mechanisms thus appear to rely on skills of the ventral trend in the left hemisphere, language is a general communicative, and cognitive, process within a complex social context. As such, it relies on the whole brain. Right hemisphere lesions lead to deficits in understanding the social implications of discourse, such as the meaning of a joke (M. Beeman, 1993; M. J. Beeman, Bowden, & Gernsbacher, 2000). The contributions from the dorsal networks may be fundamental to the communicative process, if not to specific lexical and syntactic operations. Lesions of mediodorsal frontal cortex lead to the syndrome of *transcortical motor aphasia*, in which patients fail to initiate speech and may appear mute. However, they typically show residual speech production capability upon confrontation (Goodglass, 1993).

The speech deficit with mediodorsal frontal lesions seems to reflect a loss of the impetus, the motive self-regulation of the communicative intent. Contrasting this pathology with that of ventral frontal (Broca's) lesions may provide insight into the opponent neural mechanisms regulating language production. Language articulation has evolved with strong control by the artus, the cybernetic constraint mediated within left ventral frontal networks. However, the language capacity of these networks is ineffective in producing speech unless it is balanced by the motivated impetus toward intentional communication, arising in the dorsal networks of the frontal lobe.

The child's language develops in a social context, where communication requires multiple cognitive skills. Observing the critical role of the left ventral frontal networks, we can surmise that the child maintains the feedback constraints on vocalization required by the language, just as feedback constraints are applied in the general control of action regulation. These constraints allow the articulation process to develop in line with the cultural conventions, with left ventral temporal networks providing the representation of the speech sounds and the left ventral frontal networks providing the motor articulation. The requirement for close communication between these anterior and posterior object articulation and representation networks seems to have led to the enlargement of the fiber tract connecting these regions, the arcuate fasciculus, in recent human evolution (Catani, Jones, & ffytche, 2005; Rilling et al., 2008).

At the same time, these specific articulatory skills develop within the social context, within which communication requires multiple skills in self-regulation. Studies of infants learning new words in a naturalistic context of interaction with the mother have shown that infants attend carefully to the mother's intention in speaking a new word, and use the context of her gaze or actions in interpreting the significance and meaning of the word (Baldwin, 1991, 1993). A word is not learned out of the context, such as if the mother speaks it randomly. The meaning of the word is then literally "what Mom means." The child then internalizes this as her own meaning. The brain's representations, and the implicit self assembled from them, are residuals of the social context. Self-regulation then continues to be built on the templates of internalized social transactions.

### How the Mind Grows the Brain

In this chapter, we have outlined some general psychological implications of a neurodevelopmental theory. The specific properties of dorsal and ventral control systems can be understood to underlie familiar psychological mechanisms in motivation, cognition, and personality. This is because the control of neural development, through regulating the spatiotemporal organization of neural activity, is a process of self-organization. Initially, it proceeds under general embryological guidelines, but even in the early intrauterine phases it is self-organizing, in that the fetus's own spontaneous actions shape its experiential activity dependent plasticity. In childhood, the primary motive vectors of approach (hedonic expectancies) and avoidance (anxiety and discrepancy) appear to continue the embryological activity controls of habituation and redundancy. The results of the motive vectors are manifested in the child's conceptual development, consolidated within both dorsal and ventral corticolimbic systems. Psychological function then shapes the motive process increasingly, as the child's mind operates as a more or less integrated system. The control of neural activity is then framed as the self-organization of a developing mind.

### How the Mind Evolves

Even though psychological function, and the social context that defines it, is the organizing system, it begins within the developmental process framed by the human genome. We have seen that if we are to understand the motive vectors of the neurodevelopmental process we must face specific questions about the anatomy and physiology underlying that process. How does the dorsal corticolimbic system generate its unique feedforward control properties within its unique neural circuits dominated by pyramidal neurons? How does the ventral limbic system come to integrate the redundancy of dopamine and the basal ganglia circuitry within ventral limbic and cortical networks with their unique granular cytoarchitectonics? Like all questions in biology, these only make sense in the light of evolution (Dobzhansky, 1964). Even the uniquely human features, such as hemispheric specialization or the extended and occasionally conscious self-regulation from frontothalamic circuits, must be understood in relation to the unique capacities of cognition that evolved through the counterpart of mammalian dorsal and ventral systems. Fortunately, many clues to the evolution of mammalian neural systems are readily available, within the recapitulation of the phyletic order observable in the course of embryonic morphogenesis.

Because the mind is the brain, we can therefore shift theoretical perspectives at this point and see what we can learn about the mind's developmental process by looking at how the brain grows. The previous four chapters have considered the evidence on the function of the brain, knowing that this function emerges from developing neural tissue. In the next two chapters, we extend this neurodevelopmental theory to early development, interpreting the functional systems regulating neural activity by considering their evolutionary roots, as these roots are revealed by their residuals in embryology.

# 5
## Structural Clues to Dorsal-Ventral Specialization

In Chapter 3, we outlined the functional differentiation between dorsal and ventral corticolimbic systems, seen most clearly in action regulation. In Chapter 4, we outlined how these basic motivational controls on working memory may figure in psychological capacities more generally. We now turn to specific neural mechanisms that differ between dorsal and ventral corticolimbic architectures. These mechanisms, and their anatomical substrates, may explain how neural development takes different forms depending on the relative direction from dorsal and ventral corticolimbic systems. It seems clear that the cybernetics of habituation and redundancy can only operate to shape neural networks that have evolved to respond to their primitive influences. This response involves not only specialized patterns of cortical connectivity, but cytoarchitectonics that are suited to that connectivity, and both thalamic and other subcortical specializations to regulate the specialized cortical networks. By considering these neural mechanisms in detail in this chapter, we may prepare for a review of their neuroembryological origins in Chapter 6.

Innate dispositions toward impulse or constraint are fundamental temperament variations (Derryberry & Rothbart, 1997; Derryberry & Tucker, 2006). A variety of temperamental dispositions seems to have been maintained in human evolution to allow populations to adapt to environmental conditions of opportunity or threat. Given a natural range of temperamental bias (nature), the response of family and society (nurture) may condition the child's self-regulatory motives in ways that then determine continued development. Engagement of adequate anxiety and self-control may provide the child with not only appropriate behavioral inhibition but the sensitivity to others' experience that allows effective empathy (Tucker, Luu, & Derryberry, 2005). Encouragement of behavioral initiative may allow confidence and optimism that guide later coping efforts even under challenging conditions (Tucker & Moller, 2007). In a neurodevelopmental theory, understanding these processes of child development requires understanding the neurodevelopmental mechanisms that form their physiological substrates. These are fundamentally mammalian neurodevelopmental mechanisms.

We have seen that the mammalian cortex seems to incorporate primitive subcortical control influences in paradoxical ways. It integrates a habituation bias, which is most fundamentally a control on sensory memory, to achieve fluid and efficient feedforward control of motor processes in dorsal networks. It integrates a redundancy bias, apparently evolved within the basal ganglia for structuring and sequencing motor habits, within the ventral division of mammalian cortex to provide a sustained and focused control over sensory criteria in the feedback regulation of action and cognition. By elaborating the effects of these activity controls in complex ways, evolution seems to have achieved a complexity of cybernetics in the mammalian brain that allows capacities of cognitive representation not possible in simpler brains. To optimize these complex cybernetics, mammals have evolved not only support for juvenile development through attachment and parenting, but specific cortical architectures. An effective neuropsychological theory must therefore begin by explaining how the neurodevelopmental process operates to shape different patterns of cortical architecture, cytoarchitectonic differentiation, and subcortical modulatory circuitry in the dorsal and ventral divisions of the brain.

## Connectional Architecture

A first principle for analyzing the differential function of dorsal and ventral corticolimbic systems, as reviewed in the summary of anatomical evidence in Chapter 2, is that these are segregated networks. Each of the dorsal and ventral divisions is interconnected primarily with itself, and there are only limited and specific points of interaction between them (Barbas & Pandya, 1989; Schamahmann & Pandya, 2006). Each has a base in limbic networks, with the dorsal system based in the *archicortex*, including hippocampus and cingulate gyrus, and the ventral system based in *paleocortex*, including anterior temporal, posterior orbital, and insular cortices. From these highly interconnected bases, the intrapathway connection density decreases progressively across the adjacent heteromodal, unimodal, and primary (sensory or motor) cortices. The interpathway dorsal-ventral connections are limited at the levels of both limbic and association cortices, leading to the conclusion that much of the functional processing and representation of the brain remains segregated to either a dorsal or a ventral corticolimbic system.

In macaque, there are specialized areas of association cortex that have an unusual mixture of the cytoarchitectonics of both dorsal (high density of pyramidal neurons) and ventral (high density of granular neurons) divisions. These include a region of the temporal-parietal-occipital junction in the posterior brain and a region around the principal sulcus in the frontal lobe (Barbas & Pandya, 1989; Schamahmann & Pandya, 2006). Postmortem studies of the human brain have suggested there are similar areas of mixed cytoarchitectonics, at least for the inferior parietal region (Eidelberg & Galaburda, 1984). Thus the integration of dorsal and ventral processing may be limited to a restricted set of networks in association cortex, in contrast to widespread and varied network connectivity across multiple limbic and neocortical networks within each division.

The primary dorsal-ventral integration may be organized at the limbic border of the hemisphere. This may be consistent with the more extensive interconnectivity of limbic regions generally. There is considerable ventral limbic input to the subgenual region (BA 25) of the cingulate cortex, which, as part of the cingulate cortex, is generally considered to be of archicortical (dorsal) derivation (Vogt, Vogt, & Laureys, 2006). It is interesting in this regard that Jones points out that BA 25 is one of the few cortical areas that have no clear thalamic projections (Jones, 2007). A similar dorsal-ventral integration appears to occur in the limbic networks of the posterior midline. Whereas the posterior cingulate is clearly archicortex, with extensive projections to hippocampus as well as mid and anterior cingulate cortex, there are substantial inputs from perirhinal (ventral limbic) cortex to the posterior midline and precuneus as well (Vogt et al., 2006). Thus the cingulate lobe is largely dorsal, but with considerable ventral limbic connectivity at both anterior (subgenual anterior cingulate) and posterior (precuneus and ventral posterior cingulate) extents.

Within these largely separable divisions of dorsal and ventral cortex, are there differential patterns of connectivity that help to explain their differing forms of psychological structure? The issue is locus-function specificity, whether patterns of neural connectivity can be traced to patterns of cognitive representation. Are the dorsal networks, with their more holistic and global configural representations of context, more fully interconnected over long ranges? Are the ventral networks, with their more parcellated and object-delineated representations, more restricted in their connectivity to local ranges?

## *Hemispheric Specialization: Clues to Mechanism*

Somewhat tangential evidence on this question applied to the human brain might be gained if we assume that right and left hemispheric specialization has differentially elaborated the dorsal and ventral trends, respectively (Galaburda, 1984; Liotti & Tucker, 1994). Semmes (Semmes, 1968) tracked the functional effects of focal cortical lesions in either the left or the right hemisphere. For several functional tests, including somatosensory discrimination, grip strength, and perceptual orientation, there was a close locus-function specificity for lesions in the left hemisphere, but not for lesions in the right hemisphere. Semmes concluded that a given region in the left hemisphere must be specialized for "like" elements, whereas a homologous region in the right hemisphere must be specialized for "unlike" elements. Thus at least for functional representations, the right hemisphere, which appears to elaborate the functions of the dorsal cortex particularly, does appear organized around more distributed, and widely interconnected representations of function. This would also be consistent with the bias toward limbic dominance of the direction of consolidation in the right hemisphere, implemented through an anatomical elaboration of limbic and heteromodal networks. In contrast, the left hemisphere, perhaps elaborating the ventral cortical pattern, has a functional organization that is apparently dependent on more local connectivity. This would be consistent with an emphasis on unimodal sensory and motor cortices (Goldberg & Costa, 1981).

Thus, for the human brain, in which hemispheric specialization seems closely related to cortical differentiation of dorsal and ventral divisions, it may be possible to interpret evidence of differential functional distribution over cortical regions in relation to differences in dorsal and ventral architectures. More direct evidence, relevant to mammals generally, comes from the specialized cytoarchitectonics that themselves imply specialized patterns of connectional architecture.

## Cytoarchitectonic Differentiation

The greater density of pyramidal neurons in the dorsal, archicortical regions may be directly related to the question of a greater long-range interconnectivity of the dorsal division. Pyramidal neurons are projection neurons, well suited to large scale integration. In contrast, the cytoarchitectonics of the ventral cortex include the dominance of granular cells in layer 4. These are targets of thalamic (and intercortical limbipetal) projections, suggesting a greater degree of processing of input information, possibly with greater control from local inhibitory interneurons.

In the frontal lobe, these differing pyramidal and granular cytoarchitectonic specializations seem to provide important constraints on functional organization. With the elaboration of granular layer 4 in the ventrolateral frontal lobe, the processing of sensory input (from posterior association cortex) allows object representations that serve as targets for guiding action, particularly under inhibitory, feedback guidance. This close alliance of objects in the posterior, receptive networks and anterior, expressive networks, supported by interconnection through an expanded arcuate fasciculus in humans, may be integral to the processing of language objects (words) under sequential, routinized, and hierarchic control (grammar) in action regulation (speech).

Without a granular layer in the mediodorsal frontal lobe, action regulation may need to proceed under more global control from widespread pyramidal projections, including those from superior parietal cortex. It may not be too speculative to conclude that there are processing implications that follow directly from the implications for perceptual structure inherent to these cytoarchitectonic differences. The projectional, feedforward control in dorsal cortical territories may emerge directly from the lack of a granular input layer and the corresponding emphasis on pyramidal, projectional neurons. The more constrained, reactive, and inhibited mode of processing in the ventral cortical territories may emerge directly from the elaboration of the granular layer and the enhanced local processing, inhibitory specification, and representational differentiation that result from this cytoarchitectonic specialization.

## Specialized Thalamic Modulation

The dorsal and ventral divisions of cortex seem to have different relations with subcortical circuitry, beginning with the thalamic connections that differentiate between the pyramidal and granular cytoarchitectonics, and extending to the major circuits linking the corticolimbic networks with striatal, hypothalamic, and brainstem structures.

An important clue may be the specialized connectivity between thalamus and cortex that has recently been recognized in the thalamus (Jones, 2007). The *matrix* projections from multiple thalamic nuclei connect beyond the primary cortical region associated with each nucleus, apparently providing a more global integration among differing cortical-thalamic networks. Integrating input from the layer 5 pyramidal neurons of the associated cortical regions, the matrix neurons appear to integrate as well as reciprocate multiple cortical territories. The matrix projections may be particularly important to the limbifugal, self-to-world direction of consolidation. In contrast, the *core* projections from thalamus to cortical regions are the more traditional thalamocortical connections, targeting the granular layer 4 of the cortex. The more restricted cortical pattern of these projections may be associated with greater specificity of regional sensory processing. By extension, the core projections to layer 4 may be associated with a more object-based processing mode for the corticothalamic networks, particularly important to the limbipetal, world-to-self direction of consolidation.

Whether these specializations for matrix and core patterns of corticothalamic regulation prove explanatory for dorsal and ventral functional specializations remains to be determined by evidence rather than pure speculation. However, it does seem clear that the functional operation of the cortex cannot be considered in isolation from the thalamus. Rather, the communication between cortical regions appears to be regulated by thalamic control, through mechanisms such as the matrix projections. For example, in the waking state stimulation of the cortex through transcranial magnetic stimulation results in both a local response at the site of stimulation and a spread of activity to nearby cortical regions (Massimini et al., 2005). However, when applying the same stimulation when the person was asleep, Massimini et al. observed the local response, but not the spread to other cortical regions. The implication is that the thalamocortical gating mechanism that suppresses sensory input during sleep also suppresses intercortical communication. Given the essential role of thalamic mediation in the waking state, it may be necessary to redefine the notion of direct corticocortical communication and recognize it instead as a process of corticothalamocortical communication.

If the dorsal-ventral specialization for pyramidal versus granular cytoarchitectonics implies differential control by matrix versus core thalamic projections, then there may be fundamentally different patterns of network communication in the two divisions of the cortex. The greater reliance on pyramidal matrix projections in the dorsal division may explain the holistic, configural cognitive structure supported by this unique network architecture. The direct and impulsive insertion of motive urges into the action plan within the dorsal pathway is another reflection of this architecture of widespread pyramidal interconnections, supported by the matrix regions of thalamic nuclei.

In contrast, the more restricted thalamocortical circuits associated with granular cortex and core thalamocortical projection patterns may favor local processing. With fewer pyramidal projections of the matrix regions, perhaps the thalamic nuclei target regions of the ventral cortical division more discretely, with point-to-point connectivity rather than the broad matrix pattern of cross-regional projections. The result may be discrete thalamic modulation of specific cortical zones, a regulatory pattern suitable for object perception and cognition.

## Specialized Subcortical Circuitry

The analysis of local processing within the ventral division of the hemisphere must consider regulatory influences from the basal ganglia as well as thalamus. Motor control is effected in part through surround inhibition, organized largely through circuits linking the basal ganglia to cortex, in order to specify certain elements of the movement through inhibiting others. Interestingly, surround inhibition appears to operate more strongly in the dominant (left) hemisphere, suggesting that it may be integral to the fine motor control of the dominant (right) hand (Shin, Sohn, & Hallett, 2009). In the related phenomenon of short-latency afferent inhibition, electrical stimulation of one finger results in a rapid inhibition of muscles of adjacent fingers, as well as other hand and arm muscles (Helmich, Baumer, Siebner, Bloem, & Munchau, 2005). This phenomenon is also pronounced for the left hemisphere (right hand), suggesting that it is integral to hemispheric specialization for motor control as well.

## Subcortical Control of Inhibitory Specification in Ventral Corticolimbic Networks

Although both dorsal and ventral divisions of motor cortex are engaged by basal ganglia circuits providing surround inhibition, there may be reason to consider a preferential role for these circuits in the inhibitory capacities of the ventral division of the hemisphere. Unique skills in inhibitory specification, through mechanisms analogous to the inhibitory surround, may be important to the consolidation object memory within the ventral limbic networks (Aggleton & Brown, 1999; Mishkin, 1982; Tucker, Luu, & Poulsen, 2009). The paleocortex at the limbic base of the ventral division is specialized for processing olfactory input, with extensive interconnections with the insular and caudal orbital frontal cortex as well as basal ganglia and amygdala. The amygdala is itself a complex integrative structure, with capacities for resonant processing within its internal architecture (Johnson, LeDoux, Doyere, 2009). The amygdala has extensive interconnectivity not only with ventral limbic networks, but with the basal ganglia and related thalamic nuclei as well (Groenewegen, Berendse, Wolters, & Lohman, 1990). As pointed out by Jones (2007), there is an unusual triangular circuitry linking the amygdala with the MD thalamus, amygdala with the frontal lobe, and MD thalamus with the frontal lobe. The organization of this triangular circuit appears unique to ventral networks. The magnocellular nucleus of the MD thalamus receives input from the amygdala and pyriform (olfactory) cortex, and it projects to the orbital frontal cortex specifically. In contrast, other projections from the MD thalamus, from the magnocellular and mixed regions, are to more dorsal regions of the frontal lobe, and these do not make up a triangular resonance pattern.

An important possibility is that the capacity for complex and extended resonance in the triangular circuit may create a unique form of consolidation, perhaps associated with a redundancy bias, that is integral to the specialized object-oriented processing of ventral limbic networks. Given the cytoarchitectonics of the ventral cortex emphasizing granular architecture for local processing, an important possibility is that the amygdala and ventral limbic memory controls work in close alignment with the basal ganglia in inhibitory specification, drawing on the unique consolidation architecture of ventral corticolimbic circuits, as well as on the redundancy bias of cholinergic and dopaminergic projections that target limbic regions as well as the basal ganglia. In this way, the capacity for inhibitory specification that evolved in relation to the motor control of the basal ganglia may have become elaborated in mammals for the (paradoxically opposite) task of delineating sensory objects in the granular ventral cortical networks. The lateralization of basal ganglia and ventral limbic inhibitory specification may be important not only to left hemisphere control of fine motor skills, but of perceptual differentiation and analytic cognition generally (Tucker, Frishkoff, & Luu, 2008; Tucker, Luu, et al., 2008).

## Subcortical Mechanisms of Visceral Expression in Dorsal Corticolimbic Networks

The dorsal division of cortex is closely integrated with the classical Papez circuit of the limbic system (Papez, 1937). Like Galaburda many years later, Papez emphasized the importance of Herrick's observation that, as is obvious in primitive vertebrates such as the salamander, the medial wall of the telencephalic hemisphere is closely interconnected with the hypothalamus (Herrick, 1948). In contrast, the lateral wall of the hemisphere in mammals receives the sensory projections from the dorsal thalamus. The reciprocal connections from the anterior nuclei of the thalamus to the mammillary bodies of the hypothalamus comprise key diencephalic links of the Papez circuit, with projections from the cingulate gyrus to the anterior thalamus comprising the telencephalic input. Projections from the hippocampus to the hypothalamus provide another pivotal component in Papez's classical description of a major circuit of what came to be called the limbic system.

The hippocampal, cingulate, and hypothalamic patterns of connectivity thus appear to be essential to the subcortical control of the dorsal division of the cortex. Is there a characteristic cybernetic quality to this subcortical control that could explain the apparent motive bias of the dorsal corticolimbic networks? One way to understand the cybernetic properties of dorsal cortex and its unique subcortical circuitry is in relation to the *visceromotor function* of the cingulate cortex (Neafsey, 1990; Neafsey, Terreberry, Hurley, Ruit, & Frysztak, 1993). By elaborating the visceromotor function, dorsal cortical function may be spontaneous, direct, and immediately emergent from visceral operations. This primitive control at the core of the dorsal hemisphere may be elaborated through each level of corticolimbic processing, including the more complex functions of conscious self-regulation supported by the dorsal frontal pole (Tucker, Brown, Luu, & Holmes, 2007; Tucker & Holmes, 2010). In Chapter 3 we pointed to the impairment of intentionality with seizures of the frontal pole, and the modern evidence of frontopolar projections to the rostral thalamic reticular nucleus and anterior nuclei of the thalamus, which are then reciprocally interconnected with the cingulate gyrus (Zikopoulos & Barbas, 2006). Considering the multiple levels of the dorsal circuitry in representation and control, it may be that the impulsive expression of the visceromotor function is the elemental control mode of dorsal brain cybernetics (Tucker & Luu, 2006, 2007).

The ventral limbic networks are also based in visceral functions, but in a more indirect way that may be compatible with the mechanisms of inhibitory control, with redundancy of ongoing activity, and with sensory constraints on behavior in the ventral networks. The insula is specialized for *viscerosensory functions* (Neafsey, 1990; Neafsey et al., 1993). The feedback control of action in the ventrolateral frontal lobe may be guided not only by the sensory representations in the ventral frontal division, but more generally by the cybernetic mode of constraint and restriction emergent from the insular control of the viscerosensory function (Tucker & Luu, 2006, 2007). Whereas visceromotor control is direct and emergent, applying immediate influences on behavior, viscerosensory control serves primarily to establish the presence of need or alarm states, and to engage the more extended working memory provided by the redundancy bias to organize behavior in relation to those states. Although in its elemental form the visceromotor influence is a primitive mode of control, it may be elaborated within cortical networks to become integral to the psychological function of the artus, providing the motive regulation of the behavioral restraint, inhibitory object specification, categorical perception, and differentiated form of selective attention represented within the ventral limbic system.

## Brainstem and Forebrain Neuromodulators

In the developing organism, the self-regulation of neural activity in time, such as through redundancy and habituation biases, is necessary for multiple adaptive requirements, including the sleep-wake cycle, motivational challenges and opportunities, and the effort required for cognitive operations. As described above, the concept of *arousal* in psychological theory has been considered to be a unidimensional mechanism, varying from low to high. In contrast, the neurophysiological evidence suggests that there are multiple distinct arousal and activation mechanisms, each supported by a unique neuromodulator projection system, and each with specific implications for the person's emotional state and cognitive process. Furthermore, whereas psychological theory has typically considered arousal to be independent from the *valence* of emotional state, pleasant or unpleasant, our interpretation of the major controls on cerebral arousal suggests they may have unique affective qualities, each with a specific valence. In the summary formulation of the Tucker and Williamson (1984) model, tonic activation and its redundancy bias, mediated by dopamine and acetylcholine neuromodulator systems, are associated with anxiety and vigilance for threat. Phasic arousal and its habituation bias, mediated by norepinephrine and serotonin neuromodulator systems, are associated with elation and expectancy for pleasure.

There may be a differential alignment of these affective arousal and motivation systems, supported by their brainstem and forebrain neuromodulator projection systems, with the dual corticolimbic systems. This differential alignment involves both the output or efference of the projection systems (their targeting of the brain) and the input or afference (the brain's projections to control them).

### Dorsal Corticolimbic Neuromodulation by Norepinephrine and Serotonin

The dorsal corticolimbic networks appear to be preferentially regulated by norepinephrine and serotonin controls on neural activity. The major norepinephrine projections, the *dorsal noradrenergic bundle*, proceeds from the brainstem locus coeruleus to innervate widespread regions of the forebrain, yet it is strongly biased in a *dorsal cortical* projection pattern. The projections proceed to the frontal pole then course posteriorly, with strong innervation of the cingulate cortex and dorsal association cortex, both frontal and posterior (Foote & Morrison, 1987).

The dorsal preference of these efferent norepinephrine projections is matched by the cortical control over the locus coeruleus, the brainstem origin of the norepinephrine pathways, which the most direct evidence suggests is exerted by the dorsal, but not ventral, frontal networks. Labeling of cells in monkeys in both dorsomedial and dorsolateral frontal regions, but not orbital (ventral) frontal cortex, showed uptake of the label in the region of the locus coeruleus (Arnsten & Goldman-Rakic, 1984). In addition, this tracing of frontal-to-brainstem projections also showed that only mediodorsal frontal injections resulted in label uptake around the raphe nuclei, the origins of the serotonergic projections. These findings are consistent with a model of primarily dorsal frontal control of both norepinephrine and serotonin. On the other hand, other evidence suggests there are also ventral, orbital frontal, projections to the locus coeruleus (Aston-Jones & Cohen, 2005).

### Ventral Corticolimbic Neuromodulation by Dopamine and Acetylcholine

The brainstem dopamine projections include both a nigrostriatal component, from the brainstem substantia nigra to the basal ganglia, and a mesolimbic component, from the mesencephalic dopamine nuclei to the limbic regions and frontal lobe (Williams & Goldman-Rakic, 1993). The projections to the ventral limbic and orbital frontal regions appear particularly strong, consistent with the notion that the redundancy bias of dopaminergic control is important to ventral limbic contributions to self regulation (Tucker & Luu, 2007). However, there are also major mesolimbic dopaminergic projections to the anterior cingulate cortex of the dorsal division (Vogt, 1993).

The neuromodulation of the forebrain from acetylcholine includes projections from a brainstem nucleus (pedunculopontine) and also a forebrain source, the nucleus basalis (Mesulam, Mufson, Levey, & Wainer, 1983). Although acetylcholine has been thought to be complementary in some way to the dopamine regulation of the redundancy bias (Tucker & Williamson, 1984), there is little or no preference of the cortical projections of the nucleus basalis for ventral versus dorsal regions that would support a specific ACh targeting of ventral regions (Mesulam et al., 1983).

On the other hand, the control over the nucleus basalis appears to be exerted primarily by ventral frontolimbic networks. The nucleus basalis is regulated by inputs from the orbital frontal, temporal pole, prepyriform, entorhinal, anterior insula, and medial inferotemporal regions (Mesulam & Mufson, 1984), all derivations of the paleocortex. Through their unique influence over the nucleus basalis, the ventral frontolimbic networks regulate the neuromodulation of the entire cerebral hemisphere by acetylcholine. As we saw in Chapter 2, several lines of evidence suggest that acetylcholine is perhaps the most important global control over neural plasticity and memory consolidation.

## Motivated Anatomy

In the brain's complex and varied mechanisms self-regulation, the control of the neuromodulator systems is pivotal. A few nuclei regulate widespread features of neural excitability, motivational state, and the neural plasticity of the learning process. Among the various neural mechanisms reviewed in this chapter, neuromodulation has possibly the clearest relation to the embryonic self-regulation of neural morphogenesis. Although the real cybernetic functions of neuromodulator systems, and the interactions among them, are considerably more complex than captured by the cursory outline presented here, the evidence points to a functional specialization of the dorsal division of the telencephalon for regulating, and being regulated by, norepinephrine and possibly serotonin, key agents of the habituation bias. Similarly, the ventral division is strongly regulated by dopamine, and it appears to have exclusive control over the forebrain (nucleus basalis) acetylcholine system, which taken together appear responsible for the redundancy bias. An important challenge for understanding the neural mechanisms of memory consolidation, and self-regulation generally, is to understand the joint evolution of the patterns of neural architecture with the predominant modes of neuromodulation in the dorsal and ventral divisions of the hemisphere.

We have theorized that the motive mechanisms regulating consolidation are fundamentally neurodevelopmental control processes. If so, we should be able to look to their roles in embryonic differentiation to interpret their roles in cognition. Developing structure is developing function. The next chapter will review several findings in neuroembryology that point to the developmental origins of ventral-dorsal differentiation in the mammalian cortex. Although the evidence comes from observations in embryology, interpreting this evidence turns out to require an evolutionary-developmental analysis. The features of the cortex that appear most important to human cognition are not just the uniquely human features. Rather, human intelligence has become sophisticated by elaborating the fundamental plan for dorsal and ventral functional specialization that allows mammalian memory generally. Understanding this plan turns out to require a model for how mammals evolved their unique cognitive capacities from the more elementary, and less laminar (cortical), neurocybernetics of reptiles. The developmental process of neuroembryogenesis provides striking evidence of evolutionary origins of the mammalian cortex, including the dorsal-ventral differentiation that yielded the dual mechanisms of mammalian memory.

# 6
## The Evolved Structure of Mammalian Memory

If we can understand the isomorphism between cognition and its neurodevelopmental process, then evidence on the nature of the neurodevelopmental process would provide important insight into the process of psychological development and function. New evidence on the control of neural development is emerging from studies of gene expression in mammalian and human neuroembryogenesis. In this chapter, we review several remarkable findings from this research that provide additional clues to the origins of the archicortical and paleocortical divisions of the mammalian brain. Although the primary evidence is embryological, this evidence provides interesting clues to the evolutionary transformation of the telencephalon that generated the mammalian brain and its unique memory capacity. A key event in this transformation was the emergence of the mammalian 6-layered cortex from the ancestral (reptilian and amphibian) 3-layered cortical patterns.

Once the perspective of embryogenesis is obtained, it becomes clear that an empirical effort to map the functions of the human brain without a theoretical understanding of the brain's evolved mechanisms must remain superficial. The ontogenetic process of neocortical differentiation arises through primordial mechanisms of neural development. These mechanisms first retrace the evolutionary progression to achieve neuroembryogenesis, and they then shape the ontogenetic process of learning and cognition throughout life.

We first outline several features of the embryonic development of the mammalian cortex that provide important clues to the differentiation of its dorsal and ventral divisions. Understanding this differentiation requires understanding the evolutionary origins of the 6-layered mammalian cortex from both the 3-layered reptilian pallium and the subpallial reptilian telencephalon. We therefore consider the most obvious interpretation of that evolutionary transformation.

From the perspective of this evolutionary-developmental analysis, we then reconsider the several clues to the functional differentiation of dorsal and ventral corticolimbic networks examined in Chapter 5. The specialized motive vectors regulating cognitive structure and the psychological locus of control can be seen to arise from mammalian cerebral hemispheres in which the primordial pallial architecture (3-layered pyramidal networks) have been fused with key neural mechanisms of subpallial (basal ganglia) structures, including GABAergic inhibitory modulation, direct collothalamic projections, and regulation through amygdalar controls closely linked with the basal ganglia.

We conclude this chapter with a summary of the theoretical model. The central point is that the dorsal corticolimbic networks reflect a neocortical elaboration of the pyramidal pallial architecture, whereas the ventral networks evidence a remarkable integration of pallial and subpallial circuits. We argue that an accurate theoretical model of the evolutionary roots of mammalian neocortex is essential for interpreting how the differential cybernetics of regulating neural activity in time—habituation and redundancy—operate within fundamentally different neural architectures in the dorsal and ventral divisions of the cerebral hemispheres.

## Evolution of Mammalian Embryogenesis

Although evolution is often considered as selecting for adult traits, in actuality both mutation and selection act on the embryonic mechanisms that generate those traits through the course of morphogenesis (Gould, 1977). As a result, evolution leaves its tracks in embryology, and we have only to read them. Recent research has revealed patterns of gene expression that guide the morphogenetic process, including the formation of the mammalian neocortex (Rakic, 2009). Currently, the consensus among neuroscientists is that there is no viable theory for how the mammalian neocortex evolved (Butler & Hodos, 2005). However, many of the modern findings on primate cortical connectivity by Pandya and associates (Barbas & Pandya, 1986, 1989; Pandya & Seltzer, 1982; Pandya & Barnes, 1987; Pandya & Yeterian, 1984) would seem to support the Sanides hypothesis. Sanides proposed that the neocortex evolved through successive waves of differentiation from primitive limbic cortices at two points of origin: the *paleocortex*, including insular, pyriform, and olfactory regions, and the *archicortex*, including the hippocampus and cingulate regions (Sanides, 1970). The modern connectivity evidence shows a clear division between dorsal neocortical networks, organized around the archicortical limbic base, and ventral neocortical networks, organized around the paleocortical limbic base (Pandya & Yeterian, 1984; Schmahmann & Pandya, 2007). As recognized by Pandya and associates for many years, this connectivity is consistent with the hypothesis of a differentiation of neocortex from dual, archicortical and paleocortical, points of limbic origin.

The traditional view of cortical development begins with the assumption that primary cortex (sensory or motor) is the most primitive area, and association cortex has evolved from that, with more complex forms of association cortex in more complex mammals (Kaas, 1989; Northcutt & Kaas, 1995). The Sanides model proposes the reverse, that limbic cortex was the first to evolve, and primary cortices were the last, at least in the broad progression of mammals. Clearly there was important elaboration of association and limbic areas in primate and human cortical expansion.

Among the several arguments advanced against the Sanides view (Kaas, 1988; Northcutt & Kaas, 1995), an important point is that the embryogenesis of the cortex occurs through the *radial* migration of cortical neurons from a common ventricular zone (VZ), rather than a *tangential* migration from dorsomedial (archicortex) or ventrolateral (paleocortex) points of origin, as would seem to be implied by the Sanides hypothesis. Glial cells are observed to form tracts from the VZ to the primitive cortex of the vesicle walls, and neurons migrate along these glial tracts to take up positions in the developing cortex (Rakic, 2009). In addition to this main migratory route from the VZ, recent evidence shows that neocortical neurons may originate in the subventricular zone (SVZ) of the telencephalic vesicle (forming the telencephalic cerebral hemisphere) and also the ganglionic eminence (GE; primarily associated with the basal ganglia; Rakic, 2009).

As reviewed by Rakic et al. (2009a; 2009b), the radial unit hypothesis proposes that cortical expansion in evolution is determined by the increase in the number of stem cells in the VZ and SVZ, regulated by genes that determine the rate of proliferation and that program cell death. A massive expansion of cortical columns through these mechanisms appears responsible for the expansion of the human cortex. Coincident with the last cell division cycle is the emergence of patterning centers within the VZ. These patterning centers are established by genes that regulate the expression of certain molecules, which in turn establish the positions of cells within the VZ. In this manner, the gene expression patterning centers set up cortical *protomaps*: the positions of neurons in the VZ determine their locations within the cortical plate as they migrate radially to the cortex. There are multiple regulatory genes that determine the patterning centers, and they are differentially expressed along gradients across the embryonic cerebral vesicles. These gradients may provide insight into the dual organization of the neocortex.

Once established in the cortical plate, the new cortical neurons selectively attract the appropriate afferents, including those from the thalamus. This neurotrophic guidance appears to reflect a second phase of cortical morphogenesis, following the first, gliotrophic, phase of radial migration from the VZ, SVZ, and GE (Figure 6-1).

Although the radial unit hypothesis explains much about the origins and expansion of the mammalian neocortex, it does not explain how the connectivity comes to be organized along the ventrolateral (paleocortical) and mediodorsal (archicortical) lines. Rakic (2009) points to recent observations of tangential migrations that are neurotrophic (guided by neurons rather

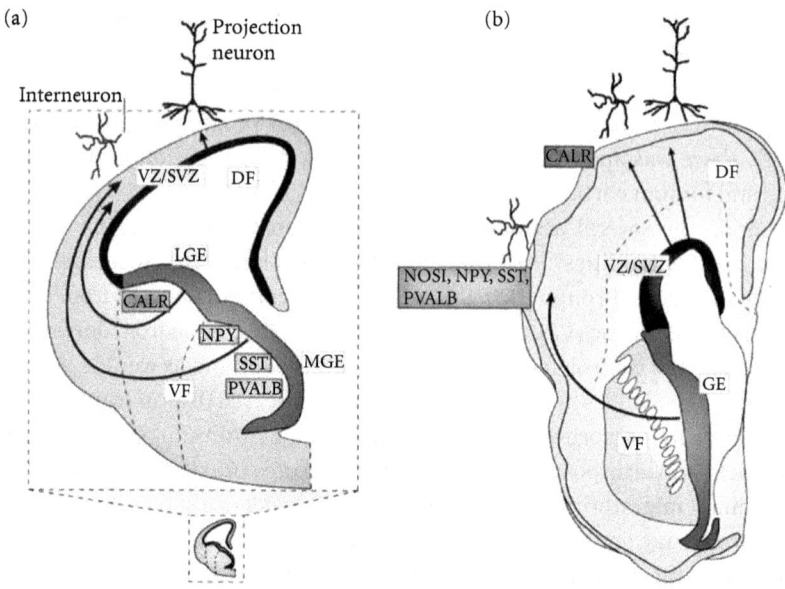

Figure 6-1. The embryonic differentiation of fetal rodent (a) and human (b) telencephalic hemispheres from dual points of origin. Migration of pyramidal projection neurons proceeds from the proliferative ventricular zone (VZ) and subventricular zone (SVZ) and progresses through the cortex in a radial (ventricular-to-surface) pattern, which in humans is marked by calretinin. Migration of interneurons proceeds from the ganglionic eminence (GE) in the ventral forebrain and is marked by parvalbumin. Reprinted with permission from Macmillan Publishers, Ltd. (Rakic, 2009).

than glia). An interesting question is whether these tangential neurotrophic migratory routes, together with the specific attraction of afferents in the second stage, may eventually prove relevant to the formation of archiocentric and paleocentric patterns of intercortical connectivity.

Although there may be little direct evidence on tangential migration tracing the dual moieties of neocortical differentiation suggested by the Sanides hypothesis, there are several forms of indirect evidence. The most surprising of these point not to pallial (primitive 3-layered cortex) but to subpallial origins of mammalian neocortex. Within the current evidence on embryogenesis, four observations may provide clues to dorsal-ventral neocortical differentiation. We outline each in the following four sections, and then consider their implications in the rest of the chapter.

### Dual Anlagen of Cortex

First, cells that make up the olfactory cortex (paleocortex) are derived from the ganglionic eminence (GE), not the VZ (De Carlos et al., 1996), whereas cells that make up the archicortex are derived exclusively from the VZ (Nowakowski & Rakic, 1979) (Figure 6-1). Given that the GE is primarily the source of basal ganglia (subpallial) differentiation, whereas the VZ is the source of pallial differentiation, this evidence suggests a new theoretical approach to the evolution of the mammalian cortex from dual subpallial (more important to paleocortex) and pallial (more important to archicortex) anlagen. The several components of this differentiation are critical to the functional organization of the mammalian neocortex. The pyramidal projection neurons are clearly of pallial origin, and emerge from the VZ/SVZ in embryogenesis. The interneurons, however, are of subpallial origin, and they emerge from the GE. Recall from Chapter 2 that the recent discovery of the calretinin matrix cells, with their diffuse cortical projections, and the parvalbumin core cells, with their specific, focal layer 4 projections, has provided new clues to the functional organization of corticothalamic networks (Jones, 2007a, 2007b, 2009). These different calcium channel forms show different origins in embryogenesis, with calretinin cells emerging from the VZ/SVZ in human embryos (Figure 6-1), whereas the parvalbumin cells emerge from the GE. This alignment suggests that the diffuse calretinin thalamocortical matrix projections reflect the primordial pallial architecture, whereas the specific and focused parvalbumin core projections reflect the specific projections to layer 4 that appeared in the mammalian cortex.

## *Dual Gradients of Neurotrophic Factors*

Second, and closely related to this question of subpallial and pallial origins, is the evidence that genes that regulate cortical protomap development (and thus the patterns guiding radial migration) appear to be expressed along dual gradients. In general, a complex set of neurotrophic factors establish gradients in both mediolateral and anteroposterior directions (Figure 6-2). However, rather than following these cardinal, orthogonal directions, many factors are organized along dual oblique vectors. One is a gradient with high concentration at the caudodorsomedial (such as Emx2) extent (with low concentration at the rostroventrolateral extent), and another that is reversed: a high rostroventrolateral (such as Pax6) density (with low concentration at the caudodorsomedial extent). These opponent trophic gradients provide important organizing vectors for telencephalic differentiation. Within the theory of cognition as a continuing neurodevelopmental process, we must consider that the developmental results of the trophic gradients may, through creating unique motive vectors, continue to organize the self-regulation of cerebral morphogenesis across the life span.

## *Lemnothalamic and Collothalamic Brainstem Primordia*

Third, cortical expansion appears to be influenced by two patterns of connectivity that reflect the organization of the dorsal thalamus: the lemnothalamic

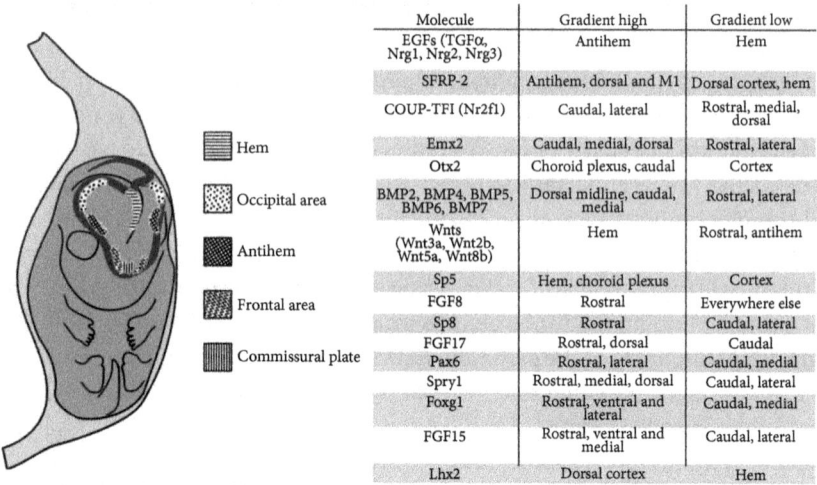

Figure 6-2. Gradients in neurotrophic factors interact to establish protomaps for neural differentiation. Redrawn from Rakic, Ayoub, Breunig, and Dominguez (2009).

and collothalamic projection systems (Butler & Molnar, 2002; Butler & Hodos, 2005). In reptiles, the lemnothalamic projections originate in the lower brainstem and target the pallium. In contrast, the collothalamic projections originate in the mesencephalic collicular networks and target what was formerly called the external striatum (now called the anterior dorsal ventricular ridge or ADVR). In mammals, the differential functions and connectivity of these brainstem and thalamic projection pathways may suggest how the evolution of dorsal and ventral divisions of the neocortex can be traced to differing subcortical primordia.

The evolution of mammalian ventrolateral neocortex may be traced to origins within the reptilian external striatum (Butler & Molnar, 2002; Figure 6-3). The embryonic differentiation of the insula, piriform cortex, and associated ventrolateral mammalian cortex from subpallial primordia is shown by mutations related to Pax-6, a neurotrophic factor that guides the differentiation of these structures. Knock-out of Pax-6 in the developing mammalian embryo blocks the differentiation of these ventral limbic structures and leads to a *pallial mound*, a mass of tissue dorsal to the striatum that Butler and Molnar observed to be strikingly similar in appearance to the reptilian external striatum (ADVR).

The collothalamic, sensory-specific projections from the thalamus thus target the external striatum in reptiles, rather than the primitive cortex. As the embryonic mutation giving rise to early mammals led to the differentiation of the external striatum into the claustrum, amygdala, and piriform and ventrolateral neocortex, the collothalamic targets were retained by these more

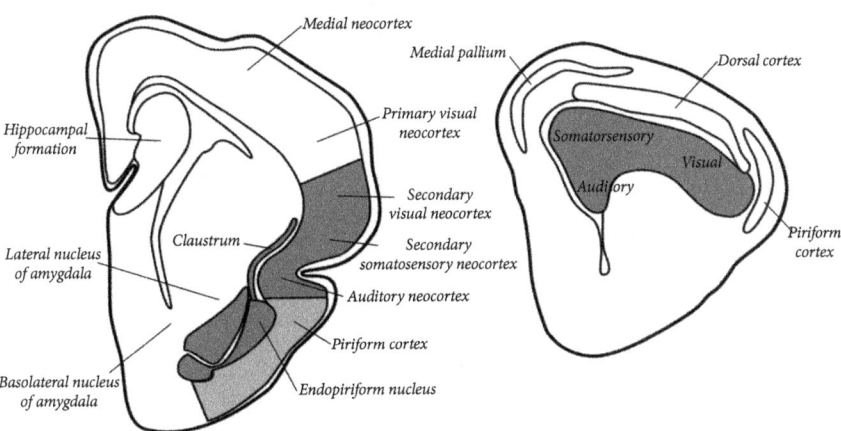

Figure 6-3. Abnormalities of genetic differentiation suggest that the reptilian external striatum (or anterior dorsal ventricular ridge; shaded, right) is the progenitor of the amygdalar-claustrum and piriform cortex, which forms the base for the mammalian ventrolateral neocortex (shaded, left). After Butler and Molnar (2002).

differentiated structures, very likely leading to the highly developed layer 4 and specific thalamic input to the granular, ventral neocortex of mammals.

## *Laminar Stages of Cortical Embryogenesis*

Fourth, cortical embryogenesis proceeds through dual stages, dependent on differing patterns of cytoarchitectonic differentiation (Figure 6-4). An initial stage is organized by pyramidal cells, with their plasticity and apparently their migration regulated by norepinephrine projections (Marin-Padilla, 1998). A later stage involves neuronal differentiation into varying cell types, regulated in part by thalamic input to layer 4. Given the evidence that these stages reflect pallial versus mixed pallial-subpallial organization patterns, this fourth observation is not entirely separable from the first three. The diffuse thalamocortical matrix projections thus appear to reflect the initial, exclusively pallial, stage of neocortical morphogenesis. Similarly, the second stage of cortical differentiation, regulated by thalamic input to layer 4, emphasizes the incorporation of specific core thalamocortical projections, apparently of collothalamic origin. These differential processes in embryogenesis may suggest how the differential cytoarchitectonics of the dorsal and ventral divisions of the mammalian neocortex can be traced to earlier versus later stages, respectively, in mammalian cortical embryogenesis. The earlier stage reflects dominance by pyramidal cells, with a thalamocortical architecture reflecting that of the amphibian

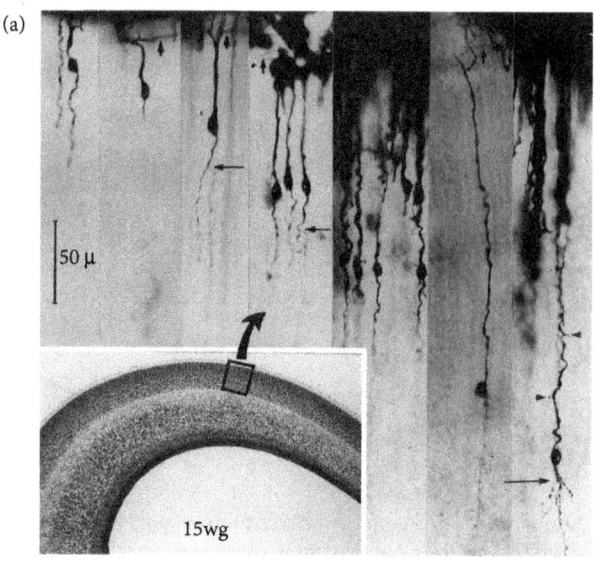
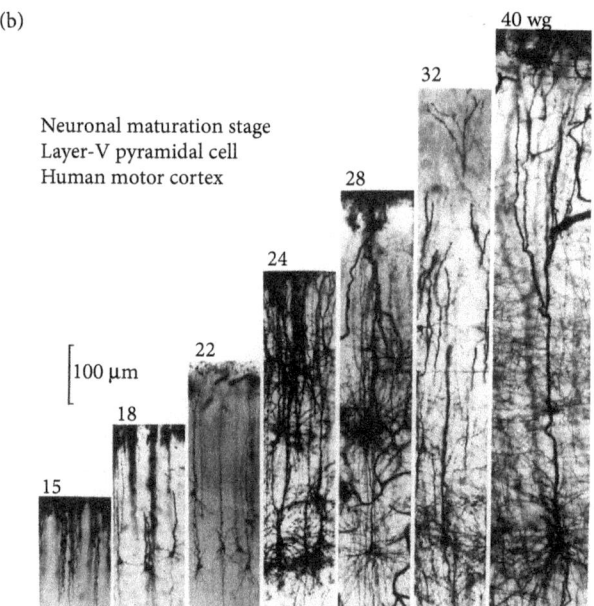

Figure 6-4. Progressive lamination of embryonic human motor cortex. **Left**: from 8 to 14 weeks the neurons that have migrated to the superficial layer (top) contact the Cajal-Retzius cells and elongate their apical dendrite, extending their cell bodies deeper into the middle layers of the cortex. Inset: embryonic human cortex shows minimal laminar differentiation at 15 weeks. **Right**: From 15 to 40 weeks there is increasing differentiation of specific neurons, effected in part through receipt of thalamic afferents, and increasing laminar stratification of the cortex. Reprinted from Marin-Padilla (1998) with permission from Elsevier.

and reptilian palliums. The later stage reflects the influence of the core specific layer 4 inputs, with an architecture reflecting the fusion of pallial with subpallial (particularly amygdala, striatum, and external striatum) elements and subcortical inputs.

## Cortical Migration From the Ganglionic Eminence and the Disappearance of the External Striatum

These neurodevelopmental clues make sense in the context of mammalian evolution. It is a key fact that the mammalian olfactory cortex has embryonic origins in the GE, the source of neurons migrating to the basal ganglia (and subpallium of reptiles). Even in the reptilian pallium, therefore, the GE origin of neuronal migration pointed to an alignment of the olfactory, piriform cortex with subpallial elements, and this alignment remains integral to mammalian neocortical embryogenesis. Even more remarkably, the GABAergic neurons of the mammalian cortex, playing key roles in the interneurons that define cortical complexity, also have their origins in the GE. These clues point to the importance of subpallial structures and neural components in the formation of the mammalian cortex. They provide a new, complementary perspective for the widespread assumption that the mammalian cortex evolved through some kind of differentiation of the pallium. It also incorporated fundamental subpallial elements.

In amphibians, the 3-layered cortex includes pyramidal cells anchored in the deep (periventricular) layer (Figure 6-5). In reptiles, there are also three layers, but the pyramidal cells extend into the middle layer. In a remarkable feat of evolutionary conservation, the mammalian hippocampus retains rudiments of both these pallial architectures. As pointed out by Ramon y Cajal, the mammalian dentate gyrus retains the amphibian structure of periventricular pyramidal cells, whereas Ammon's horn retains the reptilian structure with pyramidal cells in the middle of the 3 layers (Marin-Padilla, 1998). The human dorsal division of cortex (archicortex), with its extensive 6-layered mammalian form, relies on the hippocampus for major components of its memory consolidation. Therefore the hippocampus must achieve its contribution to memory and cognition through a network architecture integrating three evolved forms, including protomammalian (amphibian and reptilian) as well as mammalian cortical lamination and connectivity.

In amphibians and reptiles, the cortex is the pallium, which not only lacks the layer 4 (the thalamic input layer) of the mammalian neocortex, but also lacks the thalamic input. Instead, as noted above, the thalamic input in reptiles targets the external striatum (Nauta & Karten, 1970; Butler & Hodos, 2005). It may be an important fact that both evolution and embryology indicate that a major component of the thalamic input to neocortex originally targeted the

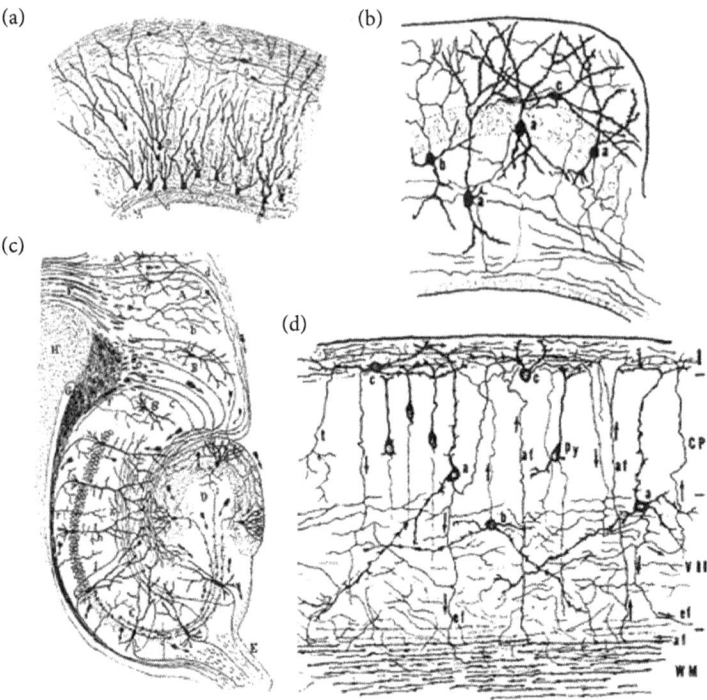

Figure 6-5. In amphibians such as the frog (A), the pyramidal cells of the pallium remain at the periventricular layer and extend processes into the neuropil. In reptiles such as the chameleon (B), the pyramidal cells migrate into the middle layer of the pallium (also called the primitive general cortex). In mammalian embryos such as that of the cat (D), the neurons migrate radially from periventricular anlagen into the superficial layer of the cortex, and extend processes down into the middle layers of the cortex that are then targeted by thalamic projections. In the mammalian hippocampus (D), Cajal observed that all three of these architectural patterns are represented, with the amphibian pattern in the dentate nucleus, the reptilian pattern in Ammon's horn, and the mammalian pattern in the entorhinal cortex interfacing with neocortex. A, B, and C are reprinted from Cajal, Ramon y, with permission from the heirs of Ramón y Cajal; D is reprinted from Marin-Padilla, 1998, with permission from Elsevier..

basal ganglia. Equally important is the realization that both evolution and embryology also point to striatal origins of the GABAergic inhibitory neurons that have become a major component of mammalian cortical architecture.

## Mammalian Variation on the Vertebrate Theme

These remarkable facts suggest that understanding the mutations of embryogenesis that led to the evolution of the mammalian brain may require insight into the functional organization of the progenitor reptilian neural architecture. The reptilian brain included a 3-layered hemispheric pallium with olfactory, septal, and hippocampal networks. This hemispheric pallium was to some degree separate from the subpallial basal ganglia that included thalamic sensory inputs to its anterior sector (the external striatum or ADVR) as well as motor outputs through the caudate, putamen, and ventral striatum and pallidum. Through these subpallial circuits, the generic (reptilian) vertebrate organization entails sensorimotor input/output circuits that are fully organized in the subcortical (subpallial) basal ganglia. These circuits include collothalamic projections to the external striatum that reflect the telencephalic extension of the collicular (mesencephalic) networks of sensorimotor integration. The collothalamic inputs represent specific sensory data, so that they are apparently well suited to forming links with specific basal ganglia motor circuits to support the stereotyped sensorimotor patterns of reptilian reflex patterns.

In contrast, the cortical neuropil of the pallium seems to support more general integrative processes. The lemnothalamic projections target more integrative nuclei of the thalamus, rather than the specific sensory nuclei targeted by collothalamic projections. As a result, the pallium both receives more integrated, multimodal inputs and integrates these within a pyramidal architecture that is suited to global integration. The pallium is the first cortical architecture to appear in the vertebrate brain, apparently realizing the representational benefits of a highly distributed, large-scale network (the primordial pallii dorsalis; Figure 6-6). Recognizing the olfactory as well as hypothalamic input to the pallium, as well as its distributed network architecture, we might speculate that the functions of the primordial vertebrate pallium may include (1) maintenance of context or mood states in relation to olfactory cues reflecting general properties of the environmental context, and (2) organization of the context model in relation to internal visceral functions. In our reptilian forebears, it would be within these pallial visceral-context (mood monitor) states that specific subpallial sensorimotor patterns (instinctive reflexes) would then be biased and modulated appropriately.

When primitive mammals first evolved, the overall structure of the telencephalon remained similar to that of other vertebrates (Figure 6-7), but there was a fairly radical shift of architecture, in which the external striatum disappeared, and the newly differentiated 6-layered neocortex took up the specific thalamic input into a newly formed layer 4, supported by extensive inhibitory GABAergic interneurons. The interpretation that seemed obvious to early brain researchers (Herrick, 1948) was that, in the morphogenetic mutations of embryogenesis that created mammals, the thalamic input projections somehow became uprooted from their striatal targets and invaded the pallium.

A more delicate interpretation, perhaps more suitable to the sensitivities of modern neuroscientists, was advanced by Nauta and Karten, who suggested that certainly not the whole structure but perhaps certain specific

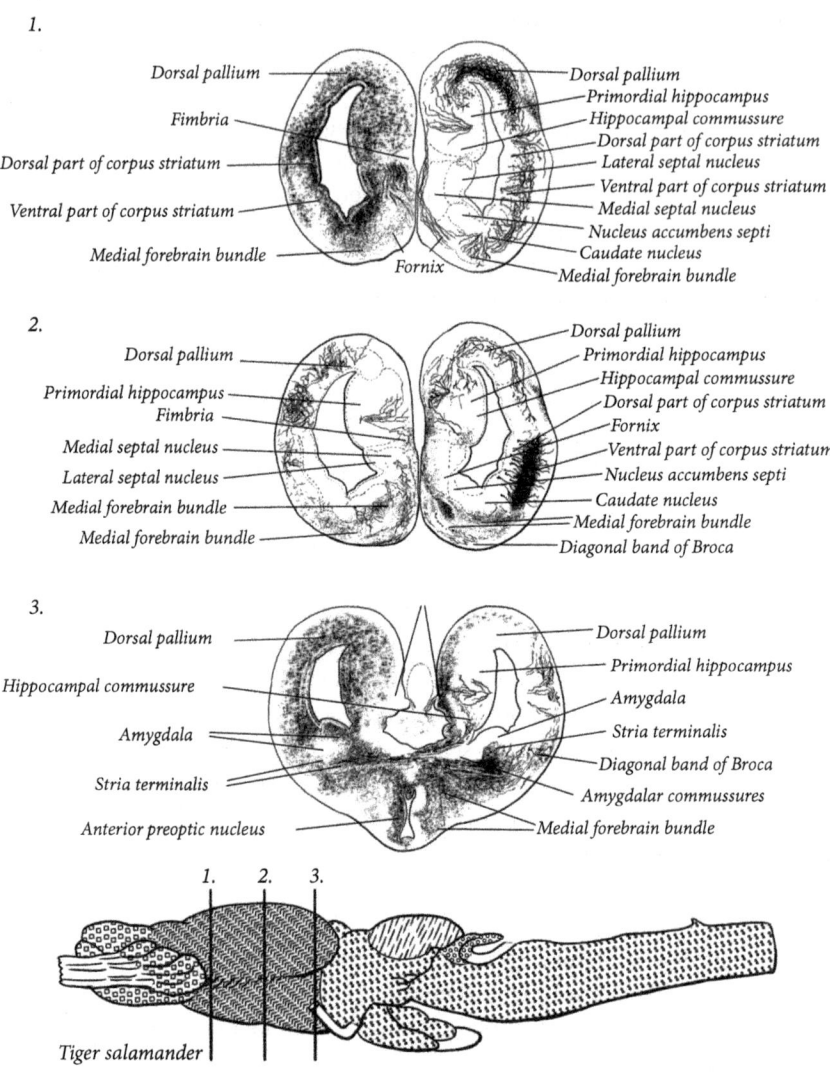

Figure 6-6. In the salamander brain, the telencephalic hemispheres form tubes around the ventricles. The dorsal pallium at the top of the tube forms what appears to be a primordium for the cerebral cortex. It is bordered by the hippocampal primordium on the medial wall and the basal ganglia (corpus striatum) and amygdala on the lateral wall. The ventral tube includes the nucleus accumbens, caudate, and medial forebrain bundle. In the transition to the mammalian hemispheric architecture, the hippocampus is drawn laterally into the temporal lobe, with the fornix extended to maintain connectivity with the septum and nucleus accumbens in their ventral and medial positions. The amygdala also undergoes a transposition to the temporal lobe; it remains connected with its visceral roots in the hypothalamus through the striaterminalis and with its somatic roots through the tail of the caudate nucleus. After Herrick (1948) and Nieuwenhuys, Ten Donkelaar, and Nicholson (1998).

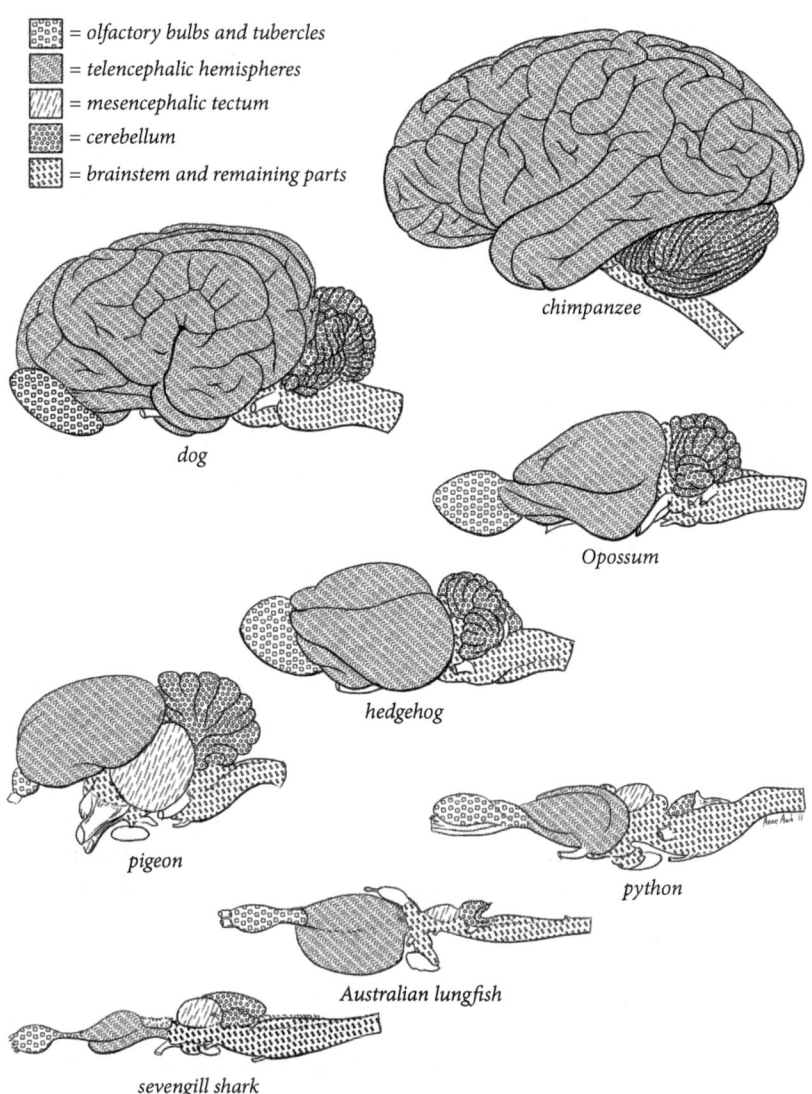

Figure 6-7. In extant species, mammals show an elaboration of the telencephalic hemispheres, particularly the cerebral cortex, that suggests the importance of the encephalization of sensation and behavior within the cerebral hemispheres. After Nieuwenhuys et al. (1998).

neurons of the reptilian external striatum (AVDR) may have become taken up into the mammalian cortex to create the novel 6-layered architecture (Nauta & Karten, 1970). The mechanism of such a change, like all evolutionary change, would be a mutation (or more likely series of mutations) of neural embryogenesis.

Thus the transition of the telencephalon from reptilian to mammalian form was such a considerable reorganization that it cannot be explained without a fairly radical interpretation. This interpretation remains beyond the theoretical horizon of current neuroscience (Butler & Hodos, 2005). The Sanides hypothesis, of differentiation of neocortex from dual limbic origins, is closely related to the question of the dual pallial and subpallial origins of mammalian cortex. The Sanides hypothesis was dismissed in the Butler and Hodos review, as well as in other neuroscience textbooks, ostensibly because of the lack of convincing evidence. Sanides developed his idea based on two sets of anatomical observations on the primate cortex. The first was the appearance of increasing cytoarchitectonic differentiation (from 3 to 6 layers, and increasing differentiation of cell types in the layers) of adjacent cortical regions, beginning at the limbic border and continuing to the lateral surface of the hemisphere. Sanides pointed out that these successive regions of increasing differentiation had the appearance of *growth rings* of the neocortex. A second observation concerned the cytoarchitectonic specialization of the dorsal (archicortical) division of the neocortex, with highly developed pyramidal layers, contrasting with that of the ventral (paleocortical) division, with a more developed granular layer 4 (Sanides, 1970).

Sanides (1970) thus developed his research observations on primate neuroanatomy into what seemed to be a novel view of the progression of neocortical evolution. Looking at the literature with this realization in mind, he then discovered that this view was not particularly new, and that there was considerable precedence in the traditional literature for this line of reasoning. In addition to the obvious reorganization of thalamic afferents from striatum to neocortex with the appearance of mammals, more fine-grained evidence on the dual origins of the pallium and neocortex had been provided by several early researchers.

Following his training in Elliot Smith's laboratory, Dart directed his students to examine the primitive general cortex (pallium) of reptiles in his native South Africa. They observed the cortex to be bounded by the hippocampal formation on the dorsal region and the pyriform cortex more ventrally (Dart, 1934) (Figure 6-8).

Examining the response of the alligator to electrical stimulation of the brain, Bagley and Richter made the interesting observation that stimulation of the dorsal cortex led to swimming movements, perhaps reflecting the links between dorsal cortex and more primitive (perhaps pallidal or mesencephalic) forms of motor control. In contrast, stimulation of the pyriform region led to sequential stepping movements (Bagley & Richter, 1924), possibly reflecting the integration of ventral regions of the pallium with the sequential motor control of the reptilian (crocodilian) basal ganglia. Patterns of motility have been important functional clues to the evolution of more differentiated neural architectures, such as in Coghill's distinction between the holistic neural control of swimming

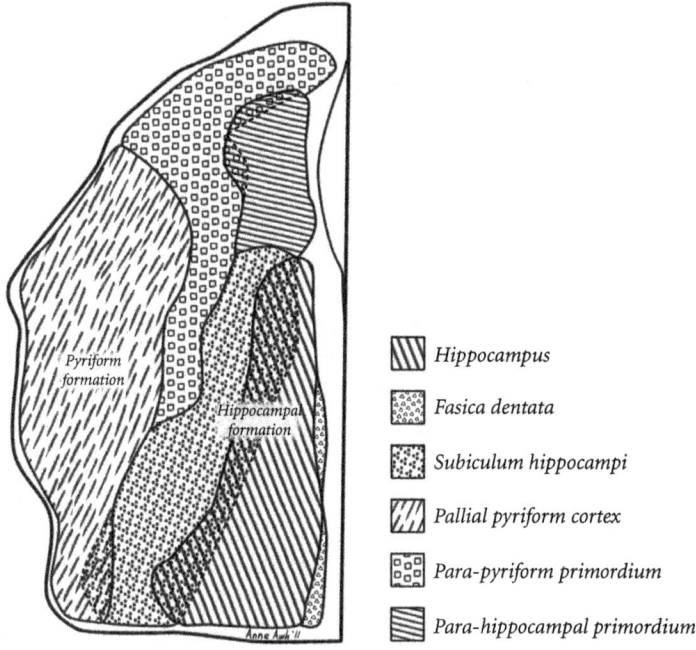

Figure 6-8. Organization of the reptilian brain (*Chameleon vulgaris*) around the hippocampal formation at the dorsal extent and the pyriform cortex at the ventral extent. This is a horizontal (axial) slice through the left hemisphere of the chameleon forebrain. Redrawn from Dart (1934).

movements versus the *partial patterns* required to articulate the inhibitory specification of tetrapodal actions (Denny-Brown, 1966; Herrick, 1948). The Bagley and Richter observations suggest that the dorsal division of the crocodilian brain retains primitive properties that may be similar to a projectional, holistic mode of control, in contrast with a more advanced, sequential, and reciprocally inhibited cybernetic capacity of the ventral division. The important clue may be that this projectional mode, whereas it can be traced to dorsal regions of the crocodilian pallium, seems to reflect an even more primitive (pallidal or paleostriatal) mode of action regulation, linked not only to the hippocampus and related cortex, but also to the ventral striatum.

To pursue the transition from the reptilian to mammalian telencephalic architecture, Abbie examined brain organization in primitive mammals, including monotremes (Abbie, 1940) and marsupials (Abbie, 1942). Abbie confirmed the dual, hippocampal and pyriform, boundaries of the cortex. Furthermore, as Sanides pointed out, Abbie anticipated the Sanides hypothesis by proposing a regular progression of differentiation of the cortex of these primitive mammals, with growth rings extending laterally from dual origins, from the hippocampal formation on the dorsal surface and from the pyriform cortex more ventrally. In the primate brain studied by Sanides, these growth rings would represent the limbic, heteromodal, unimodal, and primary sensory or motor cortices in each modality, as shown in the later 20th century by the definitive neuroanatomical studies of Pandya and his associates.

### Interpreting the Dual—Basal Ganglia and Pallial—Origins of Neocortex

The dual origins of the neocortex are seen in the embryo as VZ migration for the dorsal neocortex (archicortex) and GE (basal ganglia anlagen) migration for the ventral neocortex (paleocortex) (Figure 6-1). This is a first and fundamental developmental clue to the organization of the dorsal and ventral divisions of mammalian neocortex. It requires an evolutionary context for its interpretation. In light of both early and modern neuroanatomical observations, it seems clear that mammalian cortical evolution elaborated in some way on the 3-layered reptilian pallium, mediating between the dual hippocampal and pyriform boundaries (Sanides, 1970) (Figure 6-6). However, the incorporation of thalamic sensory afferents, extracted from the external striatum and inserted into the mammalian neocortex, remains unexplained by the idea that the neocortex simply emerged as a more differentiated pallium. Rather, a more radical interpretation is required (Nauta & Karten, 1970), in which certain neuroarchitectural features of the reptilian basal ganglia merged with the pallium to form the 6-layered neocortex. In primitive vertebrates, the telencephalic hemispheres are minimal extensions of the olfactory bulb (Figure 6-7). In more complex vertebrates such as reptiles, the uptake of thalamic projections into the pallium, primarily from nonolfactory inputs, is a major factor defining the primitive general cortex (Herrick, 1948). The subpallial striatoamygdalar circuits were the highest controls on motor function for the ancestors of mammals (Herrick, 1948), and they were closely associated with the ventral thalamic connections that helped to shape the emerging cortex's control of action regulation.

With the shift of thalamic innervation from external striatum to cerebral cortex there emerged a profound new complexity in somatic (sensorimotor) regulation, achieved through the expanded representational capacity of the new 6-layered hybrid (pallial-subpallial) neural architecture. The theoretical insight for appreciating this new capacity of mammalian function may come from understanding the complementary evolution of increasingly complex visceral (hypothalamic) control circuits (Yakovlev, 1948), providing not only elemental motive direction, but motive control over memory and cognition. In the fundamental vertebrate plan, perhaps shown most clearly in the amphibian brain (Herrick, 1948), the medial wall of the hemisphere regulates the visceral functions, through close interactions between hippocampus and septum with the hypothalamus. This is balanced by regulation of the somatic function for interaction with the outside world, through sensorimotor integration by the amygdala and basal ganglia (Isaacson, 1982). Although it is common to interpret the evolution of neocortex in relation to somatic functions primarily, the negotiation between visceral and somatic functions may have been the defining requirement for mammalian and human cortical evolution (Mesulam, 2000). In this process, the fusion of pallial and subpallial neural algorithms in dorsal and ventral moieties of neocortex may have been formative for both visceral and somatic modes of self-regulation. The differentiation of dorsal (visceromotor) and ventral (viscerosensory) modes of visceral function seems to have provided the motive regulatory basis for the unique capacities of the dorsal and ventral somatic functions of the increasingly differentiated neocortex. Memory, and cognition, emerged to span the temporal requirements of somatic coordination in service of increasingly long-term visceral goals.

## Cephalic Gradients in Genetic Regulation of Neural Morphogenesis

Thus the GE origin of olfactory cortex cells in mammalian embryogenesis may be a clue to the hybrid organization of the neocortex, and particularly its paleocortical trend, fusing aspects of the subpallial connectional architecture with the pallium to create the more articulated mammalian neocortex. The incorporation of subpallial elements appears particularly important to the ventral division of neocortex. The GE origin of olfactory cortex implies that this ventral base of mammalian neocortex shares embryologic roots with the striatum and amygdala, both of which are also derived from the GE (Aboitiz et al., 2003; Striedter et al., 1998).

The second developmental clue of neuroembryogenesis, the dual trophic gradients, can be understood in this evolutionary context. Consistent with a shared root in the neostriatum and amygdala, the embryogenesis of the diverse set of ventral structures including the insula, piriform cortex, and ventral lateral neocortex, is regulated by the same gene, Pax6. The rostroventrolateral concentration of Pax6 in the developing cortex appears to be a major factor in the differentiation of the ventral cortical division of the mammalian brain (Butler & Molnar, 2002).

In contrast, the expansion of the dorsal cortex in embryogenesis is controlled in large part by neurotrophic factors including the Emx-2 gene, which are expressed in a caudodorsomedial (dense) to rostroventrolateral (sparse) gradient. Mutant mice that lack the Emx-2 gene have reduced caudodorsomedial cortical volume, with a concomitant increase in rostroventrolateral cortical volume (Rakic, 2009). This pattern is consistent with the pallial origins of neocortex generally, but more specifically with a pallial architecture (and pallial genetic specification) of the dorsal division. In contrast, the Pax6 regulation gradient described above suggests a more remarkable integration of subpallial (external and internal striatal) neuronal populations within the neocortex (Nauta & Karten, 1970), specifically within the ventral neocortex of olfactory and pyriform (paleocortical) origins.

If we interpret the neurotrophic gradients of embryonic gene expression within the theoretical questions of the evolution of the mammalian neocortex, the evidence seems to imply that the genetic regulation of the dorsal (caudodorsomedial) neocortex reflects that of the primordial (reptilian) pallium, whereas that of the ventral (rostroventrolateral) neocortex has a more complex relation to subpallial (amygdala and basal ganglia) as well as ventral pallial (pyriform and olfactory) networks. Evidence on differing evolutionary roots of thalamic afferents to the dorsal and ventral neocortex, reviewed next, may be consistent with this implication.

## *Dual Origins of Thalamocortical Traffic*

A third observation on neuroembryology provides another key evolutionary-developmental clue to the dual configurations of cortical-subcortical organization. This is the pattern of projections from the brainstem to the forebrain, including those from the tectal (collicular) integrative centers of the upper brainstem. In simple vertebrates, the mesencephalic collicular networks provide major contributions to behavioral coordination. These collicular networks are closely integrated with the diencephalic thalamus, with the thalamus appearing to provide modulatory control over tectal function not unlike the familiar encephalization of cortical control over subcortical systems in mammals (Herrick, 1948).

In mammals, Butler and colleagues have discovered that the dorsal thalamus is organized into collothalamic and lemnothalamic divisions. Inputs to the thalamic nuclei of the collothalamic division are relayed through structures of the tectum (such as the superior and inferior colliculi) whereas inputs to the lemnothalamic division are direct. In reptiles and avians, the collothalamic division (with its striatal targets) is emphasized relative to the lemnothalamic division, whereas this pattern is reversed in early mammals (Butler & Molnar, 2002). In mammals, the nonstriatal targets of the collothalamic projections proceed to the lateral secondary and higher order sensory areas (including visual, somatosensory, and auditory cortex). These cortical targets of the collothalamic projections are believed by Butler and Molnar to have evolved from a common ventrolateral field located between the pallium and subpallium. In reptiles and avians, this field gave rise to the anterior dorsal ventricular ridge (ADVR), whereas in mammals it gave rise to the basolateral division of the amygdala, the claustrum-endopyriform nucleus, and the cortical targets (i.e., collocortex) of the collothalamic projections (Butler & Molnar, 2002).

Butler and Molnar proposed that duplication of the collocortex during evolution would give rise to the expansion of the lateral neocortical sensory cortical fields in higher mammals. Importantly, as described above, the control of this expansion appears to be regulated by the Pax6 gene, such that knock-out of this gene impairs the development of the insular and olfactory cortices, the claustroamygdalar complex, and collocortex, leaving the medial cortex (i.e., hippocampus) unaltered. Furthermore, Pax6 mutants show deficits in dopaminergic cells in the olfactory bulb (Marin & Rubenstein, 2001), implying that this gene factor regulates the actions of dopamine, a primary neuromodulator of the subpallial basal ganglia, and a proposed neuromodulator for the redundancy bias (Tucker & Luu, 2006; Tucker & Williamson, 1984).

In contrast to the concentration of collothalamic projections to the external striatum or ADVR in reptiles (and ventral neocortex in mammals), the lemnothalamic division in reptiles projects mainly to the dorsal pallium. In the sensory modalities, projections from the lateral geniculate nucleus and the ventral posterolateral and ventral posteromedial nuclei convey visual and somatosensory information, respectively, to the dorsal pallium. This pattern of thalamic afferents would be consistent with the notion that the dorsal pallium gave rise to major structures of the mammalian neocortex (Aboitiz et al., 2003), preferentially to its mediodorsal (archicortical) division.

These evolutionary considerations help to frame the understanding of the recent discovery that calretinin and parvalbumin neurons are aligned with unique thalamic projections to the cortex (Jones, 2007a, 2007b, 2009). With calretinin associated with dorsal (pallial) cortical migration from the VZ/SVZ (Figure 6-1), the calretinin mediation of the diffuse matrix projections appears to suggest a unique pattern of thalamocortical regulation for the dorsal neocortex that may be consistent with the dominance of pyramidal projection cells in the cytoarchitectonics of this division. The holistic, configural representational capacities of the dorsal neocortex, and its reliance on feedforward control (Shipp, 2005, 2007), may reflect the unique memory mechanisms of the primordial pallial mode of corticothalamic organization.

In contrast, the specificity of parvalbumin neurons in the migration from GE to neocortex (Figure 6-1) may point not only to the basal ganglia (ADVR) origins of the projections to cortical layer 4, but it may also point to the collothalamic origins (Butler & Molnar, 2002) of the specific core mode of parvalbumin thalamocortical projections. The fact that GABAergic interneurons share this GE ancestry is a remarkable and possibly convergent clue. The organization of specific sensory input in the granular (ventral) neocortex, and the refined object representation capacities that appear to emerge from inhibitory specification, seem to reflect a neocortex that has organized both circuitry and functional capacities from subpallial origins within a pallial architecture to create powerful new cybernetic forms of memory consolidation.

## Embryonic Patterns of Laminar Differentiation

A fourth observation about embryogenesis deals with the staged developmental emergence of the network architecture of the cortex (Figure 6-4). It has long been assumed that the mammalian cortex evolved from reptilian cortex through a kind of differentiation, creating 6 from 3 layers (Abbie, 1940; Herrick, 1948; Yakovlev, 1948), and this continues to make sense (Marin-Padilla, 1998). The Sanides (1970) model reinterpreted the cortical differentiation concept within the dual-trend organization framework. The embryological evidence shows that the differentiation of the cortical sheet in development proceeds through an interesting progression that includes both genetic guidance and neurotrophic self-organization. This progression may shed light on the neurodevelopmental roots of the cytoarchitectonic differences between dorsal and ventral cortical moieties and thus the mechanistic foundation for later psychological self-organization.

Development of the neocortex in mammals begins with a diffuse neuropil not unlike that of the amphibian pallium, composed of mainly input and output fibers and sparsely scattered neurons (Marin-Padilla, 1998). In amphibians, pyramidal cells extend their dendrites into this structure, but the cell bodies themselves remain near the ventricular zone. In reptiles and mammals, the cells that emerge from the VZ and SVZ migrate radially out into the neuropil, dividing it into superficial and deep layers. Once in the cortical sheet, the pyramidal neurons make contact with Cajal-Retzius (CR) cells, at which time they assume their characteristic pyramidal shape. Before that contact, they retain a primitive spindle shape. In mammals, the radial migration takes on additional complexity in later embryonic stages, as many more neurons migrate into the primitive cortical sheet to form the mammalian multilayered cortex.

Recapitulating the general progression from amphibian to reptilian to mammalian evolution, as new pyramidal cells enter the cortical plate in mammalian embryogenesis they are positioned above the older cells, first at the ventricular margin (as in amphibians), and then in the intermediate layer (as in reptiles). This inside-out or *radial* sequence of neuronal migration seems to be the mechanism by which mammalian neocortical layers are formed (Rakic, 2009). The layered organization, along with the attachment of the pyramidal neurons' apical dendrites to the CR cells, marks the first stage of neuronal differentiation, in which the cortical sheet is organized in an exclusively pyramidal network with the appearance of a primitive vertebrate pallium. The morphogenetic process is regulated by the CR cells in layer 1. Furthermore, some form of modulation from brainstem catecholaminergic projections, particularly norepinephrine (NE), appears to be critical in order to regulate the neurodevelopmental mechanisms of CR cells and the superficial layer (Marin-Padilla, 1998). This is then a fundamental neurodevelopmental role of NE in the anlagen of archicortex.

A second stage of mammalian neocortical differentiation is driven by thalamic input and involves more specific neuronal cell differentiations. Certain neurons detach their apical dendrites from the CR cells, and are then able to take on their distinctive characteristics; those that have not detached remain in the pyramidal form and become the typical pyramidal cells of the neocortex (Marin-Padilla, 1998). This second stage of differentiation leads to the formation of layer 4 granular cells, together with the GABA interneurons that are critical to local cortical circuit processing. This differentiation process appears to be driven by thalamic inputs that exert some neurotrophic influence over the target layer 4 cells (Marin-Padilla, 1998). Thus the targeting of collothalamic projections to layer 4 granular cells may be important to the second stage of differentiation. In addition, recent genetic studies of transcription factors such as Dlx2 have suggested that there is a distinct tangential migration of neurons, with subpallial (GE) origins, and with GABA as the neurotransmitter (Marin & Rubenstein, 2001). Because it is neurotrophic and organized at least in part by afferent projections, the second stage of neocortical differentiation may be important not only to thalamic afferent control, but to the connectional organization of the corticolimbic pathways in relation to both hippocampal (archicortical) and pyriform (paleocortical) divisions.

## *Emergence of Mammalian Memory Through the Collapse of the Reptilian Telencephalon*

Even through this cursory and selective review, it seems clear that the emerging evidence on transcription factors and gene expression in embryogenesis is providing new insight into the relations among multiple cellular networks reflecting the neurodevelopmental anlagen of the vertebrate neuraxis. By revealing the inherent alignment of neocortical networks and cytoarchitectural features with the primordial pallial and subpallial architectures, this embryological evidence may address the longstanding question of how the mammalian brain evolved from the ancestral forms that shared the relatively primitive neural architecture of extant reptiles and amphibians. The mammalian neocortex retains the major structure of the reptilian pallium, with subcortical communication and control channeled in large part through the dual limbic boundaries of hippocampal and pyriform networks. At the same time, the differentiation of the pallial neuropil into the 6-layered architecture of neocortex appears to have involved a radical set of mutational events. Taken together, these events represented a catastrophic collapse of the premammalian telencephalic architecture. The circuitry of the basal ganglia was fundamentally deconstructed, as the gene mutations redirected the embryonic migration of neuron precursors from the ganglionic eminence (GE)—the primordium of the subpallial basal ganglia—into the pallium. One effect was to incorporate a new class of neurons—the GABA inhibitory cells—into the cortical laminar architecture. Another effect was to redirect the collothalamic projections away from the external striatum (ADVR)—which then promptly disappeared—and into the former pallium, creating in the process the complex 6-layered neocortex of the new mammalian order.

These events were unlikely to represent a single mutation, and they were almost certainly distributed over many variations in the early protomammalian forms. Yet when complete and instantiated in the reproductive pipeline they created a profound disruption of the telencephalic architecture of the higher vertebrates, and this must have had immediate behavioral significance. The result of the breakdown of pallial and subpallial embryonic boundaries was a kind of hybrid network architecture, in which new capacities of memory representation and consolidation emerged through a synergistic fusion of both pallial (distributed networks of pyramidal projections) and subpallial (noncortical sensorimotor circuits with extensive inhibitory modulation) mechanisms of representation and control. We can speculate that this fusion allowed a novel mammalian representational capacity, capable of both broad integration (through glutamatergic pyramidal projection networks) and more local differentiation (through enhanced local circuit processing with a new density of GABAergic inhibitory interneurons).

Although the details are unclear, this fusion of network and circuit elements was almost certainly paralleled by reorganization of the diencephalic-telencephalic projections, leading to the unique mammalian forms of hypothalamic circuits with limbic structures and the core and matrix pattern of thalamocortical projections.

Importantly, the pallial-subpallial fusion was not complete, but graded across the dorsal-ventral extent of the pallial wall. The major advantages of the pallial architecture were retained by the archicortex and its pyramidal-dominant, noradrenergically modulated neocortex on the mediodorsal surface of the hemisphere. In contrast, the advantages of the subpallial architecture brought new capacities to the new hybrid network organization of the granular-dominant paleocortex of the ventrolateral hemisphere. At the same time as differing representational skills were established within the novel network architectures of the neocortex, the differing subcortical roots of the pallial and subpallial divisions of the telencephalon continued to provide unique and differential control over excitability and function of the dorsal and ventral divisions of neocortex.

The result of these mutations was that the process of memory consolidation in the mammalian neocortex could then operate on dual network substrates, regulated through different subcortical control systems. These differing control systems provide the dorsal and ventral divisions of neocortex with unique properties of arousal, network excitability, and motivational control. Taken together, these properties yield the dual, opponent and complementary, processes of the mammalian motive-memory.

In the next several sections we again consider the neural mechanisms of the dorsal and ventral cortical contributions to cognition. In Chapter 5 we saw how specific mechanisms of neural architecture and physiology may provide clues to the underlying mechanisms that are responsible for the unique contributions to cognition. Now we consider the evolutionary-developmental processes through which each of these underlying mechanisms of motivated memory arise from their specific evolutionary roots. These roots reflect the emergence of the opponent structure of the mammalian motive-memory from the fusion of pallial and subpallial architectures within the neocortex.

## Connectional Architecture

The cortex develops in relation to the dual primordia of the pallium, the hippocampus on the medial wall and the olfactory cortex on the ventrolateral wall. The multiple embryonic developmental processes differing between these primordia may be consistent with the Sanides hypothesis that these are the evolutionary origins of the mammalian cortex. However, without clearer evidence of the tangential migration of neurons from the archicortical and paleocortical sources to form the intermediate general (neo) cortex, it remains to be seen whether the linked networks of neocortex (limbic, heteromodal, unimodal, primary) can be seen as growth rings of progressive differentiation from the limbic core. Until more specific embryological evidence is obtained, the best evidence for the Sanides hypothesis is still the precise connectivity studies by Pandya and associates reviewed above (Barbas & Pandya, 1986; Pandya & Seltzer, 1982; Pandya & Yeterian, 1984, 1985). The growth rings centered on the hippocampal and piriform cores are interconnected through a highly regular web of network-to-network connections, with patterned supra- and infragranular projections, linking each to its adjacent (more or less differentiated) neighbor. This structure of the neocortex must be the structure of memory consolidation (Tucker, 1992, 2007).

In addition to the origins of neocortex in the dual hippocampal and pyriform moieties of the pallium, the evidence points to an even more remarkable differentiation. This is the incorporation of subpallial elements, both neurons and circuits, within the ventral division of cortex. In terms of connectional architecture, it is not just that dorsal and ventral divisions are each internally interconnected, with substantial intradivisional and few interdivisional connections. Rather, there are subcortical connections of the ventral division, with the thalamus, amygdala, and basal ganglia, that reflect the archaic subpallial origins of this division of cortex.

Furthermore, the embryonic patterns show another unexpected and at first confusing clue to developmental organization. Many of the genetic transcription factors, such as Em2 and Pax6, show not only a dorsal-ventral gradient, as might be predicted by the Sanides hypothesis, but a caudal-rostral skew of this gradient, leading to a caudodorsomedial pole (high density of Em2) opposed to a rostroventrolateral pole (high density of Pax6). The embryonic, and perhaps evolutionary, roots of the dorsal-ventral division of each hemisphere are thus aligned obliquely. The neurodevelopmental base of the archicortical moiety seems to be found in the caudal organization of the cortex (with an implied emphasis on somatic sensory integration). Is this why studies of connectivity of the human cortex show the posterior cingulate as the hub of cortical connectivity (Hagmann et al., 2008; Honey et al., 2009; Sporns & Honey, 2006)? In contrast, the base of the paleocortical moiety seems to be anchored in the rostral networks (with an implied emphasis on somatic motor control).

Our functional analysis of dorsal and ventral differentiation failed to anticipate this embryonic developmental architecture. In retrospect, however, the paradoxes of somatic and visceral control for sensory and motor functions may be relevant. At lower levels of the vertebrate neuraxis, the dorsal-ventral functional differentiation is for sensory versus motor functions, respectively, such as shown in the organization of the spinal cord and brainstem. The dorsal regions support sensory functions and ventral regions support motor functions. A caudal bias for the archicortex may seem incongruous, given that this is the root of the projectional *motor* control of the dorsal frontal lobe in the anterior cingulate cortex, which we have described as the impetus in action regulation and psychological function. Apparently, judging from the embryonic transcription gradients, the impetus and its direct, impulsive motivation of action arises as an elaboration of cortical networks that evolved most fundamentally from sensory anlagen.

In a similarly paradoxical fashion, the paleocortical networks at the ventral base of the cortex show a rostroventrolateral gradient that suggests a basic alignment with the motor control functions. Although this may be consistent with the close integration of paleocortical networks with the subpallial basal ganglia (and their motor functions), it still seems incongruous, given the considerable evidence that the ventral frontal networks and their feedback control mode are strongly interdigitated with *sensory* controls projecting to the elaborated layer 4. Yet this appears to be an integral component in the evolution of mammalian neocortex.

The gradients of genetic control of cortical differentiation, and the interesting paradoxes they raise, may thus suggest new perspectives on the dimensions of mammalian cerebral organization, and therefore on the neurocybernetics of complementary opposition. We have considered the basic dimensions of the neocortex as more or less orthogonal. The fundamental dimension explaining the pattern of neocortical connectivity is the core-shell dimension (Tucker, 2001), with the greatest density of network integration within the limbic core of the hemisphere, and progressively more differentiated networks within the heteromodal, unimodal, and primary sensorimotor cortices. With the possible exception of limbic networks, for each sensory and motor modality the greatest density of connectivity is clearly within rather than across modalities. The core of the hemisphere (comprising limbic networks) regulates internal, visceral functions, and the shell (comprising primary sensory and motor networks) regulates external, somatic interface with the world.

For both the core and the shell, there are then dimensions of specialization for input-output functions. The somatic shell is organized with rostral or anterior networks for motor output, linked in a kind of congruent complementarity with the caudal or posterior networks for sensory input. The visceral core seems to show its own input-output specialization, with dorsal limbic networks specialized for visceromotor control and ventral limbic networks specialized for viscerosensory control.

From this account of functional connectional architecture, we might expect embryonic gradients of neurotrophic and gliotrophic factors to be balanced neatly across the dual input-output dimensions of the visceral (core-shell) and somatic (anterior-posterior) functional divisions of the hemisphere. Instead, there are opponent trophic gradients (Em2 and Pax6) in a diagonal (caudodorsomedial vs. rostroventrolateral) dimension. Understanding this developmental pattern is an important theoretical challenge.

A caudal (caudodorsomedial) developmental origin for the dorsal moiety seems paradoxical, given the evidence that the cingulate cortex, the limbic base of archicortex, is specialized for the visceromotor function. It might seem that the dorsal limbic visceromotor function would be aligned with the rostral neocortical somatic motor function. Instead, it is aligned with the caudal, somatic sensory-integrative function. This may explain why the caudodorsomedial region of the human brain is a highly integrative center as suggested by recent functional connectivity research as well as tractography studies of connectional anatomy (Hagmann et al., 2008; Honey et al., 2009; Sporns & Honey, 2006). Apparently, neocortical evolution has somehow aligned the visceromotor function (of dorsal cingulate cortex) with the somatic sensory function (of the caudodorsomedial neocortex), both operating through the common cybernetic mode of the impetus and its habituation bias. What was fundamentally the sensory memory control of the habituation bias then is recruited to a more complex and flexible form of motor, and psychological, regulation, the impetus.

Through a complementary alignment, a rostral bias for the ventral trend must be considered in relation to the paleocortical specialization for representation of the viscerosensory function. The redundancy bias associated with the motor control of the basal ganglia seems to have been recruited to regulate the ventral division of mammalian cortex, perhaps along with the projections of the external striatum to layer 4, to support a more complex form of attentional focus required for sensory object specification. Feedback control of action regulation in the ventral frontal lobe then emerged as a more complex and powerful cybernetic mode of inhibitory specification in cognition.

## Complementary Opposition Through the Mammalian Inversion

Of course, we saw in Chapter 3 that action regulation in the mammalian neocortex involves complementary perceptual and motor systems of representation and control in both dorsal (configural-projectional) and ventral (object-feedback) moieties. But, as suggested by the diagonal pattern of the Pax6 and Em2 trophic gradients, there seems to be a primordial redirection or *inversion* of function in the evolution of the mammalian neocortex. In this functional inversion, the habituation bias, which was primordially aligned with somatic sensory control in the dorsal neuraxis, became integral in mammals not only to the visceromotor function but also to the dorsal brain's feedforward *motor* control.

A complementary opposition seems to have formed at the rostroventrolateral pole of the diagonal trophic gradient. The redundancy bias associated with the basal ganglia control of motor functions became linked to the viscerosensory function of ventral limbic networks, and then inverted its motor-sensory control function, giving rise to a form of memory consolidation that supports focused attention to *sensory* input to layer 4 in the feedback guidance of action regulation.

As we have seen from the embryological evidence, the overall organization of the neocortex seems to be achieved as a pallial cortical substrate becomes infused with subpallial circuits. To the extent that the reptilian pallium provided an integration of visceral, hypothalamic controls with the context model (integrative lemnothalamic inputs and olfactory cues), then it may make sense that the dorsal divisions of the mammalian cortex (with their strong pallial origins) would manifest a somatic sensory (caudodorsomedial) dominance. Similarly, to the extent that the reptilian strioamygdalar complex (striatum, amygdala, and external striatum) was critical to the motor reflex and pattern-generation circuits supporting reptilian action patterns, it may make sense to find the rostral dominance of the neurotrophic control (rostroventrolateral) of the ventral mammalian neocortex.

These are complex questions of the dimensional organization of the mammalian brain. Although our speculations may be incorrect, the diagonal pattern of the Emx2 and Pax6 embryonic trophic gradients suggests that a neurodevelopmental analysis requires a more complex functional interpretation of mammalian connectional architecture than would be suggested by a simple orthogonal model of visceral and somatic input-output organization, or by a simple inspection of adult mammalian neocortical anatomy. Some form of evolutionary-developmental analysis is required to understand the building blocks forming the foundation for the mammalian neocortex and its advanced memory capacity. It is this foundation that organizes in human embryonic development to guide the neurodevelopmental process of cognition throughout life.

## Cytoarchitectonic Specialization

Closely related to the differing connectional architectures of the dorsal and ventral divisions of cortex is the corresponding cytoarchitectonic specialization. The high density of pyramidal neurons in the dorsal division appears closely related to the broad pattern of interconnectivity supported by these projectional neurons. We theorized in Chapter 5 that this pattern may explain the functional capacities of the dorsal trend in holistic, configural representation and memory. In contrast, the density of granular layer 4 in the ventral trend appears related to the greater emphasis on local circuits in that division, apparently supported by a specialization for integrating GABAergic inhibitory interneurons.

Related to these architectural features is the direction of consolidation, reflecting dominance by either the limbic or primary sensory and motor neocortical networks. The strongly pyramidal pattern in the dorsal division of cortex may be consistent with the importance of the limbifugal direction of consolidation, in which the layer 5 projections to superficial (2-3) layers dominates the processing. Of course the dual directions of connections, limbifugal and limbipetal, appear to be more or less reciprocated, providing a recursive and reentrant pattern of laminar connectivity. But the dominance of pyramidal neurons in the dorsal moiety suggests that the pallial architecture is the primary template, and, consistent with Shipp's reasoning (Shipp, 2005), its pyramidal projection pattern may be consistent with a dominance of *limbifugal* traffic in the consolidation process.

The emphasis on granular layer 4 in the ventral moiety suggests a reverse, *limbipetal* direction of the dominance of corticolimbic traffic, centered on the granular input layer. In the posterior brain the granular input layer progressively models the sensory thalamic input. In the anterior brain this layer 4 gathers the posterior sensory input in order to represent the sensory objects that guide the feedback control of action regulation.

These neuronal patterns may reflect not only the derivation of dorsal and ventral divisions from mostly pallial and more evenly mixed pallial-subpallial origins, respectively, but also a sequential developmental unfolding of the primordial telencephalic architectural patterns. Recapitulating the evolution of the pallium, in the amphibian (periventricular location of the pyramidal cells) and reptilian (middle layer location) stages, the embryonic mammalian cortex demonstrates a radial migration of pyramids from the deep, periventricular to more surface locations. Migration to contact the superficial layer (Cajal-Retzius cells) then determines the architectural fate of the migrating pyramidal neurons (Marin-Padilla, 1998). With this initial stage of cortical differentiation, and with its plasticity regulated by norepinephrine, the specialized architecture of the dorsal cortex is largely determined.

In contrast, the second stage of cortical differentiation, with neuronal fate determined by disconnection from the superficial layer and by thalamic input to layer 4, may be more critical to the formation of the ventral corticolimbic architecture, with its emphasis on the granular layer. At least in broad outline, a developmental sequencing of dorsal-pyramidal and then ventral-granular differentiation may be an embryonic echo of the evolutionary process. In this process, the dorsal moiety of cortex derived from a more primitive pallial bias, whereas the ventral moiety reflects a more recent and complex integration of collothalamic input and subpallial neuronal material.

As we have suggested, both cybernetic biases of the dorsal and ventral divisions, described psychologically as impetus and the artus, may have unique cognitive advantages. However, the wholly novel architecture for the mammalian cortex seems to be the ventral moiety. It is a remarkable implication of the embryonic evidence that the ventral division is a hybrid of pallial and subpallial forms. The integration of GABAergic inhibitory circuitry with the replicated columnar network architecture of the pallium seems to have provided powerful new representational capacities. Very likely this has involved close integration with the capacity for surround inhibition supported by striatal circuits. In humans, these capacities seem to have provided unique functional skills, such as the inhibitory specification that allows categorical speech perception and the attentional differentiation of feedback control of action that allows grammar and syllogistic reasoning (Tucker, Luu, & Poulsen, 2009; Tucker, Frishkoff, & Luu, 2008). Our neurodevelopmental analysis suggests that these capacities can be traced to unique mammalian neocortical cytoarchitectonics, emergent from the collapse and fusion of the reptilian telencephalic pallial and subpallial divisions.

## Specialized Thalamic Modulation

Given its remarkable expansion in human evolution, it seems natural to focus on the cytoarchitectonics of the neocortex. Nonetheless, it is important to recognize that the function of mammalian neocortex is wholly dependent on subcortical control, particularly from the diencephalon. The classical evolutionary-developmental analysis of the vertebrate brain used the diencephalic (thalamic and hypothalamic) projections to interpret the organization of the telencephalon (Herrick, 1948). The modern neuroanatomical analysis (Jones, 2007a, 2007b), while revealing important new complexities (Sherman & Guillery, 2006), is largely consistent with the classical insights. The increasing recognition of the mechanisms of thalamic control may provide new ways of understanding the functional significance of pyramidal versus granular specialization in cortical cytoarchitectonics.

The pyramidal dominance in the dorsal neocortex reflects the retention of the pallial architecture as the primary network pattern. In mammals, the pallial pattern is elaborated of course in the 6-layered rather than 3-layered form, but with an attenuation of the layer 4 thalamic projections (and possibly of the associated inhibitory interneurons?) in the dorsal division. The absence of the granular layer 4 in the anterior cingulate cortex (Shipp, 2005) suggests that important self-regulatory functions may be achieved by a neocortex with a strongly pyramidal dominant cytoarchitecture, reminiscent of the pallial primordium. The dominance of lemnothalamic inputs to the dorsal neocortex may be consistent with the retention of global contextual information from the subcortex (rather than discrete sensory inputs) within the massively parallel pyramidal architecture of the dorsal brain. Furthermore, the dominance of calretinin matrix cells in the embryogenesis of the human dorsal neocortex from the VZ and SVZ must establish a corresponding dominance of diffuse, interregional patterns of projections that figures importantly in the cognition and memory of the archicortical division of neocortex.

The dominance of the granular layer 4 in the ventral regions of neocortex has long been a defining feature anatomically (generating the term *frontal granular cortex*). An evolutionary-developmental analysis may suggest both the origins and the significance of this cellular dominance. The core parvalbumin thalamic input, formerly restricted to the subpallial basal ganglia (external striatum) in premammalian vertebrates, now targets layer 4 of the mammalian neocortex. Although thalamic input of course remains critical to sensory and motor function of the dorsal as well as the ventral neocortical divisions, the specialization of the granular layer in the ventral brain may be a clue to a corresponding functional specialization for certain forms of input processing, perhaps integrating local inhibitory cortical circuits. The particular importance of collothalamic projections to the ventral brain is a further indication of how the ventral hemispheric networks have elaborated subcortical circuits that in this case reflect roots in mesencephalic (collicular) integrative functions. The unique properties of the ventral division of the hemisphere are not due solely to features of cortex, but to the combined corticothalamic network architecture that includes an elaboration of the specific projection patterns of the core parvalbumin projection cells. The fact that inhibitory GABA interneurons migrate into neocortex from the ganglionic eminence during cortical embryogenesis suggests that the complex elaboration of inhibitory interneurons in the mammalian cortex (perhaps most dense ventrally?) must also be traced to subpallial origins.

## Specialized Subcortical Circuitry

With its apparent elaboration of the pallial architecture, the evolution of the dorsal neocortex would have relied on the septal-hippocampal circuitry of the pallium. We can look for the motive drive of consolidation in the dorsal hemisphere, therefore, in the basis of this circuitry, including its connections with the hypothalamus and nucleus accumbens.

As we have seen, the granular paleocortical architecture is the preferential target of the collothalamic projections, reflecting roots of the ventral division of the hemisphere in the collicular integrative networks of the mesencephalon. Furthermore, the strong amygdalar connections of ventral limbic networks also appear to reflect developmental roots of the ventral hemisphere in the reptilian subpallial architecture. The unique resonance within the amygdala (Schafe, Nader, Blair, & LeDoux, 2001), and the apparent tight reentrance within the amygdala-MD-frontal triangular circuit (Jones, 2007b), may support a form of consolidation that is particularly important to feedback control and motivational focus of attention. These resonant memory circuits may be remnants of subpallial (strioamygdalar) control circuits in the ventral moiety that have no direct parallel in the memory operations of the archicortical networks of the dorsal hemisphere.

Just as the lemnothalamic projections target the dorsal thalamus preferentially, there appears to be an association between the ventral limbic cortical networks and collothalamic inputs from the thalamus and ventral thalamus, including the thalamic reticular nucleus (TRN). This association may explain the preferential orbital and ventrolateral frontal control over the TRN (Yingling & Skinner, 1976), and thus over the TRN's inhibitory control over sensory and motor regions of the thalamus. In contrast, the more dorsal regions of the frontal pole (BA9), including dorsal regions of BA10, exert control over the rostral TRN which in turn connects to the limbic nuclei of the thalamus (with their anterior and posterior cingulate and precuneus projections) (Zikopoulos & Barbas, 2006). Through these dual frontopolar networks, human consciousness may be supported by hierarchic representation and control skills of differentiated selective attention (ventrally) and motivated, goal-oriented intentionality (dorsally) (Tucker, Brown, Luu, & Holmes, 2007; Tucker & Holmes, 2010).

In considering how the consolidation process of the neocortex is regulated across each corticolimbic pathway, we have pointed to the recent evidence on the functional and anatomical differentiation of the core (parvalbumin) and matrix (calbindin) projections to neocortex emanating from each major thalamic nucleus (Jones, 2007a, 2007b). It seems likely that the more widespread pyramidal connectivity of the calbindin matrix projections would be important in the first stage of embryonic cortical differentiation. These projections appear to be emphasized within the pyramidal-dominant dorsal division of the neocortex. In contrast, the core parvalbumin projections to granular layer 4 would be expected to be emphasized in the second stage of embryogenesis, with the cellular differentiation of nonpyramidal cells. These core-thalamic projections would appear to dominate within the ventral neocortex, given its more extensive granular layer.

If this model of differential neurodevelopment is accurate, it may provide clues to the differing forms of adaptive resonance and thus consolidation that are supported in dorsal and ventral divisions. It may be that with its neurodevelopmental bias emphasizing the projectional pyramidal network pattern of the pallium, the dorsal neocortex has elaborated on the matrix projections from the thalamus that support interregional integration. In contrast, with its elaboration of the granular layer, the ventral neocortex may have evolved more complex patterns of local cortical processing that elaborate the core parvalbumin sensory projections from the thalamus. Although the core and matrix projections appear to complement each other functionally, and thus co-occur in each thalamic nucleus (Jones, 2007b), it is an important theoretical possibility that the several differentiations between dorsal and ventral divisions (cytoarchitectonic, connectional patterns, subcortical substrates) imply differential thalamic contributions to consolidation, relying on greater influence from matrix projections in the dorsal division versus a more balanced matrix-core pattern in the ventral brain. The embryonic neurodevelopmental evidence (Marin-Padilla, 1998) provides interesting clues to the developmental mechanisms for this differential thalamic specialization, with the pyramidal dorsal cortical architecture organized as the fundamental, protomammalian pallial base for the cortex, and the granular (hybrid pallial-subpallial) ventral cortical architecture elaborated in part through core thalamic projections to layer 4.

## Brainstem and Forebrain Neuromodulators

These several features of structural differentiation of the dorsal and ventral divisions, including connectional anatomy, cytoarchitectonics, and subcortical controls, appear to have evolved to provide cortical networks that are tuned to control by specialized neuromodulator systems, producing unique cybernetics of neural control over time. Given the cumulative nature of the neurodevelopmental process, the most primitive controls on neural activity, such as habituation and redundancy, may be the most important. The brainstem and forebrain neuromodulator projection systems control not only motive arousal, but also the emotional quality of experience and the adaptive direction of cognitive processing. It may be an important theoretical realization that these elemental vectors of self-regulation operate first to regulate neural embryogenesis, and then to regulate learning and behavioral integration throughout life (Brown, 2011; Trevarthen, 1985). The continuity of neuromodulation is a fundamental basis for the neurodevelopmental analysis of cognition.

Norepinephrine (NE) appears essential to regulating the neural tone and plasticity of the pyramidal neurons as they form the initial pallial architecture of the neocortex through contacting the Cajal-Retzius cells with their apical dendrites (Marin-Padilla, 1998). In the juvenile mammal, NE and serotonin (5-HT) remain critical to the experience-dependent neural plasticity of learning, as evidenced by the impairment of ocular dominance columns in the visual system when they are absent (Singer, 1987). Although in relation to postnatal development we would typically consider these neuromodulators to be concerned with regulating arousal, mood, or attention, the interesting concept for neurodevelopmental theory is that the role of neuromodulation in neural morphogenesis is the same role played in the neural plasticity of cognition. If so, the evidence on consolidation suggests that the neuromodulation of neural plasticity must occur not just in the period of perceptual-motor interaction with the environment, but during the extended, ongoing process of background consolidation, including sleep.

The embryologic evidence of NE modulation of pyramidal cell organization of the initial, pallial phase of neocortical organization provides an explanation for the preferential role of NE in regulating the function of the dorsal hemisphere. The projections from the major dorsal NE bundle are specific if not exclusive to the dorsal division of the hemisphere in adult mammals (Foote & Morrison, 1987; Morrison & Foote, 1986). Cortical control over the brainstem NE locus coeruleus is exerted by the mediodorsal, but not ventrolateral, regions of the frontal lobe (Arnsten & Goldman-Rakic, 1984). This specialization can be understood in relation to the self-regulatory mechanisms of the fundamentally pallial architecture of the dorsal division of the neocortex.

The frontal control over the brainstem nuclei for serotonergic projections, the dorsal raphe nuclei, is also exerted by mediodorsal but not ventrolateral frontal lobe (Arnsten & Goldman-Rakic, 1984), implying a parallel specificity of this neural mechanism to the dorsal brain, at least in terms of the cortical control over the brainstem projections.

In contrast with these apparent pallial-derived modes of self-regulation, the ventral neocortex appears to draw from control systems that evolved within subpallial circuits, in the basal ganglia. Dopamine (DA) and acetylcholine (ACh) appear to be restricted to the subpallial basal ganglia in reptiles. In many studies, chemical markers of these substances were used as anatomical indicators of the subpallial basal ganglia in reptiles (MacLean, 1949, 1986; Pribram & MacLean, 1953). Furthermore, the basal ganglia continue to manifest a high density of DA and ACh in mammals. The prominence of ventral limbic control over the forebrain nucleus basalis ACh (Mesulam & Mufson, 1984; Mesulam, Mufson, Levey, & Wainer, 1983) may be consistent with the importance of subpallial control systems to the functions of the ventral hemisphere. Importantly, the ventral limbic (anterior temporal, orbital frontal) control over the nucleus basalis ACh projections would result in a pivotal modulation of consolidation for the entire hemisphere by the ventral frontolimbic networks.

The mesolimbic DA system projects to dorsal as well as ventral regions of the frontal lobe, including the anterior cingulate cortex. However, we may speculate that constancy and focus of function with DA modulation of the redundancy bias is particularly important to ventral limbic consolidation mechanisms, including the unique triangular circuits of the amygdala projections to both MD thalamus and frontal cortex, complemented by the MD-frontal projections. As dorsal frontal networks of action regulation are tuned by this DA influence, they would shift from a native habituation mode (impulse) toward modulation by redundancy (constraint), although how this works at the mechanistic level is unclear. Our reasoning has been that DA modulation of redundancy is particularly integrated with the ventral limbic cytoarchitectonics (granular cytoarchitectonics, dominance of GABA); how DA control of the agranular dorsal frontal lobe (Shipp, 2005) would operate differently from this is an important question.

## Evolved Roots of Self-Regulation

Thus there are several intriguing clues for understanding how both neural structure and process emerge in embryonic development. There may be direct implications of these embryonic trends for understanding the neural mechanisms of the cognitive process. We may reason from the specific structures of cortical architecture to explain the cognitive representational skills that emerge from those structures (Semmes, 1968). The differentiated neural mechanisms of the dorsal and ventral divisions of the mammalian neocortex may be better understood in light of their evolutionary origins, as suggested by the recapitulation of those origins in the process of embryogenesis. Fundamentally, the theoretical understanding of the cognitive process requires analysis of the patterned neural architecture that supports motivational control of memory consolidation. Scientifically, this analysis may be gained through an evolutionary-developmental theory. Philosophically, it is interesting to realize that the ongoing process of cognition in personal experience reflects an unbroken continuity of neurodevelopmental mechanisms that organize the brain through a recapitulation of the major formative stages of vertebrate neural evolution.

### *Pallial Anlagen of the Dorsal Hemisphere*

The entire neocortex appears to be built on the pyramidal 3-layer pallial scaffolding of the telencephalic vessicle, with the hippocampal networks anchoring network functions dorsally and the olfactory pyriform networks anchoring them ventrally. Nonetheless, the several features of neural architecture reviewed above suggest that the dorsal division of the mammalian neocortical hemisphere has retained and elaborated this pallial form, apparently in a complementary opposition to the ventral division of the hemisphere, which evolved its own specialized processing capacity through embryonic mutations that incorporated unique neuronal materials and connections of the subpallial basal ganglia.

The dorsal capacity for configural organization of cognition, such as in spatial memory, may emerge from several features of the dorsal networks that reflect elaboration of the primordial pallial architecture. As Cajal pointed out, the pallial architecture is retained at the base of the dorsal networks, in both amphibian (periventricular pyramidal layer) and reptilian (intermediate pyramidal layer) forms within the dentate and CA regions of the hippocampus, respectively. Given the central role of the hippocampus in memory function within dorsal networks, it must be that the archaic pallial architecture confers certain advantages in reentrant neurophysiological excitement that continue to organize consolidation of this division of neocortex.

A related feature is the cytoarchitectonic dominance of pyramidal cells, which provide interregional projections. This broad pattern of connectivity may be integral to the organization of more global, integrative, and configural representations within the dorsal networks. However, in addition to the cross-modal interconnection between networks of a similar level of differentiation (for example, heteromodal visual to heteromodal auditory), the intramodal adjacent network connections (heteromodal visual to unimodal visual) are critical for supporting the processing of consolidation. In the dorsal pathway, this consolidation processing may be supported by the extensive pyramidal projection neurons, in this case organized within the matrix thalamocortical architecture, allowing the layer 5 pyramidal projections to layer 1 in multiple regions to support widespread integration through network resonance and oscillatory coupling (Jones, 2009). The facile cross-modal processing in dorsal networks appears to reflect a projectional or feedforward bias in dorsal network organization, in which representational patterns are readily propagated across cortical networks generally, as well as propagated within each corticolimbic (sensory or motor) pathway specifically. This rapid propagation appears to be suited to configural representations of multiple sensory elements in the perceptual processing of the posterior brain.

In the organization of action, the rapid propagation of network resonance in the dorsal cortical architecture seems consistent with a feedforward form of control, in which motive resonance at the limbic core establishes goals that are rapidly translated into action-goal urges that are then actualized into patterns of action, rapidly and holistically. The projectional mode appears closely related to the limbifugal direction of consolidation, rapidly elaborating the visceromotor function of the limbic base within the reentrant consolidation across multiple levels of the dorsal cortex. These several neural mechanisms may be essential to the expectant cognition underlying approach learning, as evidenced by findings such as the integral role of hedonic expectancy to approach learning seen in Chapter 1.

The embryonic modulation of the initial pyramidal architecture of corticogenesis (in the pallial form) by NE appears critical to the organization of the superficial layer 1 and the 3-layered pyramidal primordium. This form of control begins by setting the essential physiological tone for the layer 1 (C-R cell) regulation of the pyramidal cell alignment and differentiation. We theorize that NE modulation exerts an essentially similar influence on the plasticity of neocortical architecture throughout development, such as is exemplified by the neural plasticity of ocular dominance columns (Singer, 1987), and such as may guide the expectant, feedforward control of the psychological process of the dorsal division of the neocortex in cognitive development.

The rapid habituation of the NE response appears to be integral to the unique form of cybernetic function of this neuromodulator system (Tucker & Williamson, 1984). The habituation bias supports response to change but not constancy. In the perceptual domain, this control influence can be described as *extraversion*, in that ongoing arousal depends on orienting to changing events in the environment. In the action domain, this control influence leads to impulsiveness, in the sense of rapid discharge of motives into actions. This form of action regulation is consistent with our emphasis on limbifugal dominance in consolidation, and with the projectional, ballistic, feedforward control described for the mediodorsal motor control networks by Goldberg (1985).

How can this habituation bias—and its facilitation of change—be reconciled with the slow learning system of the dorsal limbic regions described by Gabriel and associates (Kubota & Gabriel, 1995)? This slow learning system and its context model are central to the theoretical model of learning as action regulation (Luu & Tucker, 2003a, 2003b; Tucker & Luu, 2007). The answer may be that the dorsal networks support a form of learning in which any given contents habituate rapidly enough to have only a limited influence on the consolidation process. The result is that the context model, while continually refreshed by limbifugal urges, is only minimally altered by the flow of external events, and it continues to shape cognition and behavior primarily through anticipatory feedforward control. Thus, in one of the several paradoxes in the control of neural activity over time, the slow learning process is achieved through a neural control system that changes quickly under a habituation bias (thereby consolidating only gradually).

Thus several unique properties of the dorsal, archicortical division of the mammalian brain may be understood to arise from the evolutionary emergence of its network properties from the primitive pallium of the premammalian telencephalon. Remarkably, important features of neural plasticity in the dorsal hemisphere of mature animals appear to reflect a kind of continuation of the process of embryogenesis, through which NE modulation of pyramidal networks allows broad, configural representations to be framed under feedforward control. What begins as an embryonic neurodevelopmental mechanism for organizing the structure of cortical networks continues as a psychological neurodevelopmental mechanism for organizing the cognition and structure of cortical networks. Although our theoretical characterization of these mechanisms is certainly crude at this point, there may be an important foundational principle implied by the initial pattern of this evidence. We may form the outlines of a neurodevelopmental theory of cognition, in which the functional control of neural plasticity in learning in postnatal life reflects a continuation of the mechanisms of embryonic neural morphogenesis.

### Subpallial Anlagen of the Ventral Hemisphere

In the mutational events we have described as the collapse of the reptilian telencephalon, the entire pallium seems to have become fused with subpallial neuronal elements and connections, including GABAergic interneurons and collothalamic projections to the new layer 4. Although affecting the neocortex broadly, these events dominated the neural architecture of the ventral hemisphere particularly, leading to novel and powerful cybernetic properties. The fundamental features of mammalian cognition can be seen to result from this fusion. Furthermore, human evolution seems to have continued to elaborate the cybernetic advantages of the new ventral neocortex, elaborating this hybrid (pallial-subpallial) architecture across large expanses of the hemisphere, and yielding important contributions to structured capacities in object-symbolic cognition such as language and mathematics.

Based on the evidence reviewed above, the granular cortical layer 4 appears to reflect an incorporation of formerly subpallial thalamic projections, reflecting the rudiments of the external striatum, in a remarkable interdigitation with the pyramidal pallial architecture. In the subpallial form, the striatothalamic circuitry appears to have integrated sensory data in direct forms of motor pattern generation. GABAergic inhibitory control is integral to this circuitry of the basal ganglia, and in mammals it may remain important to the neural circuits of the neocortex that now process the redirected thalamic (primarily collothalamic) targets.

What is the effect of this fused or hybrid architecture? We can theorize along three lines of progressively complex reasoning to characterize the unique cognitive properties of the 6-layered neocortex achieved by its interdigitation of pallial and subpallial elements. The first considers some form of sensory elaboration, through local circuits that provide extensive processing of the thalamic input. The second is closely related: an increased capacity for sensory guidance in the feedback control of action provided by the granular architecture of the ventral frontal lobe. The third is a new form of consolidation provided by the integration of the distributed pallial cortical architecture with the migrant neural elements of the basal ganglia (including GABAergic interneurons and collothalamic projections). This new form of consolidation includes a kind of reentrance that achieves increased inhibitory specification of discrete cognitive elements.

A first consequence of the incorporation of subpallial elements within the ventral division may be the extended processing of sensory input in the expanded granular layer, apparently emphasizing collothalamic projections. This seems to have allowed the ventral neocortex to achieve a high degree of complexity in representing sensory features. As we have seen, there is then the paradoxical development in which the redundancy bias supports a *sensory* process, together with the dominance of sensory constraints in the limbipetal dominance of consolidation. This new form of opponent complementarity may be a reflection of the oblique axis of telencephalic organization produced through the major neurotrophic gradients along the caudodorsomedial to rostroventrolateral gradient. The result for the ventral division of the hemisphere is a sustained, focused, and differentiated pattern of perception achieved by the cybernetic mode that evolved initially for sequential control of the somatic motor system in the basal ganglia. In contrast with the broad configural representations of the dorsal networks, the more focal processing supported by the cybernetics of the redundancy bias may allow sensory patterns to be held in memory as criterial objects.

A second consequence of the integration of subpallial elements into the ventral neocortex may be the support of more complex forms of action regulation. The invasion of the granular layer into the ventral frontal lobe seems to have been integral to a new capacity for representational differentiation to support the sensory guidance of actions. In contrast, without a granular layer, the ACC of the dorsal frontal lobe cannot integrate complex sensory data for guiding action, and must therefore rely on the ballistic actualization of limbic urges. Because it has an extensive granular layer, the ventral frontal hemisphere can be said to be a kind of perceptual cortex, integrating multiple sensory as well as visceral inputs to provide feedback guidance to the action plan.

Elaborated at the rostroventrolateral extent of the trophic gradients, the redundancy bias seems to have evolved originally for constancy and sequencing of motor control. Yet in the ventral neocortex this control mode becomes paradoxically important to a highly developed processing of sensory inputs, in the ventral layer 4. Then again, the focused perceptual capacity and object memory thereby created seems to have been recruited to the task of cognitive guidance of action regulation, with the restrictive feedback control linked to sensory guidance of the artus.

In addition to a general emphasis on input processing, and new capacities for feedback cybernetics in action regulation, there may be a third level of explanation for the specific mechanisms afforded by the subpallial circuitry incorporated within the ventral neocortex. The subpallial mechanisms may have provided a new capacity for *inhibitory specification* of discrete representational elements in the learning process. This inhibitory specification may parallel the specification of the local neocortical architecture for the thalamic input. As outlined in Chapter 4, the capacity for object perception can be explained in terms of inhibitory specification, as the defining features of the object are grouped through an inhibitory surround that suppresses the nonobject elements of contextual embedding.

As we have seen in Chapter 5, an inhibitory surround has been observed in the motor system, as selected actions are facilitated through inhibitory suppression of irrelevant and interfering actions. The integration of the inhibition and sequencing capacities of the basal ganglia (Gurney, Prescott, & Redgrave, 2001; Redgrave, Prescott, & Gurney, 1999) may be particularly important to the ventral cortex's control of action regulation. Through an evolutionary-developmental analysis, we may understand how the mammalian neocortical mutations allowed the basal ganglia mode of control to be recruited to support the ventral cortical network capacity for inhibitory specification. Although the specifics are unclear in our knowledge, the incorporation of GABAergic interneurons into the mammalian neocortical architecture seems relevant to the issue of inhibitory control, in relation not only to GABAergic modulation of the basal ganglia, but also to the capacities for local circuit processing that they appear to provide to the neocortex. To the extent that the collothalamic projections were organized in relation to striatal targets in the subpallial external striatum, we might speculate that their new cortical targets would reflect control through GABAergic interneurons in a way analogous to striatal circuits.

If it is true that the ventral neocortex reflects a particular incorporation of circuits and inhibitory capacities of the striatum, this would be compatible with a particular influence of the striatal neuromodulators, DA and ACh, on the neurophysiology and cognition of the ventral division (Tucker & Luu, 2006, 2007). The original formulation is that these neuromodulators provide a redundancy bias (Tucker & Williamson, 1984), through which a few cognitive or sensorimotor elements are given strong and continued representation in working memory. The redundancy bias itself can be seen as a kind of inhibitory control, in that the normal flow of behavioral change is restricted to a limited attentional focus to the currently active cell assembly. This primitive cybernetic mode may have been elaborated within the tetrapodal sequencing controls that evolved in the reptilian basal ganglia, and then again within the inhibitory specification within mammalian ventral limbic networks that incorporated subpallial elements.

It is thus an interesting possibility that several key components of the neural circuitry of the basal ganglia, including GABA interneurons, collothalamic projections to an expanded layer 4, and DA and ACh neuromodulation, were incorporated into the neocortex with the telencephalic collapse. The confluence of these elements seems to have been particularly important to the ventral division. The result may be a highly developed capacity for inhibitory specification within the fused ventral mammalian network, through which perceptual objects are extracted from the surround, and through which discrete actions are sequenced under inhibitory control, such that they can be guided by feedback control in relation to discrete sensory criteria.

Although we have emphasized the more elementary cognition of the ventral division of the hemisphere supporting perceptual objects and feedback control of actions, the capacity for inhibitory specification may be integral complex capacities of human intelligence, including symbolic representation, analytic cognition, and the structured organization of critical reasoning (Tucker, Luu, & Poulsen, 2009). The incorporation of subpallial circuits of inhibitory control may have led to a specialized mode of consolidation within the ventral division of the hemisphere. The triangular circuit, linking amygdala and anterior temporal networks to MD thalamus and frontal networks, provides a mechanism of consolidation that is not only aligned with the granular architecture of the ventral frontal lobe, but may be modulated particularly by the mesolimbic DA projection system. Similarly, the anterior temporal, insular, and orbital frontal control of the nucleus basalis ACh neuromodulator projection system may extend the ventral limbic capacity for inhibitory specification to influence the consolidation operations within the entire cerebral hemisphere.

# 7
## Self-Organizing Ontogenesis on the Phyletic Frame

In this final chapter, we consider how a general understanding of the structure and process of the mammalian brain may be important to interpreting the unique features of the human brain. The explanation comes from an evolutionary-developmental analysis, as it is informed by the embryological evidence. When we bring this framework to psychological theory, we find not only new ways of thinking about familiar questions, but opportunities to recognize unexamined preconceptions. The preconceptions include assumptions about the separateness of arousal, motivation, and the control of cognition. Components of the mind that can be easily isolated in an academic psychology may be seen to reflect common underlying mechanisms in a biological analysis in which psychological function is achieved by neurodevelopmental mechanisms.

As we have considered the formative mechanisms in neural development, we have gained insight into elementary control systems regulating the activity of neural networks that are the physical basis of the cognitive process. Neural control systems shape synaptic activity and neural plasticity within the patterns of vertebrate neural architecture. The neurotrophic gradients frame the vertebrate patterns of neural architecture, and the brainstem and forebrain neuromodulator systems then operate as the root vectors of self-regulation through the course of ontogenetic differentiation. The cortical-limbic-diencephalic networks of the forebrain have evolved to optimize the cybernetic utility of the unique forms of activity control of the neuromodulator systems, creating vertically integrated mechanisms of arousal, memory consolidation, and cognitive representation.

Through an evolutionary-developmental analysis, we thus discover that the motive process for psychological organization is a simple continuation of the motive process for neuroembryonic growth and differentiation. In psychological development this motive process simultaneously shapes arousal, the neural plasticity of learning, and the ongoing consolidation of cognition. As we studied the neuroembryological evidence, we found that the dual divisions of the mammalian cortex can be traced to dual roots in pallial and subpallial self-regulatory architectures. The specific ways that connectivity implies function in the mammalian brain are revealed by the evolved patterns of these cortical architectures. The mechanisms emergent from primitive subcortical roots are critical to the operations of the mammalian cortex in the consolidation of memory. These mechanisms are simultaneously the mechanisms of cognition, organizing the mind of the person, and the mechanisms of neural development, organizing the connectional architecture of that person's brain.

## Cognition Is the Neurodevelopmental Process

We recognize that the evolutionary analysis of embryological development has not been the most popular way of understanding human cognition. Even if it is the case that neurodevelopmental mechanisms are responsible for the self-regulation of cognition, it might be expected that these would only be the mechanisms that remain operative after birth. They would not include embryonic mechanisms generally, such as the gradients of trophic factors that guide the early stages of morphogenesis. Nonetheless, many of the clues to understanding the neural basis of cognition point to evolutionary-developmental roots. A careful theoretical analysis of these roots may suggest new perspectives on how function emerges from, and then shapes, structure in the mammalian brain.

More simply, we may come to realize that cognition *is* the neurodevelopmental process. Cognition may be the brain's continuing ontogenetic unfolding, in which the structure of neural connectivity emerges continuously from the motivated cybernetics of memory consolidation. Considered in this way, cognition is not an isolated activity of the mind, restricted to the discrete episodes of conscious thought that we might assume through naive introspection. Rather, in this more biological sense cognition is not only the process of conscious thought; it is also the continuous background process that effects memory consolidation, as experiences are digested in unconscious mental activity, and as the adaptive significance of events is realized through the integrative mechanisms of sleep.

### *The Dynamic Momentum of Connections*

If cognition is the neurodevelopmental process, a single theoretical framework will explain both neuroembryology and psychology. The scientific analysis of psychology can then be carried out in relation to the literal neural substrates of the dynamic, continually unfolding architecture of the brain. With developmental principles of distributed representation, we can reason from connectivity to understand function.

In this reasoning, we must realize that the brain's connectivity is both dynamic and continuous. The negotiation between stability and plasticity is dynamic, meaning that earlier patterns of cognition and behavior may be modified by significant new experiences, and by significant operations of self-regulation. At the same time, the context for ongoing development is the organism's history. The brain's organization is continuous, such that each act of cognition is shaped by the cumulative residuals of developmental history.

Scientific theory of the brain and mind must therefore appreciate the momentum of development that frames each moment of the cognitive process. The theory required for both neural systems analysis and psychological analysis is developmental theory, in which the reasoning of the theory frames the current operations of the brain and mind within the full context of the organism's developmental history. Although developmental theory has been important in certain classical traditions in psychology, particularly in psychoanalysis (Bowlby, 1997; Kohut, 1978; Mahler, 1968), it seems to have been difficult to integrate with the experimental studies of academic work, even when the focus is child development (Cicchetti & Tucker, 1994). In our theoretical analysis of neural mechanisms of cognition, developmental reasoning leads to the interpretation of each operation of ontogenesis (each act of cognition) in terms of the developmental context. This context is manifested in each individual's pattern of neural connections, formed over a lifetime, yet continually consolidated into new forms through the ongoing process of cognition.

Although an integrative theoretical framework will continue to prove challenging, there must be important scientific progress that will soon result from the increasing recognition that neural connectivity implies function concretely and directly. As new tools of studying cerebral connectivity become available for humans (Wedeen et al., 2008), we can test connectionist reasoning against anatomical data, with increasingly precise resolution. The ability to examine hemodynamic fluctuations with fMRI is providing clues to functional networks of the brain (Honey et al., 2009; Lacruz, Garcia Seoane, Valentin, Selway, & Alarcon, 2007; Protzner & McIntosh, 2008). New evidence suggests that these networks of functional connections show increasing differentiation during development (Dosenbach et al., 2010).

Although new neuroimaging tools are maturing rapidly, at the time of this writing the concepts of cortical architecture remain fairly primitive (Hagmann et al., 2008), and the concepts of functional networks are still quite preliminary (Greicius, Supekar, Menon, & Dougherty, 2008). The best evidence on the anatomical architecture of the human brain remains the detailed primate studies of Pandya and his associates, as outlined in Chapter 2, even with the obvious requirement to reconsider this architecture in light of the unique mutations in recent human evolution (Schmahmann & Pandya, 2006). The analysis of functional networks must match the highly organized anatomy of the brain.

The theoretical challenge for a neurodevelopmental theory is then to identify the elementary neural mechanisms that are explanatory for regulating neural plasticity in development, with the insight that these must explain the major patterns of human experience and behavior. In our analysis of the dorsal and ventral corticolimbic systems in human self-regulation, we have emphasized that elementary motivational controls are essential not only to shaping neural activity in time, but to framing the modes of expectancy that are integral to characteristic patterns of both cognitive structure and the interpersonal locus of control in personality. It may be useful in this final chapter to review the principles from this analysis and to consider their implications for understanding the neurodevelopmental process.

## Principles of Neurodevelopmental Theory

Neural plasticity is dependent on the elementary controls on neural activity in time, interacting with the control of connectivity patterns forming the spatial structure of neural networks, and thus the representational space of cognition. We have modeled the elementary controls on neural activity as redundancy and habituation. The brainstem and forebrain neuromodulator systems shape the elemental cybernetic processes of neural development. The circuitry and network architecture of the brain appear to have evolved within the functional parameters of these constant, essential regulatory influences.

As described in Chapter 1, mammals appear to have evolved modes of regulating neural plasticity that have become integral to representational memory. Because of the mammalian capacity for memory, and because memory is motivated to regulate action, mammalian learning is always expectant. In reptiles, motives lead to actions, but in mammals they lead to expectancies that then shape more complex planning of actions-in-context. Mammalian learning appears to be mediated through a hedonic goal-oriented expectancy in one mode (tuned by a habituation bias) and a reactive, yet anticipatory, externally directed feedback control in the other (tuned by a redundancy bias). Arousal and activation involve thus not only shape neural plasticity in a general sense; they tune the balance of representational stability and plasticity through specific cybernetic mechanisms for controlling neural activity in time.

These controls on connections in time entail a primitive strategy for allocating neural connectivity, and thus representational space. As the habituation bias limits activity in time it thereby increases the scope of neural connections that are active in a given interval in time. As the redundancy bias sustains neural activity in time, it restricts the scope of connections that are engaged over a given interval. Even in a simpler mammal, such as the rat, the evidence from learning research shows that the controls on activation and arousal are cognitive mechanisms, strategically tuning expectancies for future events.

The effect of tuning expectancies is to shape network connectivity. As we saw in Chapter 2, the nested corticolimbic networks of the mammalian brain are able to regulate their own processes of memory consolidation, through network communications that are excited by the resonant responses of the hemisphere's limbic core. With limbic resonance tuned in large part by the circuitry of the thalamus and hypothalamus, this tuning ensures that the cortex and basal ganglia are continuously regulated by the brainstem and forebrain neuromodulators that guide the redundancies and habituations of the consolidation process. The integrated corticolimbic system of the mammalian brain self-regulates its consolidation through the highly differentiated network patterns of the thalamus, with inhibitory interactions playing at least as large a role as excitatory ones (Buzsaki, 2006).

Although memory consolidation is typically considered as one process, the evidence on differential capacities of dorsal and ventral memory circuits suggests that it is actually two. As we saw in Chapter 3, the dorsal and ventral corticolimbic memory systems apply different and complementary cybernetic biases on the control, and consolidation, of behavior. The dorsal networks operate in a projectional mode, as motivated urges of limbic visceromotor processes actualize themselves in the impulses of behavior. This form of cognition seems to emphasize consolidation in the limbifugal direction, as syncretic representations at the limbic base shape the processing across multiple (heteromodal, unimodal, primary) networks of the neocortex.

The ventral corticolimbic networks seem to operate in a reactive mode, as the implicit values of limbic viscerosensory response guide the selective differentiation of target criteria and the feedback control of actions in relation to the distance from those criteria. With the dominance of the limbipetal direction of consolidation, the cognition of the ventral division is shaped by a consolidation process in which the patterns in sensory and motor cortices strongly constrain the emergent patterns of cognition and action regulation.

Thus the expectant and reactive forms of learning we examined in Chapter 1 are found to reflect the operation of dual motive-cognitive systems, each integrating brainstem, diencephalic, limbic, and neocortical influences uniquely. The control of neural plasticity, and therefore the control of the neurodevelopmental process, is found to emerge through a process of self-organization in mammals, as the patterns of memory determine the neurophysiological mechanisms of consolidation, and these in turn reshape the connectional architecture of the growing brain and mind.

In Chapter 4 we considered how the dual modes of neurodevelopmental self-organization may have general implications for psychological process and structure, including not only the executive functions in cognition but the locus of control of the individual within the social environment. These psychological implications reflect the inherent affective qualities of activation and arousal. The affective qualities shape the organization of intelligence, providing the motive biases integral to hypothesis and criticism. They also frame the pathologies of the self and cognition in psychopathology. At least in broad outline, we may see how an adequate theory of neural development would be a powerful theory of psychology.

Recognizing the fundamental nature of primitive motive controls, and the cognitive and neural structures they generate, we examined clues to the specific neural mechanisms of the dorsal and ventral corticolimbic systems in Chapter 5. These clues included not only differential properties of dorsal and ventral cortical architectures, but also the apparent coevolution of cortical patterns with specific subcortical motive biases. The patterns of cortical connectivity and cytoarchitectonics in the dorsal and ventral networks seem to have evolved as unique network architectures that optimized the cognitive controls applied by the habituation and redundancy biases.

To gain a fresh perspective for interpreting these clues, and for understanding some of their inherent paradoxes, we turned in Chapter 6 to the neuroembryological evidence, both classical and recent. The origins of the dorsal and ventral corticolimbic systems, as seen through an evolutionary-developmental analysis of embryological mechanisms, can be traced to the emergence of the mammalian cortex from its reptilian pallial and subpallial progenitors. The dorsal and ventral divisions of the mammalian telencephalon can be seen to have generated unique cognitive and motivational biases because of their differential dependences on pallial (dorsal) and mixed pallial-subpallial (ventral) modes of operation. Both representational capacity and network stability arise through the opponent balance of these primitive modes, elaborated within a complex web of telencephalic networks. In psychological terms, we can now understand the opponent complementarity of the impetus and the artus. The unique cybernetics of each mode can be elaborated and expressed freely because the brain as a whole is balanced by the inherent opposition of its complement.

## *Neurocybernetic Complementarity in Representation and Control*

The majority of this theoretical analysis applies to mammals in general. Yet an evolutionary-developmental analysis of human uniqueness may be framed most clearly within the complex matrix of multiple balanced representation and control dimensions of the mammalian brain. The requirements for regulating information input and output creates important functional specializations of neural networks—with integral opponent balances—at both the somatic and the visceral domains of the cerebral hemisphere's network organization.

The somatic input/output specialization is well known, with motor control organized in frontal or rostral networks and perceptual integration in posterior or caudal networks. Our understanding of how this specialization supports function has become more complex in recent years, as we appreciate the reciprocal balance that has evolved between sensing and acting. Perceiving the actions of another is not just a posterior brain operation; it involves mirroring and analysis-by-synthesis within frontal premotor networks (Rizzolatti, Fadiga, Fogassi, & Gallese, 1999). Similarly, organizing one's own actions is not just a motor function; it requires integrated, perceptual concepts of the actions organized within posterior parietal networks (Goodglass, 1993; Jeannerod, 1994; Pisella et al., 2009). The differing rostrocaudal biases in language self-regulation in aphasia provide an important example of balanced self-regulation of cognition. There is impoverished production but intact semantic constraints in Broca's aphasia, and fluent but semantically loose productions in Wernicke's aphasia (Goodglass, 1993). In Chapter 3 we described the relation between anterior and posterior networks in action regulation as a *congruent* complementarity, in that the capacities of the receptive networks seem closely matched to those of the expressive networks, whether specialized in the dorsal or the ventral mode. However, the expressive and receptive network representations in language are not only complementary; they are in important respects *opponent*. Each is balanced by the other, and each is biased against the other. When the bias is unbalanced by brain damage, we see the classical deficits in cognitive self-regulation of the Broca's and Wernicke's aphasias.

Within both anterior and posterior regions of the cortex, there is also a kind of complementary opposition between the somatic domain, in primary sensory and motor cortex, and the visceral domain, in limbic cortex. The somatic domain captures the detailed requirements for interface with the world. The visceral domain monitors the requirements for internal homeostasis, including the base species needs for social attachment, sex, and territory. Cognition, and the memory consolidation that achieves the differentiation of neural architecture, can be seen as an arbitration between the visceral and somatic domains, through a kind of complementary opposition between diffuse visceral urge and articulated reality constraint. The microgenetic analysis of clinical memory disorders provides instructive illustrations of the levels of cognition that are unmasked by brain lesions (Brown, 1988). In the dense amnesia following a brain lesion, the consolidation process is fully ineffective in achieving cognition. As the amnesia clears, and the primitive anlage of cognition emerges, the patient often shows confabulation, apparently reflecting the motivational process as it begins to structure cognition without constraint by facts. Effective cognition must arbitrate between the complementary opposition of the visceral and somatic domains, creating concepts that weave the opportunities charged by needs with the constraints imposed by the demands of reality.

A similar functional organization, through balanced complementary opposition, may operate within the visceral domain itself, and this may explain the dual organization of the limbic system. The inherent opposition and dynamic balance between visceromotor and viscerosensory control may explain the division of the hemisphere between dorsal and ventral cortical-thalamic-limbic systems. The direct and impulsive control mode of the visceromotor function, supported by the habituation bias, may be elaborated throughout the dorsal division of the hemisphere, for both perception and action. This is a direct expression, through the dominance of the limbic influence, manifested by a dominance of the limbifugal direction of consolidation.

In contrast, there is a more indirect, constrained control mode of cybernetics integral to the viscerosensory function, through which the redundancy bias maintains working memory so that actions meet the demands of the viscerosensory criteria. Through this cybernetic mode, consolidation in the limbipetal direction allows both somatic input (perceptual) and somatic output (motor) functions to dominate the representations of the ventral hemisphere. Framed by the limbipetal direction of consolidation, memory is organized in relation to the constraints at the somatic interface.

### *Cybernetics of Spatiotemporal Scope*

Thus balanced systems of input/output functions are organized for the both the somatic and visceral domains of hemispheric function. Although the somatic input-output functions are obviously important, it also seems clear that the visceral, limbic regulation is critically important to the consolidation process. We think that the functional and structural specializations of the dorsal and ventral moieties of the cortex have evolved to elaborate the complex cognitive cybernetics of the mammalian brain from the habituation and redundancy biases of the primitive vertebrate brain. In the dorsal brain, the multiple functional capacities, including projectional control of action, holistic organization of cognitive structure, and the paradoxical psychology and locus of control of extraversion, can be seen as emergent from a control on neural activity that habituates rapidly, thus expanding the scope of representational capacity in working memory over time. In the ventral brain, the capacities of feedback constraints on action, analysis of perceptual space into discrete objects, and the paradoxical psychology and locus of control of introversion, can be seen to emerge from a control that sustains neural activity through a redundancy bias, thereby restricting the scope of functional connectivity over time.

## Uniquely Human Mutants

These dimensions of opponent balance in representation and control are generic for the mammalian brain. Yet they form an essential basis for human cognition. Both the motive basis of self-regulation and the cognitive basis of conceptual structure are explained in important ways by the functional differentiation of visceral-motive systems between the dorsal and ventral divisions of the hemispheres, and by their unique and complex relations with somatic motor and sensory systems of the cortex. These motive-cognitive relations may be understood through a comprehensive evolutionary-developmental analysis of the structure and process of mammalian memory.

Traditionally, scientific rationales for human uniqueness have been unidimensional. The natural tendency seems to be to seize on the one salient feature of human evolution, whether it is an unusual vocal tract, a particular lobe of the brain, or a specific linguistic skill, and conclude that this is how we humans became what we are. As we have seen through several levels of analysis, a more comprehensive theory may be gained by considering how human evolution has elaborated complex skills from the basic mammalian cybernetics.

Nonetheless, there are many unique features of the human brain, and these present interesting challenges for further theoretical work. Each of these suggest ways of extending the more basic evolutionary-developmental analysis presented here.

### Hemispheric Specialization

The fact that humans have evolved differential cognitive and emotional capacities in the left and right hemispheres has been apparent for many decades. Although lateral asymmetries have appeared many times in evolution, and can be seen in a variety of extant species (MacNeilage, 1986), human hemispheric specialization has been integral to the evolution of intelligence, creating the unique neural architecture of language. This is a significant fact for understanding human nature, and yet its implications have been lost on psychology and neuroscience. Even philosophers seem to have missed the point. Instead, we understand hemispheric specialization in the popular culture through the efforts of advertisers, as we are reminded that decisions such as in buying a car have a different basis in the left and right sides of the brain.

Fortunately, the descriptions of the left and right sides of the brain in advertising have been more or less consistent with the scientific evidence. The left hemisphere does contribute analytic skills to perception and cognition (Hecaen, 1962; Robertson & Lamb, 1991; Tyler & Tucker, 1982), and it allows rational, sequential thought through the medium of language (Brown, 1988; Goodglass, 1993; Tucker, 1981). The right hemisphere does support holistic, intuitive thought (Beeman, 1993; Bogen & Bogen, 1969), and its function is strongly modulated by emotional responses (Borod, 1992, 2000; Tucker, 1981). Although there is little solid evidence on the evolutionary process, or even the genetic basis, underlying hemispheric specialization, it seems clear that the differentiation of function between the two sides of the brain must have been a major factor in the recent evolution of human intelligence.

If it is indeed the case that left and right hemispheric specialization reflects an elaboration of ventral and dorsal corticolimbic systems (Galaburda, 1984; Liotti & Tucker, 1994), then there may be another dimension of opponent complementarity to understand for a theory of human brain function. As we have seen for relations among other dimensions of cerebral specialization (such as visceral vs. somatic and sensory vs. motor), the dimension of hemispheric specialization is not orthogonal to, but aligned in complex ways with, other dimensions. Understanding human intelligence may therefore require insight into multiple interlocking yet asymmetric dimensions of opponent complementarity.

As we have seen, there is a fundamental alignment of embryonic trophic factors in caudodorsomedial to rostroventrolateral, or opposite, gradients. This would seem to reflect the evolution of the dorsal cortical networks from somatic sensory systems (as, perhaps paradoxically, they have become regulated by the visceromotor cingulate limbic networks) and the corresponding evolution of ventral cortical networks from somatic, basal ganglia, motor systems (as they have become regulated by the viscerosensory insular-amygdalar networks). This alignment would imply the remarkable evolutionary strategy of opposing visceral with somatic input/output controls. The caudodorsomedial pole reflects the somatic *sensory* center dorsally, and it is balanced by motive control from cingulate viscero*motor* networks. The rostroventrolateral pole reflects the somatic *motor* nexus ventrally, and it is balanced by motive control from insular viscero*sensory* networks. It seems that the crossing of visceral and somatic input/output controls in this way appears to confer important stability, perhaps through increased cybernetic complexity, in the mammalian corticolimbic process of representation and control. The evolution of human hemispheric specialization may have extended the advantages of opponent complementarity, through allowing an entire hemisphere to elaborate one mode of self-regulation, balanced to be sure by the opposite mode elaborated within the opposite hemisphere.

Of course, there are both dorsal and ventral corticolimbic divisions within each hemisphere. Yet one appears dominant on each side. The right hemisphere's holistic cognitive skill appears to elaborate on the configural memory, and very likely the projectional control mode, of the dorsal corticolimbic division. A limbic dominance of the consolidation process would be consistent with the right hemisphere's skill in emotional communication, through both auditory (tone of voice) and visual (facial expression) channels (Borod, 2000). In contrast, the left hemisphere's analytic cognitive skill appears to draw on the object memory, and very likely the feedback constraint, of the ventral corticolimbic division. The somatic dominance of consolidation may be consistent with a form of cognition such as language that becomes closely reality-oriented and less limbic-constrained, and with a specialization of unimodal networks in the left hemisphere (E. Goldberg & Costa, 1981).

The gross anatomical asymmetries of the brain may fit with differential elaboration of dorsal and ventral networks (Galaburda & Geschwind, 1981; Galaburda, LeMay, Kemper, & Geschwind, 1978; Geschwind & Levitsky, 1968), with the larger regions in the right hemisphere (such as the dorsolateral frontal region) pointing to elaboration of dorsal networks, and larger regions in left hemisphere (such as in the temporal lobe) pointing to elaboration of ventral networks.

The classical observations on differential effects of hemispheric lesions on sensory and motor skills may suggest that hemispheric development does not simply elaborate dorsal versus ventral corticolimbic network functions in a simple alignment, but extends the more general "diagonal" (caudodorsomedial to rostroventrolateral) functional dimension. Left hemisphere lesions may impair motor control more than similar right hemisphere lesions (Denny-Brown, 1966; Goodglass, 1993). Certainly verbal cognition is important to clinical assessment of apraxia because the motor commands are given verbally. Yet the left-lateralization of skilled movements may be more general than just handling linguistic commands. Although the frequency of motor deficits with left hemisphere lesions has sometimes been explained as a trivial result of right-handedness, the phenomena of right-handedness itself points to the fundamental nature of left hemisphere control of skilled movements (Shin, Sohn, & Hallett, 2009).

Similarly, one of the first observations of unique cognitive functions of the right hemisphere emphasized the *imperception* that occurs with right hemisphere lesions (Jackson, 1879). The integration of perceptual space within the right hemisphere (drawing on the configural representational skills of dorsal networks particularly) has long been evidenced by the disproportionate left unilateral neglect from right hemisphere lesions (Heilman & Van Den Able, 1979). There is a corresponding right neglect from left parietal lesions (Posner, Walker, Friedrich, & Rafal, 1987), but it is more transient and less clinically significant than left neglect with right parietal lesions. The right hemisphere therefore plays a key role in maintaining the representation of the sensory context, perhaps consistent with a specialization for the dorsal networks and their developmental roots in the caudodorsomedial gradient of cerebral development.

Hemispheric specialization is thus a unique development in human evolution that may nonetheless be understood as an elaboration of the complex web of interlocking functional specializations that occurred with mammalian neocortical evolution. If it is the case that the right hemisphere becomes specialized to elaborate the network architecture of the dorsal, pallial, pattern, this may be consistent with both the lesion evidence (Semmes, 1968) and the EEG coherence evidence (Thatcher, Krause, & Rhybyk, 1986; Tucker, Roth, & Bair, 1986) suggesting the right hemisphere has a more distributed network architecture than does the left. It is interesting to think that this involves the preferential organization of dorsal corticothalamic relations around the diffuse calretinin matrix projections of the thalamus (Jones, 2009). In contrast, if it is the case that the left hemisphere's skill in analytic perception, object memory, and linguistic function reflect an elaboration of the unique patterns of control and representation in the ventral neocortical division, this would reflect an equally remarkable continuity in neural evolution, through which forms of intelligence that have been considered as uniquely human may be understood to emerge from the inhibitory cybernetics that evolved first within subpallial neural systems. The focal representational organization of the left hemisphere (Semmes, 1968) may then reflect the specialization for the parvalbumin core thalamocortical projections.

In interpreting embryonic mechanisms, it is reasonable to begin with the genetic origins of hemispheric specialization, common to all humans. However, it seems clear that there are also complex epigenetic processes shaping an individual's pattern of hemispheric specialization, as the extended neoteny and continued plasticity of the large human cortex allows functional patterns to organize themselves within the individual's corticothalamic connectional architecture. Computational models of neural networks have shown how self-organizing maps arise through competitive interaction of differentiated inputs for the common space of connections (Kohonen & Honkela, 2007). The left and right hemispheres operate through a dynamic, competitive interaction, as shown by the reciprocal inhibition underlying attentional neglect and extinction (Birch, Belmont, & Karp, 1967), or by the maladaptive suppression of the right hemisphere's rudimentary language skills by a damaged left hemisphere (Heilman, 1979). Emphasizing the primitive evolutionary roots of hemispheric specialization should not detract from recognizing the dynamic, self-organizing nature of the functional balance that arises in each child's development through hemispheric patterns of complementary opposition. To the extent that the frontopolar organization of multiple networks in self-monitoring becomes integral to the neurodevelopmental process of epigenetic hemispheric specialization (Tucker, Brown, Luu, & Holmes, 2007; Tucker & Holmes, 2010), the dorsal and ventral frontopolar networks may provide differential frontothalamic mechanisms for the conscious participation in right and left hemispheric cognition.

### *Mutations at the Hemispheric Core*

As with all evolutionary changes, human neural adaptations must involve mutations of the embryological process. One curious specialization of the human brain is the high density of spindle cells (Von Economo neurons) in frontolimbic networks. An increasing density of spindle cells correlates closely with the increasing hominoid cognitive sophistication, from monkeys to chimps to bonobos to humans (Nimchinsky et al., 1999). This progression suggests that spindle cells may have had a formative role in human evolution (Allman, Watson, Tetreault, & Hakeem, 2005).

From a naive recapitulation view, in which ontogeny retraces the phyletic order, we would expect the uniquely human features to reflect terminal additions, modifications of the most recently evolved regions of mammalian neocortex. Yet the spindle neurons are prominent in the most *primitive* of mammalian cortical regions, in the agranular anterior cingulate cortex (archicortex) and the dysgranular insular cortex (paleocortex) (Allman et al., 2005). Furthermore, rather than a highly differentiated neuronal type, these cells take the fusiform or spindle shape of the early embryonic stage of the pyramidal cell (Marin-Padilla, 1998).

Certainly, a heterochronic mutation, retarding the development of neural differentiation, could lead to a primitive form, such as a spindle-shaped pyramidal cell. Yet in this case the regression of this cell type to a primitive form appears to have been integral to the evolution of complex human cognitive capacities. As we saw in Chapter 4, the anterior cingulate and insular networks are found to show primary activations in many neuroimaging studies of cognition. It is as if some primitive quality has been rediscovered by a regressive mutation and evolutionary selection that serves well the extended plasticity of neotenous human limbic networks. Advanced capacities are achieved paradoxically through retardation of the maturational process. The functional role of the spindle cells and their connections may be an important clue to the more general theoretical and empirical challenge of understanding how the extended immaturity of human limbic development, through the juvenile period into adulthood, has evolved in concert with the complexity of human neocortical function (Tucker, 1992).

## Diencephalic Counterpoint

As a component of the functional networks of cognition, the mammalian cortex is inseparable from the thalamus. Any concepts of cortical function must be built around thalamic mechanisms, including the thalamic reticular nucleus as well as the core and matrix thalamocortical projections (Jones, 2007a, 2007b). One of the most impressive advances in computational modeling of the brain has been the incorporation of thalamocortical mechanisms in the adaptive resonance model of cortical learning (Grossberg & Versace, 2008). The predictions from this model may be testable through recently developed models of neural synchronization of cortical sources (Buzsaki, 2006; Buzsaki, Kaila, & Raichle, 2007; Luu, Tucker, & Makeig, 2004), as we begin to recognize how cortical synchronizations could reflect underlying thalamic mechanisms (Jones, 2009).

The human neurodevelopmental process may draw on species-specific evolution in the architecture of the thalamus as well as the cortex. Because their functions are so entwined, the cortex and the thalamus clearly have evolved in concert. The embryonic migration of GABAergic cells from the telencephalic ganglionic eminence to the thalamus, through the *corpus giganticothalamicum*, is a uniquely human mutation. A reasonable speculation is that it reflects a developmental requirement to expand the inhibitory neural capacity of the thalamus to deal with the massive human cortex (Rakic, 2009). Thalamic inhibitory capacity may be necessary because the highly developed networks of the human cortex rely so heavily on neural inhibition (Buzsaki, 1996, 2006).

It is of course remarkable that the origin of the GABAergic migration to the diencephalon is the ganglionic eminence, the *telencephalic* embryonic source of the basal ganglia. In this mutation, ontogenesis reverses the phyletic order. The elaboration of inhibitory control may be one of many adaptations in the thalamus to reflect the coevolution of diencephalic mechanisms with the expanding telencephalic architecture. As we traced the multiple neural mechanisms that appear to underlie the differentiation of dorsal and ventral corticolimbic systems in Chapter 5, it was clear that communication with the thalamus is critical to each of these, whether it is the differential dorsal-ventral targeting of lemnothalamic and collothalamic projections, or what we speculate is a differential core-versus-matrix specialization for the granular-versus-pyramidal cytoarchitectonics. Even the complex human cortex is only one component of the vertically integrated neuraxis.

Inhibitory control is such a critical facet of complex neural networks that a unique human mutation of GABA neuron migration must be considered as a valuable clue. As we reviewed the embryological origins of the mammalian cortex in the last chapter, the migration of GABAergic neurons from the basal ganglia into the pallial mantle is itself an astounding development, with important implications for the new cybernetic properties that accrued to the mammalian cortex. Finding a unique GABAergic population migrating from the anlage of the basal ganglia into the embryonic human thalamus is therefore highly suggestive: what we learn about thalamic regulation of the cortex of mammals in the coming years will then allow clearer interpretation of the specific mutations of human diencephalic anatomy.

## *Conscious Learning*

More detailed conceptual models of the integrated function of the cortex and thalamus may be required for interpreting one of the most unique features of recent human brain evolution, the enlargement of the frontal pole (Gilbert et al., 2006; Semendeferi, Armstrong, Schleicher, Zilles, & Van Hoesen, 2001). Recent studies of human conscious self-monitoring have suggested that those persons who are most aware of their own cognitive processes, as shown by judging their accuracy in a difficult perception task, have a larger frontal pole than those who are less aware (Fleming, Weil, Nagy, Dolan, & Rees, 2010). As shown in macaques (Zikopoulos & Barbas, 2006), the primate frontal pole has a detailed pattern of connectivity with the thalamic reticular nucleus (TRN) and thereby with the interconnected thalamic nuclei. These connections allow the frontal pole to exert control over the balanced thalamocortical mechanisms regulating both the state of consciousness and the differentiated patterning of selective attention. If our interpretations from Chapter 3 are correct, the dorsal and ventral divisions of the frontal pole have differential roles in voluntary intention and selective attention, respectively (Tucker et al., 2007; Tucker

& Holmes, 2010). These capacities may form concrete cybernetics of human consciousness.

The roles of dorsal and ventral divisions of the frontal pole, in goal-oriented intention versus constrained attentional selection, seem to be consistent with the differing connectional architectures, cytoarchitectonics, and neuromodulator controls of the dorsal and ventral divisions of the hemisphere more generally. As a result, there may be implications for understanding the unique capacities that humans have evolved for frontal regulation of attention and cognition. Although spontaneous consolidation of memory through intrinsic limbic-cortical connectivity may be an ongoing, continuous neurodevelopmental process, a more deliberate form of network consolidation may be achieved through frontopolar engagement of the TRN-thalamus-cortical circuitry regulating conscious intentionality and differentiated attention (Tucker & Holmes, 2010). The hierarchic architecture of the frontopolar-thalamic circuitry may be especially important in conscious self-control.

For the dorsal regions of the frontal pole, there is a preferential recruitment of the rostral pole of the TRN and its connectivity with the anterior thalamic nuclei (AD, AM, AV) and then the cingulate gyrus. This circuitry of Papez may be critical to self-regulation of the state of alertness and the hedonic goal-oriented impetus to action, both of which may combined to support the intentional control of cognition and behavior. To the extent that the right hemisphere becomes specialized to elaborate the cognition of the dorsal corticolimbic networks, intentional control may be integral to its holistic and affectively charged cognition.

For the more ventrolateral frontal networks, there is a preferential recruitment of the more caudal and lateral regions of the TRN and the associated connectivity with sensory nuclei of the thalamus, providing more direct control over thalamic gating of selective attention (Zikopoulos & Barbas, 2006). This gating would be complementary to the ventrolateral frontal lobe's sensory feedback guidance of action regulation. With these specific mechanisms, the ventral division of the frontal pole would be able to provide a focused, differentiated attentional control over the contents of consciousness. This form of inhibitory specification of attention may be integral to the left hemisphere's analytic cognition. If consciousness reflects differential contributions of the dorsal and ventral divisions of the cognitive process, then perhaps individual differences in conscious self-monitoring (Fleming et al., 2010)—and expansion of the relevant dorsal and ventral networks of the frontal poles—will be found to comprise differential modes awareness. These modes may arise from primitive roots in action regulation, organizing the impulse to action versus its constraint and feedback guidance.

The enlargement of the frontal pole in human evolution may thus provide an interesting clue to the mechanisms of conscious self-regulation. Human consciousness requires both the extended buffer that allows working memory and the capacity for self-awareness that allows deliberate control of the cognitive process. The frontal pole has a unique connectivity with the TRN and therefore thalamocortical projections on the one hand—and both dorsal and ventral limbic networks on the other—that together lead to the capacity for *representation of the regulatory function*. This is a capacity for concepts of motives, leading to an awareness of the motive process rather than the primitive operation of motives in their native unconscious form. With this more complex form of self-regulation, learning becomes both conscious and deliberate, organized and monitored through self-aware cognition.

Given the dual components of frontothalamic representation, it may be useful to recognize dual components of conscious learning, with the impetus supporting the intentional control of behavior, and the artus supporting the attentional differentiation of experience. With the integral role of motives in the cognitive process, effective cognition may be optimized by an awareness of the motive process. This is not just a monitoring of goals, or a consciousness of the results of cognition, but an ongoing awareness of the motive controls—the anxiety of the artus and the elation of the impetus—that are the continuous experiential engines of the neurodevelopmental process of cognition.

### *Radical, Protracted Neoteny*

The radical neoteny of human development, extended through several decades at least, is an important clue to the uniqueness of human neural development. The fact that mutations of developmental timing affect the entire hierarchy of vertical integration, from brainstem to cortex, reminds us that the human brain results from a unique experiment of nature in the entire morphogenetic process, not just in one anatomical aberration. The human child is highly immature and dependent on social support prior to puberty (Tucker, 1989). Even after the metamorphosis of puberty (Tucker & Moller, 2007), maturation is a relative process, allowing most individuals to make the transition to the parent role while retaining the neotenous neural plasticity required for continued cognitive development. Through our radical neoteny, morphogenesis remains incomplete, such that complex cognition becomes the dynamic product of the neurodevelopmental process.

Only through a juvenile period that extends over two decades can humans achieve their complex intelligence. The recent evolution of symbolic capacity has been seen as the hallmark of human advances (Deacon, 1997; Wynn & Coolidge, 2008), and it may have reflected contributions from several of the unique features of the human brain reviewed above. But as we examine the foundations of human cognition in the motive vectors regulating consolidation in the pallial (archicortical) and hybrid pallial-subpallial (paleocortical) forms, these are generic features of the mammalian brain. Perhaps the most important quality of human development is not any unique cognitive form, but an elaboration of the generic mammalian form through maturation that is retarded over decades, unmasking the raw mechanisms of the neurodevelopmental process. Through radical neoteny, we maintain the infantile, larval form. We are the amphibians of mammals, extending the early stage of life in the larval form to acquire the culture, more or less, then transforming into the adult form in the metamorphosis of puberty and the individuation of an adult self (Tucker & Moller, 2007).

In their raw, primitive forms, the impetus and the artus operate in the child's mind with the capacity to create extended, hierarchic, and sophisticated forms of representation. Because they are not fixed into the adaptive behavior patterns of simpler adult mammals, the motive vectors also remain plastic over the course of child development, vulnerable to becoming exaggerated in the familiar human psychopathologies (Barbas, 1995).

The impetus operates as an extension not only of visceral urges, but also of an egocentric perspective that may degrade to operations of a narcissistic self. But when organized within a sophisticated conceptual structure, the impetus provides the child with increasing intentional control of cognition that allows a deliberate agency in thought and action.

The artus operates from an opposite, allocentric attentional perspective. The object capacity of cognition may be exaggerated by anxiety and hostility, leading to an alienation of self from world when this perspective is exaggerated, organizing anxious obsessions and paranoid fixations in the more degraded forms. But in more sophisticated forms the object capacity of the artus supports objectivity, as the allocentric perspective allows differentiated perception of properties of the world that are minimally constrained by the child's inherent egocentrism.

Because human development is maintained for so long in the neotenous form, the evolutionary-developmental roots of embryogenesis may be particularly important to shaping the neurodevelopmental process.

## The Context of Evolution

Just as the ontogenetic history of the organism constrains the current cognitive process, the phylogenetic history of the species constrains the broad outlines of the neurodevelopmental process. Human development begins through retracing primordial vertebrate forms, and then organizes the mammalian cortical architecture in a uniquely human pattern. The spontaneous activity of the fetus serves to self-regulate the activity-dependent specification of neural tissue.

In prescientific times, spontaneous fetal activity was thought to be the beginning of life, the *quickening*. Within a neurodevelopmental theory, the self-regulation of neural activity may continue the process of self-definition as the brain is shaped by birth into a particular culture, where neotenous human neural plasticity makes for remarkable adaptations of intelligence. We can now understand that the child's psychological growth operates through the same mechanisms of neural activity and synaptic differentiation that shape neural development in embryogenesis. The neurodevelopmental process continues in each act of cognition, as motives operate with their inherent cybernetic biases to shape the structure of experience and behavior.

## Nested Frames of Evolution and Development

It may be that the psychological analysis of child development can be informed in concrete detail by the biological context, evolution. In fact, we have found it difficult to develop a developmental theory of human brain function without understanding the operational principles of the mammalian brain. As we have seen, those principles are difficult to grasp until we understand them as reworking the more venerable developmental patterns of the reptilian brain. Even as embryonic mutations are the engine of evolution, they operate within the conservative program of the genome. Particularly conserved are the fundamental mechanisms of morphogenesis, such that not only the trophic gradients but the dynamic controls on neural activation and arousal—with the unique cybernetic biases that they bring to the task of synaptic differentiation—have continued to guide the developmental process across each transition of vertebrate speciation. As a result, the specific ways that motives influence human cognitive structure can be understood in relation to their primordial origins, as they are genetically determined by the conserved patterns of the neurodevelopmental process. These patterns may be revealed through an evolutionary-developmental study of the embryogenesis of mammalian neural architecture.

The human embryo organizes its brain broadly within the mammalian genetic pattern, specified in detail by the human genome. With the quickening, the neurodevelopmental process engages behavioral self-organization, as the fetus's motive vectors of habituation and redundancy achieve the differentiation of embryonic connectional anatomy through active influences of actions and the sensations they create. Self-organization begins with the venerable self-regulatory strategies of the brainstem and then reworks the major phyletic transformations of the midbrain, diencephalon, and multiple variations of the telencephalon. Understanding the algorithms of self-regulation in cognition may be informed by studying the root vectors of developmental organization, as in the incorporation of pallial and subpallial architectures within the neocortex.

At birth, the process continues, only now it appears to us as an experiential process, with psychological rather than anatomical results. But the process is the same. There is no point at which ontogenesis stops, and psychological function begins. Cognition arises through anatomical differentiation. Psychological growth, and cognition, is the engine of neural growth.

Because of this identity of neural structure and cognitive function, the concept of *orthogenesis*, that development is similar on many levels, may be reconsidered (Werner, 1957). Werner applied the evolutionary-developmental reasoning of 19th-century biology to understand psychological development. This reasoning provides insights that we should not lose in the 21st century. The child's development can best be understood through understanding its evolutionary context. So, too, with cognition.

In his theoretical analysis of the way that human brain lesions disrupt the developmental process of cognition, Brown (Brown, 1977, 1987, 1988) faced the problem of relating familiar cognitive processes, such as language perception or route-finding, to the evolved structure of the human brain, with its rhombencephalic, mesencephalic, diencephalic, and telencephalic levels, each nested within the vestiges of its anlagen. Brown applied Werner's concept of *microgenesis* to create a dynamic view of human brain function organized across the multiple levels of the vertebrate neuraxis. Through microgenesis, a thought begins as a motive-arousal impulse in the rostral brainstem. The thought is elaborated through motive controls in the hypothalamus, amygdala, and striatum, and gains semantic form in limbic and association networks. Only then is it articulated in preverbal form in frontal premotor regions. Considered in Brown's microgenetic analysis, each act of cognition recapitulates the phyletic order. This approach addresses the brain's anatomy directly, and it provides an account of vertical integration in developmental terms.

Microgenesis is a powerful theoretical metaphor for understanding the nested frames of neural development. Yet, in a more literal analysis, cognition may be related more simply to one developmental process, ontogenesis. The nested frames, of phylogenesis, ontogenesis, and microgenesis, are only apparent; for any child they reflect different regulatory scales of the single ontogenetic process. Because evolution has proceeded through embryonic mutations, an individual's ontogenesis recapitulates the evolutionary program, in the broad outlines as retained for that specific genome. Ontogenesis begins with these broad outlines, then continues to organize neural activity and structure through the motive vectors that begin in neuroembryology. Because of this, psychological development, whether considered over decades or over a few seconds of cognition, remains an ontogenetic process, shaping neural networks continuously in waking and sleep. Cognition is ontogenesis, considered in the present moment.

### *Motive Control of Consolidation*

Among the developmentalists, it was probably only psychoanalysts who considered each act of cognition as the product of a cumulative ontogenetic process. Cognitive psychologists typically consider each instance of cognition as a de novo operation, an information processing task that arises from some general capability, rather than from the continuous organization of a historical self. If cognition is an ongoing neurodevelopmental process, then each developmental phase (and each idea) is fully constrained by prior phases of the ontogenetic process.

A concrete guide to reasoning about the cumulative process of cognition is provided by connectionist models. The stability-plasticity dilemma of distributed networks causes the capacity of new learning to be constrained by prior learning (Grossberg, 1980). Freud described the embedding of new cognition within the frame of prior experience as *transference*, reflecting the transfer of latent unconscious mental contents into the current mental process (Freud, 1953). Because of transference, the psychoanalyst could understand fundamental developmental issues through observing the patient's behavior in the present moment.

Through understanding principles of distributed representation, we can appreciate that personal history is indeed implicit in current cognition, but no specific transference is required. New learning is invariably embedded within an existing representational network. Each cognitive process is a developmental event, an act of the historical self. Furthermore, each cognitive process is a transformational event; as the representation is consolidated, the self is then changed. The degree of change depends on the negotiation between assimilation and accommodation, effecting the consolidation of cognition. These are realizations that may seem strange to our everyday psychological reasoning, yet they follow directly from the understanding of cognition as a neurodevelopmental process. Thought shapes the literal anatomical structure of the brain, and the self.

Humans are remarkable in the continuation of neural plasticity beyond the juvenile period. Radical neoteny seems to apply to the motivational networks particularly, allowing extended plasticity of the primate limbic networks well into adulthood. The relative rigidity of the sensory and motor neocortices explains the loss of learning capacity for sensory and motor skills with human maturation, such as the inability to learn to speak (or hear) a foreign language with native fluency after puberty (Tucker, Frishkoff, & Luu, 2008). The neurodevelopmental process of cognition remains vibrant and flexible in adults, primarily because the limbic networks, and to a lesser extend the adjacent association cortices, retain their juvenile capacity for plasticity (Barbas, 1995). Complex knowledge is organized through extensive consolidation within the limbic-associational networks, with consolidation motivated in the way Freud described as the *motive-memory* (Freud, 1895).

As we have studied the mechanisms of the neurodevelopmental process, we have seen that these are inherently motivational. The spontaneous process of neural activity in the embryo arises through motivational vectors. Throughout ontogenesis these vectors differ for the corticolimbic architectures in the dorsal and ventral divisions of the brain. The redundancy bias of primitive memory function in the ventral telencephalon seems to have evolved in concert with the basal ganglia, amygdala, and associated ventral limbic networks. The inhibitory controls from GABAergic mechanisms seem to be particularly important to these same networks, as suggested by the embryonic migration of GABAergic neurons from the subpallial ganglionic eminence. When cognitive consolidation is modulated by the redundancy bias, neural development is shaped by a unique cybernetic mode, a way of controlling neural activity that is intrinsic to certain forms of network differentiation. The motive process is then interlocked with specific patterns of network representation. Process yields structure. The constancy of focused attention supported by the redundancy bias is well-suited to the ongoing guidance of behavior by viscerosensory constraints, and by the representation of object memory criteria in the ventrolateral frontal networks.

When framed by the artus, cognition and neural development thus give a certain form to the ongoing structure of the self. This is a mode in which memory is readily modified, under control not only by internal viscerosensory evaluations, but also by external somatic sensory criteria that are charged by those evaluations. The psychological executive process is selective attention, engaging both the focus of the redundancy bias and the differentiation of perception and action through core thalamic and ventral somatic sensory representations of object memory. The artus is thus the mode of accommodation (Piaget, 1936/1992), as the contents of the mind are made reactive to both homeostatic and environmental constraints.

A complementary mode of motivated learning occurs when cognition is dominated by the impetus, operating under feedforward control in dorsal corticolimbic pathways, and modulated by the unique cybernetics of the habituation bias. Perceptual scope is broad, as the capacity of working memory is readily cleared. Action regulation in this mode is impulsive, as the cybernetics of the habituation bias are suited only to launch behavior under projectional, hedonic urges. The corticothalamic architecture is well adapted to this control mode, and the hedonic value of goal representations and integral monitoring of agency provide a substrate of intentionality for directing behavior. The executive process of intentionality is not well integrated in modern psychological theory, but it should be. This is a mode in which cognition asserts itself on the world, through assimilation (Piaget, 1936/1992). In this cybernetic mode, egocentric expectancies are effective in aligning personal values with ongoing environmental events, and the context model effectively guides the informational exchange with the world.

The Piagetian concepts of assimilation and accommodation are useful for a neurodevelopmental theory because they are consistent with cognition as a process of self-organization, changing the self through each act of incorporating external information (or not). By recognizing the inherent motivational biases of the activity controls underlying assimilation (habituation) and accommodation (redundancy), we can understand what Piaget did not: that cognition is intrinsically motivated, formed through the inherent cybernetics of motive controls on neural activity.

The motivational base of cognition is inherently primitive, aligned with visceral functions at the limbic core. Yet the limbic networks hold complex cognitive representations themselves, and they remain highly plastic throughout human development. As a result, the motives that regulate consolidation are themselves transformed by the developmental process, and may progress from egocentric maintenance of the narcissistic, childish self (Kohut, 1978) toward support of more complex values reflecting both socialization and personal decisions (Maslow, 1968).

## Complementary Opposition in the Mammalian Cybernetic Inversion

Cognition is typically considered to be internal to the mind. However, careful inspection of each concept shows that it mediates the relation of the mind to the world (Harvey, Hunt, & Schroder, 1961). The dual motive vectors of the corticolimbic networks shift the locus of control in cognition, between internal and external direction, in ways that are complex and even paradoxical. The paradoxes of internality and externality may be only apparent, however, and they may be resolved as we recognize the differential effects of the cybernetic modes of habituation and redundancy over different intervals of time.

Insight into the paradoxes of internality and externality can be gained through an evolutionary analysis, by considering the ways that the mammalian cortex provided complementary, second-order elaborations of the more elemental cybernetic biases of redundancy and habituation that operated in premammalian forms. These somewhat complex theoretical issues may be important to understanding the motive cybernetics of each aspect of neural development and cognition, including not only the cognitive process but also the structure of representational capacity. The primitive motive vectors determine how the locus of control is balanced between internal and external causality.

The controls on the neural activity of sensation and action may be easiest to characterize in their premammalian forms. The dorsal-ventral division of function in the vertebrate brain, as seen in the brainstem and spinal cord, separates somatic sensation in dorsal pathways and somatic action in ventral pathways. Visceral regulation is interposed between these functions. In the mammalian telencephalon, an important clue is the diagonal trophic gradient, with Pax6 and Em2 varying in opposite directions from the rostroventrolateral to the caudodorsomedial extent of the developing cerebrum. This clue from morphogenesis suggests that the dorsal division of the cortex emerges from sensory (caudal) anlagen, while the ventral division emerges from roots in motor (rostral) systems.

In the reptilian telencephalon, somatic action is controlled by the basal ganglia, with the sequence control required for tetrapodal locomotion provided by the complex inhibitory circuitry that evolved in the basal ganglia (Redgrave, Prescott, & Gurney, 1999). Activation control for this circuitry was provided by dopamine and acetylcholine modulation in what we have called, after Pribram (Pribram, 1971), the redundancy bias. In terms of the structure of behavior, the redundancy bias can be seen to achieve its spatiotemporal strategy in motor control in this primitive form. By sustaining action in one muscle group while inhibiting that in others there is inhibitory specification, a short-term cybernetic constancy that achieves action selection. Furthermore, the redundancy bias elevates the threshold for change in the action pattern, such that only the most dominant action can be expressed, elevated to the focus of action by the inhibitory surround. With important contributions from the GABAergic inhibitory control that evolved within the basal ganglia circuitry, the redundancy bias provides an integral cybernetic influence to structured behavior that became integral to the cerebral cortex as the subpallial elements were incorporated within the new 6-layered cortical network architecture.

This control influence of redundancy, described as tonic activation, was implicit within the concept of motor readiness in the Pribram and McGuinness theory (Pribram & McGuinness, 1975). The concept of a redundancy bias was elaborated in somewhat more explicit cybernetic terms in the Tucker and Williamson account (Tucker & Williamson, 1984). In both formulations, the fundamental control of tonic activation was to support motor readiness, priming for action, with cognitive regulatory influences emergent from that primitive basis. In the present theoretical analysis, we have seen that the redundancy bias is integral to the mammalian ventral limbic networks, where, perhaps paradoxically, it supports not only motor readiness but also the sustained neural activity to support the *sensory* criteria for object specification and the feedback control of action. This is a kind of inversion of function, in which a specific form of motive cybernetics is redirected from its subcortical purpose to an opposite role, in sensory specification and constancy, in regulating the distributed representations of the mammalian cortex.

Remarkably, the embryological evidence provides clues to this mutational redirection. The subpallial (basal ganglia) elements of GABAergic inhibitory neurons and collothalamic inputs were incorporated into the pallium to create the 6-layered neocortex, including the granular sensory input layer 4 that dominates the cytoarchitectonics of the ventral division. We have seen how the cybernetics of the redundancy bias, interdigitated with capacities of inhibitory control, appear to be integral to the memory consolidation of the ventral division, supporting object specification, attentional selection, and feedback control of action. These functional skills elaborate a kind of cognition that is keyed on sensory data, and they are supported in the fused pallial-subpallial ventral cortical network architecture by the cybernetic mode of the redundancy bias, a subcortical control that in its primordial form evolved as a motor regulatory device.

Another inversion of functional control in the mammalian cortex can be seen in the redirection of the primitive cybernetics of phasic arousal with its integral habituation bias. This is fundamentally the control mechanism of vertebrate sensory systems, located in the dorsal division of the neuraxis, allowing elementary memory capacity to be readily distributed to capture the novel sensory elements of the dynamic environmental context. At least judging from the dorsal targets of the noradrenergic and serotonergic neuromodulators, the habituation bias seems to be integral to the control of the vertebrate pallium. The pallium was the first form of distributed cortical network architecture. The neurodevelopmental regulation of its pyramidal cell architecture requires norepinephrine in morphogenesis (Aboitiz, et al., 2003) and presumably in ongoing function. In mammals, the pallial architecture seems to have been retained as the dominant connectional pattern of the dorsal division of neocortex, with its dominance of pyramidal architectonics. Remarkably however, although the primitive sensory control of the habituation bias appears dominant in this division, the dorsal cortex in mammals becomes highly important to *motor* control, with the mediodorsal frontal cortex supporting a projectional control of actions that allows mammals flexible and fluid motor capacities that far exceed the mechanical and stereotyped actions of reptiles. Furthermore, in primates the pyramidal motor system evolved to bypass the basal ganglia altogether (at least in final motor output) and provide direct cortical control over the brainstem motor centers.

Thus the evolution of the 6-layered neocortex in mammals involved a kind of paradoxical specialization of the motive control of consolidation in the dorsal and ventral moieties that we describe as the *mammalian cybernetic inversion*. What was formerly the exclusive cybernetic mode of motor control, the redundancy bias, was redirected to provide a new complexity, and very likely a new stability, in the neurodevelopmental and cognitive process within the ventral cortex, through emphasizing sensory constraints rather than elemental motor control. What was formerly the exclusive cybernetic mode of sensory systems, the habituation bias, became redirected with the collapse of the vertebrate telencephalon and the mutations forming the neocortex, such that it then provided new capacities in not only the representational capacities of a distributed cortex, but the fluid, projectional control of action in the dorsal frontolimbic networks. Because this projectional control was synchronized with the caudodorsomedial networks of sensory integration, it could directly manifest the visceromotor motive direction from limbic networks while remaining embedded within the perceptual context model.

### Cybernetics of Memory in Time

The several features of dorsal-ventral specialization for cognitive structure and process that we have considered in the previous chapters can be traced to the unique integration of motive controls with cortical representational structure resulting from the mammalian inversion. The complexity of cognitive capacity resulting from this event also appears to have extended to the regulation of the locus of control. In the premammalian telencephalon, the redundancy bias was associated with motor readiness of the basal ganglia, which was engaged strongly in threat or strong need states, leading to a cessation of exploratory attention and a narrow, internal focus of attention and intention. This is the primary mode of internality. In contrast, the habituation bias supported the broad representation of the context in perceptual memory, acquired through exploration of novel events. This is the primary mode of externality.

Through the mammalian cybernetic inversion, these primary modes were inverted by a cortex that elaborated new properties of representation from the primitive cybernetics, along with new more complex biases on the locus of control. These fused modes of motive control and cognitive representation were examined as principles of action regulation in Chapter 3, and we learned of their neuroembryological origins in Chapter 6. As a result of the mammalian cybernetic inversion, the impetus of dorsal cortical networks—first evolved as a sensory control—now operates under *internal* control, as urges are shaped into actions with primary motivation from the visceromotor function of cingulate cortex. In this process, the habituation bias supports the direct emergence of impulses, at the same time that it ensures stable coping through rapid attenuation of positive hedonic responses to pleasurable events. Intentional approach to goals in the world is motivated and controlled by the impetus.

Similarly, in the fused mode resulting from the mammalian inversion, it is the artus—formerly the motor control mode of the basal ganglia—that regulates action under *external* control, as the sensory representations in ventrolateral frontal regions hold the criterial targets, sustained by the ventral limbic viscerosensory response to those targets. The redundancy bias of the artus then fits well with the requirement to sustain working memory to consolidate a new cognitive context model under conditions when the implicit hedonic context model has failed (such as under threats or strong motive states). Attentional avoidance of threats in the world is motivated and controlled by the artus.

In this complex and inverted form of mammalian self-regulation, cognition then shifts between assimilation and accommodation, in what can be described as a cognitive process of primitive learning. In the mode of assimilation, the internal hedonic context model is adequately predictive of motivationally significant environmental events, and new information can just be assimilated through self-regulation by the impetus, without internal change. In the mode of accommodation, however, external events are perceived that are foreign to the context model, and the self (or at least the current hedonic context model) must change, through self-regulation by the artus. These properties, making mammalian learning unique from that of reptiles, result directly from the mammalian inversion. Redundancy is redirected to regulate anticipatory sensory (and viscerosensory) guidance. Habituation and its integral hedonic expectancy are redirected to regulate the motor (and visceromotor) process.

## Inverting the Locus of Control

The locus of control described in the previous section follows the immediate dominance of internal and external information associated with the dual modes of cognitive control, in a way that is consistent with shifts in the viscerosomatic (internal-external) vector of consolidation. Yet over longer intervals of time, paradoxically, the locus of control in mammalian and human cognition becomes constrained by the fundamental, and premammalian, cybernetic bias. It is important to understand these somewhat complex relations of memory in time.

The influence of the impetus on dorsal pathways leads to internal control, as characterized in Goldberg's initial model of mediodorsal frontal control of action (G. Goldberg, 1985). Egocentric motives are discharged into action. However, over a more extended behavioral interval, the dominance of cognition by the impetus leads to *extraversion*. Under this control mode and its habituation bias, the contents of working memory are rapidly attenuated, and then changed, in response to orienting to novel environmental events. Cognition over time is then directed by the flux of the environment—or by immediate internal (visceral) urges—rather than by any continuity of internal representations. This is the fundamental psychological orientation of the extravert.

The psychological characteristics of extraversion are thus based in an openness to external stimulation, yet the extravert's primitive internal control of experience and behavior is manifested in the impulsiveness that leads the extravert to act without careful attention to environmental constraints, or to the needs of others. Even in more extended and adaptive cognitive mediation of behavior, the extravert shows strong intentionality in service of egocentric hedonic goals. In more balanced, intelligent personalities, this self-regulatory mode supports confidence, optimism, and decisiveness. When exaggerated and unbalanced, it leads to impulsiveness, narcissism, and psychopathy (Shapiro, 1965).

The reliance of the impetus on dorsal frontolimbic networks is also shown from a different perspective by the *pseudodepression syndrome* manifested by persons with lesions of these networks. Without the hedonic motive initiative of the impetus, the patients appear clinically depressed (Blumer & Benson, 1975). Thus, even though the complex features of representation in the dorsal brain reflect an inversion of the habituation bias, leading to fluid action regulation and the direct control of behavior by internal urges, the more complex psychological locus of control in the dorsal brain over time reflects the more primitive cybernetic mode of the habituation bias, shifting the constituent domains of the mind toward the extravert's dependence on mirroring the external flux.

In a parallel and similarly paradoxical fashion, the corticolimbic systems organizing the artus are, in their primary operation in mammals, biased toward external, sensory, control of behavior. This is the powerful fusion of motive counterpoint with cortical representation created by the mammalian functional inversion in the evolution of the 6-layered neocortex. The cybernetic mode of motor readiness—redundancy—provides powerful capacities when recruited by a mammalian cortex that elaborates sensory objects. Yet when dominant over time the cybernetic mode of the artus leads to *introversion*, the primordial locus of control of the redundancy bias. The focus on cognitive accommodation supported by the redundancy bias has a primitive basis in aversive motives of motor control—anxiety and hostility—that inherently disengage the individual from hedonic attachment to the context, even as they maintain attention to that context. The direct effect of the redundancy bias on working memory is to sustain the internal control over cognition (Tucker & Williamson, 1984). Yet by recognizing the representational counterpoint and complementarity of temporal span created by the mammalian inversion, we can now understand why the introvert manifests a kind of internal control over cognition that is at the same time occupied by external threats. This is the allocentric vector of attention, from world to self (Jeannerod, 1994). The effect is to cause the introvert to be sensitive, or in extreme cases hypersensitive, to the regard of others (Shapiro, 1965).

Through these somewhat paradoxical dynamics of representation and control, the personalities that become dominated by the artus reflect the anxiety associated with ventral limbic and orbital frontal activity (Tucker & Derryberry, 1992), manifesting cognitive styles with the restricted impulses, focused attention, and critical attitudes of the obsessive-compulsive and paranoid disorders (Shapiro, 1965). They are sensitive, and may be oversensitive, to the social context. The opposite self-regulatory deficit is shown by persons who are dominated by the impetus, because they have lost the complementary opposition of the artus. As we saw in Chapter 4, orbital frontal lesions, causing the loss of the balancing opponent influence of the artus, result in the *pseudopsychopathic syndrome* (Blumer & Benson, 1975). The psychopathic, manic, and histrionic personalities are fundamentally egocentric and defective in their social sensitivity and social propriety.

The regulation of intention, attention, and the locus of control thus emerge as integral consequences of the primitive control modes of the mammalian brain. These control modes have different effects on the locus of control in their immediate operation from the effects they have over longer intervals. Primitive controls on neural activity (habituation and redundancy) seem to have been put to new purposes in the mammalian inversion, creating representational complexity for integrating visceral with somatic domains of traffic between self and world. In their differential influences over the span of memory in time, the primitive neuromodulatory influences of redundancy and habituation can be understood as modes of introversion and extraversion, where the control of internality and externality involve multidimensional and, at first, paradoxical implications. Although we have mostly analyzed these implications for the locus of control in the elementary neurocybernetic terms that are appropriate to the motivation and attentional control of mammals generally, we have also seen that there are implications for the organization of human personality. Because human development requires an intensely social context, the negotiation of introversion and extraversion as modes of relation in the social context becomes a critical factor in successful negotiation of the neurodevelopmental process.

## *Self-Organizing Ontogenesis*

The process of morphogenesis plays out in each individual through the program of the genome as it is guided by various mechanisms of trophic factors, neuromodulators, and the active interactions of glia and neurons themselves. For the organized neural networks of the fetal brain, the network architecture is organized through the control of neural activity, and therefore the control of activity-dependent specification, by the intrinsic network dynamics. The activity determining network dynamics in neural morphogenesis in utero appears to be the same that continues to regulate cerebral organization in psychological development, in the process of memory consolidation.

Launched into their initial workings by genetic control, the motive vectors of neural activity eventually come under psychological control. The child is a self-organizing system, with motive processes that are engaged in the negotiation between internal needs and environmental contingencies. Because human self-organization in childhood occurs in a social context, the motive regulation of the locus of control is a critical component of the human ontogenetic process. The shifts in orientation with affect state and motive bias—from extraversion and engagement of social contact toward introversion and independence or alienation—become powerful determinants in the organization of the self. We have seen that motive self-regulation starts in primitive or reactive modes, but gains the capacity for more effortful forms as cognitive development provides increasing abstract representational capacity. Cognition then increasingly determines the mediation between internal needs and external demands.

At certain points in childhood, we become conscious, more or less. Some implicit self-awareness emerges in the intelligent three-year old. The school-age child gains increasing consciousness of mental process through the disciplines of monitoring schoolwork and the challenges of understanding peer interactions. The adolescent becomes acutely self-aware at about the time of increasingly abstract intellectual skills. Conscious self-direction may remain a relatively fragile capacity throughout life, in contrast with the minimally conscious urges, habits, and transferences of developmental history that determine most behavior. Yet, even if it is only the implicit developmental inertia constraining the person's present choices, we can say that the self-regulation of ontogenesis becomes increasingly psychological, and individuated. Of course, because it continues to operate on the phyletic biological substrate, human ontogenesis, and the cognition that is its present reflection, emerge only through the neurodevelopmental process. But this process is regulated by the demands of individuation within a unique social context, such that the brain function of each child becomes increasingly differentiated as an experiment of nature within a unique historical era.

Through a neurodevelopmental analysis, we can recognize the continuity of ontogenesis, from the evolutionary-developmental foundations of neuroembryological differentiation to the social and psychological foundations of individuation and self-actualization. There are clear advantages of recognizing this continuity for scientific theory. Theories become highly constrained if they must be consistent with the facts of both neural and psychological development. There may also be interesting philosophical insights from a neurodevelopmental analysis, as we come to understand how the familiar motive processes of self-regulation reflect the venerable organizing vectors of mammalian neural evolution.

# References

## Chapter 1. Neurodevelopmental Mechanisms of Learning

Amsel, A., & Stanton, M. (1980). Ontogeny and phylogeny of paradoxical reward effects. *Advances in the study of behavior, 11*, 227–274.

Berridge, K. C. (2009). "Liking" and "wanting" food rewards: Brain substrates and roles in eating disorders. *Physiology and Behavior, 97*(5), 537–550.

Bitterman, M. E. (1968). Reversal learning and forgetting. *Science, 160*(823), 100.

Black, J. E., & Greenough, W. T. (1986). Developmental approaches to the memory processes. In J. L. Martinez Jr. & R. P. Kesner (Eds.), *Learning and memory: A biological view* (pp. 55–81). Orlando, FL: Academic Press.

Cannon, W. B. (1915). *Bodily changes in pain, hunger, fear, and rage.* New York: Appleton.

Cannon, W. B. (1927). The James-Lange theory of emotions: A critical examination and an alternative theory. *American Journal of Psychology, 39*, 106–124.

Carlsson, A. (1988.). The current status of the dopamine hypothesis of schizophrenia. *Neuropsychopharmacology, 1*, 179–186.

Chein, J. M., & Schneider, W. (2005). Neuroimaging studies of practice-related change: fMRI and meta-analytic evidence of a domain-general control network for learning. *Brain Research: Cognitive Brain Research, 25*(3), 607–623.

Coolidge, F., & Wynn, T. (2005). Working memory, its executive functions, and the emergence of modern thinking. *Cambridge Archaeological Journal, 15*(1), 5–26.

Cooper, J. R., Bloom, F. E., & Roth, R. H. (1974). *The biochemical basis of neuropharmacology.* New York: Oxford University Press.

Csikszentmihalyi, M., & Rathunde, K. (1993). The measurement of flow in everyday life: Towards a theory of emergent motivation. In J. E. Jacobs (Ed.), *Nebraska Symposium on Motivation* (Vol. 40, pp. 60). Lincoln: University of Nebraska Press.

Descarries, L., & Lapierre, Y. (1973). Norepinephrine and axon terminals in the cerebral cortex of the rat. *Brain Research, 51*, 141–160.

Fair, D., Dosenbach, N., Church, J., Cohen, A., Brahmbhatt, S., Miezin, F., et al. (2007). Development of distinct control networks through segregation and integration. *Proceedings of the National Academy of Sciences of the United States of America, 104*(33), 13507–13512.

Fibiger, H. C., & Phillips, A. G. (1986). Reward, motivation, cognition: Psychobiology of mesotelencephalic dopamine systems. In F. E. Bloom (Ed.), *Handbook of physiology. Section 1: The nervous system. Volume IV. Intrinsic regulatory systems of the brain* (pp. 647–675). Bethesda, MD: American Physiological Society.

Freud, S. (1895). Project for a scientific psychology. In J. Strachey (Ed.), *The standard edition of the complete psychological works of Sigmund Freud* (Vol. 1, pp. 295–344). London: Hogarth Press.

Gabriel, M. (1990). Functions of anterior and posterior cingulate cortex during avoidance learning in rabbits. *Progress in Brain Research, 85,* 467–483.

Gabriel, M., Kubota, Y., Sparenborg, S., Straube, K., & Vogt, B. A. (1991). Effects of cingulate cortical lesions on avoidance learning and training-induced unit activity in rabbits. *Experimental Brain Research, 86*(3), 585–600.

Gabriel, M., & Sparenborg, S. (1986). Anterior thalamic discriminative neuronal responses enhanced during learning in rabbits with subicular and cingulate cortical lesions. *Brain Research, 384*(1), 195–198.

Gabriel, M., Sparenborg, S., & Kubota, Y. (1989). Anterior and medial thalamic lesions, discriminative avoidance learning, and cingulate cortical neuronal activity in rabbits. *Experimental Brain Research, 76*(2), 441–457.

Gabriel, M., Taylor, C., & Burhans, L. (2003). In utero cocaine, discriminative avoidance learning with low-salient stimuli and learning-related neuronal activity in rabbits (*Oryctolagus cuniculus*). *Behavioral Neuroscience, 117*(5), 912–926.

Greenough, W. T. (1975). Experiential modification of the developing brain. *American Scientist, 63*(1), 37–46.

Greenough, W. T., & Black, J. E. (1992). Induction of brain structure by experience: Substrates for cognitive development. In M. Gunnar & C. Nelson (Eds.), *Minnesota Symposium on Child Psychology: Vol. 24. Developmental behavioral neuroscience.* (pp. 155–200). Hillsdale, NJ: Erlbaum.

Grossberg, S. (1970). Neural pattern discrimination. *Journal of Theoretical Biology, 27*(2), 291–337.

Grossberg, S. (1980). How does a brain build a cognitive code? *Psychological Review, 87,* 1–51.

Grossberg, S. (1984). Some psychophysiological and pharmacological correlates of a developmental, cognitive and motivational theory. In R. Karrer, J. Cohen, & P. Tueting (Eds.), *Annals of the New York Academy of Science: Vol. 425. Brain and information: Event-related potentials* ( pp. 54–82). New York: New York Academy of Sciences.

Hamburger, V., Balaban, M., Oppenheim, R., & Wenger, E. (1965). Periodic motility of normal and spinal chick embryos between 8 and 17 days of incubation. *Journal of Experimental Zoology, 159*(1), 1–13.

Hamidi, M., Tononi, G., & Postle, B. R. (2008). Evaluating frontal and parietal contributions to spatial working memory with repetitive transcranial magnetic stimulation. *Brain Research, 1230,* 202–210.

Hebb, D. O. (1949). *The organization of behavior.* New York: Wiley.

Heims, S. J. (1991). *The cybernetics group.* Cambridge, MA: MIT Press.

Hendler, J. A. (1995). Types of planning: can artificial intelligence yield insights into prefrontal function? In J. Grafman, K. J. Holyoak, & F. Boller (Eds.), *Annals of the*

*New York Academy of Sciences: Vol. 769. Structure and functions of the human prefrontal cortex* (pp. 265–276). New York: New York Academy of Sciences.

Herrick, C. J. (1948). *The brain of the tiger salamander.* Chicago: University of Chicago Press.

James, W. (1884). What is emotion? *Mind, 4,* 118–204.

Kamin, L. J. (1965). Temporal and intensity characteristics of the conditioned stimulus. In W. F. Prokasy (Ed.), *Classical conditioning* (pp. 118–147). New York: Appleton-Century-Crofts.

Kringelbach, M. L., & Berridge, K. C. (2009). Towards a functional neuroanatomy of pleasure and happiness. *Trends in Cognitive Sciences, 13*(11), 479–487.

Kringelbach, M. L., & Berridge, K. C. (2010). The functional neuroanatomy of pleasure and happiness. *Discovery Medicine, 9*(49), 579–587.

Krechevsky, I. (1932). "Hypotheses" in rats. *Psychological Review, 39,* 516–532.

Lenartowicz, A., & McIntosh, A. (2005). The role of anterior cingulate cortex in working memory is shaped by functional connectivity. *Journal of Cognitive Neuroscience, 17*(7), 1026–1042.

Lindsley, D. B. (1957). Psychophysiology and motivation. In M. R. Jones (Ed.), *Nebraska Symposium on Motivation* (Vol. 5). Lincoln: University of Nebraska.

Lindsley, D. B. (1960). Attention, consciousness, sleep, and wakefulness. In J. Field, H. W. Magoun, & V. E. Hall (Eds.), *Handbook of physiology: III. Neurophysiology.* Washington, DC: American Physiological Society.

Luu, P., & Pederson, S. (2004). The anterior cingulate cortex: Regulating actions in context. In M. I. Posner (Ed.), *Cognitive neuroscience of attention* (pp. 232–244). New York: Guilford.

Luu, P., & Tucker, D. M. (2003). Self-regulation by the medial frontal cortex: Limbic representation of motive set-points. In M. Beauregard (Ed.), *Consciousness, emotional self-regulation and the brain* (pp. 123–161). Amsterdam: John Benjamins.

Malmo, R. B. (1959). Activation: A neuropsychological dimension. *Psychological Review, 66*(6), 367–386.

Marin-Padilla, M. (1998). Cajal-Retzius cells and the development of the neocortex. *Trends in Neurosciences, 21*(2), 64–71.

Markham, J. A., & Greenough, W. T. (2004). Experience-driven brain plasticity: beyond the synapse. *Neuron Glia Biology, 1*(4), 351–363.

McCulloch, W. S., & Pitts, W. (1990). A logical calculus of the ideas immanent in nervous activity. *Bulletin of Mathematical Biology, 52*(1–2), 99–115.

Mesulam, M. (1981). A cortical network for directed attention and unilateral neglect. *Annals of Neurology, 10*(4), 309–325.

Miller, B. T., Deouell, L. Y., Dam, C., Knight, R. T., & D'Esposito, M. (2008). Spatiotemporal dynamics of neural mechanisms underlying component operations in working memory. *Brain Research, 1206,* 61–75.

Miller, E. K., & Cohen, J. D. (2001). An integrative theory of prefrontal cortex function. *Annual Review of Neuroscience, 24,* 167–202.

Moruzzi, G., & Magoun, H. W. (1949). Brain stem reticular formation and activation of the EEG. *Electroencephalography and Clinical Neurophysiology, 1,* 445–473.

Muller, N. G., & Knight, R. T. (2006). The functional neuroanatomy of working memory: contributions of human brain lesion studies. *Neuroscience, 139*(1), 51–58.

O'Reilly R, C., & Frank, M. J. (2006). Making working memory work: A computational model of learning in the prefrontal cortex and basal ganglia. *Neural Computation, 18*(2), 283–328.

Papini, M. R. (2002). Pattern and process in the evolution of learning. *Psychological Review, 109*(1), 186–201.

Papini, M. R. (2003). Comparative psychology of surprising nonreward. *Brain, Behavior and Evolution, 62*(2), 83–95.

Paus, T. (2001). Primate anterior cingulate cortex: where motor control, drive and cognition interface. *Nature Reviews: Neuroscience, 2*(6), 417–424.

Piaget, J. (1936/1992). *The origins of intelligence in children*. New York: International Universities Press.

Poremba, A., & Gabriel, M. (1997). Amygdalar lesions block discriminative avoidance learning and cingulothalamic training-induced neuronal plasticity in rabbits. *J Neurosci, 17*(13), 5237–5244.

Posner, M. I. (1978). *Chronometric explorations of mind*. Hillsdale, NJ: Erlbaum.

Posner, M. I., & Petersen, S. E. (1990). The attention system of the human brain. *Annual Review of Neuroscience, 13*, 25–42.

Posner, M. I., & Rothbart, M. K. (1998). Attention, self-regulation and consciousness. *Philosophical Transactions of the Royal Society of London: B. Biological Sciences, 353*(1377), 1915–1927.

Postle, B. R., Ferrarelli, F., Hamidi, M., Feredoes, E., Massimini, M., Peterson, M., et al. (2006). Repetitive transcranial magnetic stimulation dissociates working memory manipulation from retention functions in the prefrontal, but not posterior parietal, cortex. *Journal of Cognitive Neuroscience, 18*(10), 1712–1722.

Pribram, K. H., & Gill, M. M. (1976). *Freud's "project" re-assessed*. New York: Basic Books.

Pribram, K. H., & McGuinness, D. (1975). Arousal, activation, and effort in the control of attention. *Psychological Review, 82*, 6–149.

Price, J. L. (1999). Prefrontal cortical networks related to visceral function and mood. *Annals of the New York Academy of Sciences, 877*, 383–396.

Rakic, P. (2009). Evolution of the neocortex: A perspective from developmental biology. *Nature Reviews: Neuroscience, 10*(10), 724–735.

Rakic, P., & Singer, W. (Eds.). (1988). *Neurobiology of neocortex*. New York: Wiley.

Rescorla, R. A. (1980). Simultaneous and successive associations in sensory preconditioning. *Journal of Experimental Psychology: Animal Behavior Processes, 6*(3), 207–216.

Rescorla, R. A., & Wagner, A. R. (1972). A theory of Pavlovian conditioning: Variations in the effectiveness of reinforcement and nonreinforcement. In A. H. Black & W. F. Prokasy (Eds.), *Classical conditioning: II. Current research and theory* (Vol. 65–99). New York: Appleton-Century-Crofts.

Rosenblatt, F. (1958). The perceptron: A probabilistic model for information storage and organization in the brain. *Psychological Review, 65*, 386–408.

Rumelhart, D. E., & McClelland, J. L. (1986). *Parallel distributed processing: Explorations in the microstructure of cognition: Vol. I. Foundations.* Cambridge, MA: MIT Press.

Schacter, F., & Singer, J. E. (1962). Cognitive social and physiological determinants of emotional states. *Psychological Review, 69,* 379–399.

Schildkraut, J. (1965). The catecholamine hypothesis of affective disorders: A review of supporting evidence. *American Journal of Psychiatry, 122,* 509–522.

Schneider, W., & Chein, J. (2003). Controlled and automatic processing: Behavior, theory, and biological mechanisms. *Cognitive Science, 27,* 525–559.

Sejnowski, T. J. (2002). *Thalamocortical assemblies in the sleeping and alert brain.* Paper presented at the Oscillatory Dynamics Conference at Rancho Sante Fe.

Singer, W. (1987). Activity-dependent self-organization of synaptic connections as a substrate of learning. In J. P. Changeux & M. Konishi (Eds.), *The neural and molecular basis of learning* (pp. 301–336). New York: Wiley.

Smith, G. D., & Sherman, S. M. (2002). Detectability of excitatory versus inhibitory drive in an integrate-and-fire-or-burst thalamocortical relay neuron model. *J Neurosci, 22*(23), 10242–10250. doi: 22/23/10242 [pii]

Swanson, L. A. (2003). *Brain architecture: Understanding the basic plan.* New York: Oxford University Press.

Tennant, W. A., & Bitterman, M. E. (1975). Blocking and overshadowing in two species of fish. *Journal of Experimental Psychology: Animal Behavior Processes, 1*(1), 22–29.

Trevarthen, C. (1985). Neuroembryology and the development of perceptual mechanisms. In F. Falkner & J. M. Tanner (Eds.), *Human growth, a comprehensive treatise: Volume 2. Posnatal growth, neurobiology.* New York: Plenum.

Tucker, D. M., Derryberry, D., & Luu, P. (2000). Anatomy and physiology of human emotion: Vertical integration of brainstem, limbic, and cortical systems. In J. Borod (Ed.), *Handbook of the neuropsychology of emotion* (pp. 56–79). New York: Oxford.

Tucker, D. M., & Desmond, R. E., Jr. (1998). Aging and the plasticity of the self. In K. W. Shaie & M. P. Laughton (Eds.), *Annual Review of Gerontology and Geriatrics: Vol. 17. Focus on emotion and adult development.* (pp. 266–281). New York: Springer.

Tucker, D. M., & Luu, P. (2006). Adaptive binding. In H. Zimmer, A. Mecklinger, & U. Lindenberger (Eds.), *Binding in human memory: A neurocognitive approach* (pp. 85–108). New York: Oxford University Press.

Tucker, D. M., & Luu, P. (2007). Neurophysiology of motivated learning: Adaptive mechanisms of cognitive bias in depression. *Cognitive Therapy and Research, 31,* 189–209.

Tucker, D. M., & Williamson, P. A. (1984). Asymmetric neural control systems in human self-regulation. *Psychological Review, 91*(2), 185–215.

Vogt, B. A., Rosene, D. L., & Pandya, D. N. (1979). Thalamic and cortical afferents differentiate anterior from posterior cingulate cortex in the monkey. *Science, 204*(4389), 205–207.

von de Malsburg, C., & Singer, W. (1988). Principles of cortical network organization. In P. Rakic & W. Singer (Eds.), *Neurobiology of neocortex* (pp. 69–99). New York: Wiley.

West, R. W., & Greenough, W. T. (1972). Effect of environmental complexity on cortical synapses of rats: Preliminary results. *Behavioral Biology, 7*(2), 279–284.

Wiener, N. (1961). *Cybernetics; or Control and communication in the animal and the machine.* Cambridge, MA: MIT Press.

Wise, R. A., & Rompre, P. P. (1989). Brain dopamine and reward. *Annual Review of Psychology, 40,* 191–225.

## Chapter 2. Consolidating Memory

Adamec, R. E. (1990). Amygdala kindling and anxiety in the rat. *Neuroreport, 1*(3–4), 255–258.

Adamec, R. E. (1993). Partial limbic kindling—brain, behavior, and the benzodiazepine receptor. *Physiology and Behavior, 54*(3), 531–545.

Adamec, R. E. (2000). Evidence that long-lasting potentiation in limbic circuits mediating defensive behaviour in the right hemisphere underlies pharmacological stressor (FG-7142) induced lasting increases in anxiety-like behaviour: Role of benzodiazepine receptors. *Journal of Psychopharmacology, 14*(4), 307–322.

Aggleton, J. R., & Brown, M. W. (1999). Episodic memory, amnesia, and the hippocampal-anterior thalamic axis. *Behavioral and Brain Sciences, 22,* 425–489.

Alheid, G. F., & Heimer, L. (1988). New perspectives in basal forebrain organization of special relevance for neuropsychiatric disorders: The striatopallidal, amygdaloid, and corticopetal components of substantia innominata. *Neuroscience, 27*(1), 1–39.

Barbas, H., & Pandya, D. N. (1984). Topography of commissural fibers of the prefrontal cortex in the rhesus monkey. *Experimental Brain Research, 55*(1), 187–191.

Barbas, H., & Pandya, D. N. (1987). Architecture and frontal cortical connections of the premotor cortex (area 6) in the rhesus monkey. *Journal of Comparative Neurology, 256*(2), 211–228.

Barbas, H., & Pandya, D. N. (1989). Architecture and intrinsic connections of the prefrontal cortex in the rhesus monkey. *Journal of Comparative Neurology, 286,* 353–375.

Barbas, H., & Rempel-Clower, N. (1997). Cortical structure predicts the pattern of corticocortical connections. *Cerebral Cortex, 7*(7), 635–646.

Berridge, C. W. (2006). Neural substrates of psychostimulant-induced arousal. *Neuropsychopharmacology, 31*(11), 2332–2340.

Berridge, K. C. (2009). "Liking" and "wanting" food rewards: Brain substrates and roles in eating disorders. *Physiology and Behavior, 97*(5), 537–550.

Berridge, K. C., & Kringelbach, M. L. (2008). Affective neuroscience of pleasure: Reward in humans and animals. *Psychopharmacology Series-Berlin, 199*(3), 457–480.

Borod, J. C. (1993). Cerebral mechanisms underlying facial, prosodic, and lexical emotional expression: A review of neuropsychological studies and methodological issues. *Neuropsychology, 7,* 445–463.

Borod, J. C. (2000). *The neuropsychology of emotion.* New York: Oxford.

Brand, M., & Markowitsch, H. J. (2006). Memory processes and the orbitofrontal cortex. In D. H. Zald & S. L. Rauch (Eds.), *The orbitofrontal cortex* (pp. 285–306). New York: Oxford.

Brown, J. (2011). Theoretic Note: The relation of embryology to linguistic and cognitive process. *Journal of Psycholinguistic Research, 40*, 189–194.

Brown, J. W. (1979). Language representation in the brain. In I. H. D. Steklis & M. J. Raleigh (Eds.), *Neurobiology of social communication in primates* (pp. 133–195). New York: Academic Press.

Brown, J. W. (1988). *The life of the mind: Selected papers*. Hillsdale, NJ: Erlbaum.

Brown, J. W. (1989). The nature of voluntary action. *Brain and Cognition, 10*, 105–120.

Butler, A. B., & Molnar, Z. (2002). Development and evolution of the collopallium in amniotes: A new hypothesis of field homology. *Brain Research Bulletin, 57*(3–4), 475–479.

Cajal, S. R. (1899). *Comparative study of the sensory areas of the human cortex*. Worchester, MA: Clark University.

Chrobak, J. J., & Buzsaki, G. (1998). Operational dynamics in the hippocampal-entorhinal axis. *Neuroscience and Biobehavioral Reviews, 22*(2), 303–310.

Coyle, J. T., Price, D.L., & DeLong, M.R. (1983). Alzheimer's disease: A disorder of cortical cholinergic innervation. *Science, 219*, 1184–1190.

Damasio, A. R. (1989). Time-locked multiregional retroactivation: A systems-level proposal for the neural substrates of recall and recognition. *Cognition, 33*, 25–62.

Denny-Brown, D. (1966). *The cerebral control of movement*. Springfield, IL: Charles C. Thomas.

Desimone, R., & Ungerleider, L. G. (1989). Neural mechanisms of visual processing in monkeys. In F. Boller & J. Grafman (Eds.), *Handbook of neuropsychology* (Vol. 2, pp. 267–299). Amsterdam: Elsevier.

Doran, S. M., Van Dongen, H. P., & Dinges, D. F. (2001). Sustained attention performance during sleep deprivation: Evidence of state instability. *Archives Italiennes de Biologie, 139*(3), 253–267.

Fleming, J. F. R., & Crosby, E. C. (1955). The parietal lobe as an additional motor area: The motor effects of electrical stimulation and ablation of cortical areas 5 and 7 in monkeys. *Journal of Comparative Neurology, 103*, 485–512.

Freud, S. (1895). Project for a scientific psychology. In J. Strachey (Ed.), *The standard edition of the complete psychological works of Sigmund Freud* (Vol. 1, pp. 295–344). London: Hogarth Press.

Galaburda, A. M., LeMay, M., Kemper, T. L., & Geschwind, N. (1978). Right-left asymmetrics in the brain. *Science, 199*(4331), 852–856.

Galaburda, A. M., Sanides, F., & Geschwind, N. (1978). Human brain: Cytoarchitectonic left-right asymmetries in the temporal speech region. *Archives of Neurology, 35*(12), 812–817.

Glick, S. D., Ross, D. A., & Hough, L. B. (1982). Lateral asymmetry of neurotransmitters in human brain. *Brain Research, 234*, 53–63.

Gold, P. E., & Zornetzer, S. F. (1983). The mnemon and its juices: Neural modulation of memory processes. *Behavioral and Neural Biology, 38*, 151–189.

Goldberg, E., & Costa, L. D. (1981). Hemisphere differences in the acquisition and use of descriptive systems. *Brain and Language, 14,* 144–173.

Goldman-Rakic, P. S., Bates, J. F., & Chafee, M. V. (1992). The prefrontal cortex and internally generated motor acts. *Current Opinion in Neurobiology, 2,* 830–835.

Grossberg, S. (1980). How does a brain build a cognitive code? *Psychological Review, 87,* 1–51.

Grossberg, S. (1984). Some psychophysiological and pharmacological correlates of a developmental, cognitive and motivational theory. In R. Karrer, J. Cohen, & P. Tueting (Eds.), *Annals of the New York Academy of Sciences: Vol. 425. Brain and information: Event-related potentials* ( pp. 54–82). New York: New York Academy of Sciences.

Grossberg, S., & Versace, M. (2008). Spikes, synchrony, and attentive learning by laminar thalamocortical circuits. *Brain Research, 1218,* 278–312.

Harkness, K. L., & Tucker, D. M. (2000). Motivation of neural plasticity: Neural mechanisms in the self-organization of depression. In M. D. Lewis & I. Granic (Eds.), *Emotion, development, and self-organization* (pp. 186–208). New York: Cambridge University Press.

Hendler, J. A. (1995). Types of planning: Can artificial intelligence yield insights into prefrontal function? In J. Grafman, K. J. Holyoak, & F. Boller (Eds.), *Annals of the New York Academy of Sciences: Vol. 769. Structure and functions of the human prefrontal cortex* (pp. 265–276). New York: New York Academy of Sciences.

Hinton, G. E. (2010). Learning to represent visual input. *Philosophical Transactions of the Royal Society of London: B. Biological Sciences, 365*(1537), 177–184.

Jackson, J. H. (1931). The evolution and dissolution of the nervous system. *Selected writings of John Hughlings Jackson* (Vol. 2, pp. 45–75). London: Hodder and Stoughton.

Janowsky, J. S., Laxer, K. D., & Rushmer, D. S. (1980). Classical conditioning of kindled seizures. *Epilepsia, 21,* 393–398.

Jones, E. G. (2007a). *The thalamus* (Vol. 1). Cambridge, UK: Cambridge University Press.

Jones, E. G. (2007b). *The thalamus* (Vol. 2). Cambridge, UK: Cambridge University Press.

Jones, E. G. (2009). Synchrony in the interconnected circuitry of the thalamus and cerebral cortex. *Annals of the New York Academy of Sciences, 1157,* 10–23.

Kaas, J. H. (1988). Development of cortical sensory maps. In P. Rakic & W. Singer (Eds.), *Neurobiology of neocortex* (pp. 101–113). New York: John Wiley.

Kaas, J. H. (1989). Why does the brain have so many visual areas? *Journal of Cognitive Neuroscience, 1*(2), 121–135.

Kamin, L. J. (1965). Temporal and intensity characteristics of the conditioned stimulus. In W. F. Prokasy (Ed.), *Classical conditioning* (pp. 118–147). New York: Appleton-Century-Crofts.

Knutson, B., Fong, G. W., Adams, C. M., Varner, J. L., & Hommer, D. (2001). Dissociation of reward anticipation and outcome with event-related fMRI. *Neuroreport, 12*(17), 3683–3687.

Kringelbach, M. L., & Berridge, K. C. (2010). The functional neuroanatomy of pleasure and happiness. *Discovery Medicine, 9*(49), 579–587.

Lindsley, D. B. (1951). Emotion. In S. S. Stevens (Ed.), *Handbook of experimental psychology* (pp. 473–516). New York: Wiley.

Lindsley, D. B. (1957). Psychophysiology and motivation. In M. R. Jones (Ed.), *Nebraska Symposium on Motivation* (Vol. 5). Lincoln: University of Nebraska.

Luu, P., & Tucker, D. M. (2003a). Self-regulation and the executive functions: Electrophysiological clues. In A. Zani & A. M. Preverbio (Eds.), *The cognitive electrophysiology of mind and brain* (pp. 199–223). San Diego: Academic Press.

Luu, P., & Tucker, D. M. (2003b). Self-regulation by the medial frontal cortex: Limbic representation of motive set-points. In M. Beauregard (Ed.), *Consciousness, emotional self-regulation and the brain* (pp. 123–161). Amsterdam: John Benjamins.

Malmo, R. B. (1959). Activation: A neuropsychological dimension. *Psychological Review, 66*(6), 367–386.

Massimini, M., Ferrarelli, F., Huber, R., Esser, S. K., Singh, H., & Tononi, G. (2005). Breakdown of cortical effective connectivity during sleep. *Science, 309*(5744), 2228–2232.

Mesulam, M. M. (2000). Behavioral neuroanatomy: Large-scale networks, association, cortex, frontal syndromes, the limbic system, and hemispheric specializations. In M. M. Mesulam (Ed.), *Principles of behavioral and cognitive neurology* (pp. 1–120). Oxford: Oxford University Press.

Mesulam, M. M., & Mufson, E. J. (1984). Neural inputs into the nucleus basalis of the substantia innominata (Ch4) in the rhesus monkey. *Brain, 107*, 253–274.

Mesulam, M. M., Mufson, E. J., Levey, A. I., & Wainer, B. H. (1983). Cholinergic innervation of cortex by the basal forebrain: Cytochemistry and cortical connections of the septal area, diagonal band nuclei, nucleus basalis (substantia innominata), and hypothalamus in the rhesus monkey. *Journal of Comparative Neurology, 214*, 170–197.

Minzenberg, M. J., & Carter, C. S. (2008). Modafinil: A review of neurochemical actions and effects on cognition. *Neuropsychopharmacology, 33*(7), 1477–1502.

Mishkin, M. (1982). A memory system in the monkey. *Philosophical Transactions of the Royal Society of London: B. Biological Sciences, 298*(1089), 83–95.

Monrad-Krohn, G. H. (1924). On the dissociation of voluntary and emotional innervation in facial paresis of central origin. *Brain, 47*, 22–35.

Morrison, R. S., & Dempsey, E. W. (1942). A study of thalamocortical relations. *American Journal of Physiology, 135*, 281–292.

Morrison, R. S., & Dempsey, E. W. (1943). Mechanism of thalamocortical augmentation and repetition. *American Journal of Physiology, 138*, 297–308.

Mountcastle, V. B., Lynch, J. C., Gorgopoulos, A., Sakata, H., & Acuna, C. (1975). Posterior parietal association cortex of the monkey: Command functions for operations within extrapersonal space. *Journal of Neurophysiology, 38*, 871–908.

Myslobodsky, M. S., Mintz, M., Lerner, T., & Mostofsky, D. I. (1983). Amygdala kindling in the classical conditioning paradigm. *Epilepsia, 24*(3), 275–283.

Nader, K., & Einarsson, E. O. (2010). Memory reconsolidation: an update [Review]. *Annals of the New York Academy of Sciences, 1191*, 27–41.

Nauta, W. J. H. (1971). The problem of the frontal lobe: A reinterpretation. *Journal of Psychiatric Research, 8,* 167–187.

Nauta, W. J. H. (1986). Circuitous connections linking cerebral cortex, limbic system, and corpus striatum. In B. K. Doane & K. E. Livingston (Eds.), *The limbic system: Functional organization and clinical disorders* (pp. 43–54). New York: Raven Press.

Nauta, W. J. H., & Haymaker, W. (1969). Hypothalamic nuclei and fiber connections. In W. Haymaker, E. Anderson, & W. J. H. Nauta (Eds.), *The hypothalamus* (pp. 136–209). Springfield: Charles C. Thomas.

Neafsey, E. J. (1990). Prefrontal cortical control of the autonomic nervous system: Anatomical and physiological observations. In H. B. M. Uylings, C. G. Van Eden, J. P. C. De Bruin, M. A. Corner, & M. G. P. Feenstra (Eds.), *The prefrontal cortex: Its structure, function and pathology* (pp. 147–166). New York: Elsevier.

Neafsey, E. J., Hurley-Gius, K. M., & Arvanitis, D. (1986). The topographical organization of neurons in the rat medial frontal, insular and olfactory cortex projecting to the solitary nucleus, olfactory bulb, periaqueductal gray and superior colliculus. *Brain Research, 377,* 261–271.

Neafsey, E. J., Terreberry, R. R., Hurley, K. M., Ruit, K. G., & Frysztak, R. J. (1993). Anterior cingulate cortex in rodents: Connections, visceral control functions, and implications for emotion. In B. A. Vogt & M. Gabriel (Eds.), *Neurobiology of the cingulate cortex and limbic thalamus* (pp. 206–223). Boston: Birkhauser.

Northcutt, G. R., & Kaas, J. H. (1995). The emergence and evolution of the mammalian neocortex. *Trends in Neurosciences, 18,* 373–379.

Pandya, D. N., & Barnes, C. L. (1987). Architecture and connections of the frontal lobe. In E. Perecman (Ed.), *The frontal lobes revisited* (pp. 41–72). New York: IRBN.

Pandya, D. N., & Seltzer, B. (1982). Association areas of the cerebral cortex. *Trends in Neurosciences, 5,* 386–390.

Pandya, D. N., & Yeterian, E. H. (1984). Proposed neural circuitry for spatial memory in the primate brain. *Neuropsychologia, 22*(2), 109–122.

Pandya, D. N., & Yeterian, E. H. (1985). Architecture and connections of cortical association areas. In A. Peters & E. G. Jones (Eds.), *Cerebral cortex: Volume 4. Association and auditory cortices* (pp. 3–61). New York: Plenum Press.

Papini, M. R. (2003). Comparative psychology of surprising nonreward. *Brain, Behavior and Evolution, 62*(2), 83–95.

Pribram, K. H. (1981). Emotions. In S. K. Filskov & T. J. Boll (Eds.), *Handbook of clinical neuropsychology* (pp. 102–134). New York: Wiley.

Pribram, K. H., & McGuinness, D. (1975). Arousal, activation, and effort in the control of attention. *Psychological Review, 82,* 6–149.

Price, J. L. (1999). Prefrontal cortical networks related to visceral function and mood. *Annals of the New York Academy of Sciences, 877,* 383–396.

Redgrave, P., Prescott, T. J., & Gurney, K. (1999). The basal ganglia: A vertebrate solution to the selection problem? *Neuroscience, 89*(4), 1009–1023.

Rizzolatti, G., & Fadiga, L. (1998). Grasping objects and grasping action meanings: The dual role of monkey rostroventral premotor cortex (area F5). *Novartis Foundation Symposium, 218,* 81–95.

Rizzolatti, G., Fadiga, L., Fogassi, L., & Gallese, V. (1999). Resonance behaviors and mirror neurons. *Archives Italiennes de Biologie, 137*(2–3), 85–100.

Rothbart, M. K., Taylor, S. B., & Tucker, D. M. (1989). Right-sided facial asymmetry in infant emotional expression. *Neuropsychologia, 27*(5), 675–687.

Rumelhart, D. E., & McClelland, J. L. (1986). *Parallel distributed processing: Explorations in the microstructure of cognition: Vol. 1. Foundations.* Cambridge, MA: MIT Press.

Sackeim, H. A., Gur, R. C., & Saucy, M. C. (1978). Emotions are expressed more intensely on the left side of the face. *Science, 202,* 434–436.

Sanides, F. (1970). Functional architecture of motor and sensory cortices in primates in the light of a new concept of neocortex evolution. In C. R. Noback & W. Montagna (Eds.), *The primate brain: Advances in primatology* (Vol. 1, pp. 137–208). New York: Appleton-Century-Crofts.

Sanides, F. (1975). Comparative neurology of the temporal lobe in primates including man with reference to speech. *Brain and Language, 2*(4), 396–419.

Schildkraut, J. (1965). The catecholamine hypothesis of affective disorders: A review of supporting evidence. *American Journal of Psychiatry, 122,* 509–522.

Schultz, W., Tremblay, L., & Hollerman, J. R. (1998). Reward prediction in primate basal ganglia and frontal cortex. *Neuropharmacology, 37*(4–5), 421–429.

Sereno, M. I. (1998). Brain mapping in animals and humans. *Current Opinion in Neurobiology, 8*(2), 188–194.

Shepard, R. N. (1984). Ecological constraints on internal representation: Resonant kinematics of perceiving, imagining, thinking, and dreaming. *Psychological Review, 91*(4), 417–447.

Shin, H. W., Sohn, Y. H., & Hallett, M. (2009). Hemispheric asymmetry of surround inhibition in the human motor system. *Clinical Neurophysiology, 120*(4), 816–819.

Shipp, S. (2005). The importance of being agranular: A comparative account of visual and motor cortex. *Philosophical Transactions of the Royal Society of London: B. Biological Sciences, 360*(1456), 797–814.

Squire, L. R. (1986a). Mechanisms of memory. *Science, 232,* 1612–1619.

Squire, L. R. (1986b). Memory functions as affected by electroconvulsive therapy. *Annals of the New York Academy of Sciences, 462,* 307–314.

Squire, L. R. (1987). *Memory and brain.* New York: Oxford University Press.

Squire, L. R. (1998). Memory systems. *Comptes Rendus de l'Academie des Sciences: Serie III. Sciences de la Vie-Life Sciences, 321*(2–3), 153–156.

Steriade, M. (2003). *Neuronal substrates of sleep and epilepsy.* New York: Cambridge University Press.

Steriade, M., Jones, E. G., & Llinas, R. R. (1990). *Thalamic oscillations and signaling.* New York: Wiley.

Steriade, M., & McCarley, R. W. (1990). *Brainstem control of wakefulness and sleep.* New York: Plenum.

Tucker, D. M. (1981). Lateral brain function, emotion, and conceptualization. *Psychological Bulletin, 89*(1), 19–46.

Tucker, D. M. (1992). Developing emotions and cortical networks. In M. Gunnar & C. Nelson (Eds.), *Developmental behavioral neuroscience: Minnesota symposium on child psychology* (Vol. 24, pp. 75–127). Hillsdale: Erlbaum.

Tucker, D. M. (2001). Motivated anatomy: A core-and-shell model of corticolimbic architecture. In G. Gainotti (Ed.), *Handbook of neuropsychology, 2nd edition: Volume 5. Emotional behavior and its disorders* (pp. 125–160). Amsterdam: Elsevier.

Tucker, D. M. (2002). Embodied meaning: An evolutionary-developmental analysis of adaptive semantics. In T. Givon & B. Malle (Eds.), *The evolution of language out of pre-language* (pp. 51–82). Amsterdam: J. Benjamins.

Tucker, D. M. (2007). *Mind from body: Experience from neural structure*. New York: Oxford University Press.

Tucker, D. M., & Luu, P. (2006). Adaptive binding. In H. Zimmer, A. Mecklinger, & U. Lindenberger (Eds.), *Binding in human memory: A neurocognitive approach* (pp. 85–108). New York: Oxford University Press.

Tucker, D. M., & Luu, P. (2007). Neurophysiology of motivated learning: Adaptive mechanisms of cognitive bias in depression. *Cognitive Therapy and Research, 31,* 189–209.

Tucker, D. M., Luu, P., & Poulsen, C. (2009). Neural mechanisms of recursive processing in cognitive and linguistic complexity. In T. Givon & M. Shibatani (Eds.), *Syntactic complexity: Diachrony, acquisition, neuro-cognition, evolution* (pp. 461–490). Amsterdam: John Benjamins.

Tucker, D. M., & Williamson, P. A. (1984). Asymmetric neural control systems in human self-regulation. *Psychological Review, 91*(2), 185–215.

Ungerleider, L. G., & Mishkin, M. (1982). Two cortical visual systems. In D. J. Ingle, R. J. W. Mansfield, & M. A. Goodale (Eds.), *The analysis of visual behavior* (pp. 549–586). Cambridge, MA: MIT Press.

Walker, M. P. (2009). The role of sleep in cognition and emotion. *Annals of the New York Academy of Sciences, 1156,* 168–197.

Werner, H. (1957). *The comparative psychology of mental development*. New York: Harper.

Wilson, F. A. W., O Scalaidhe, S. P., & Goldman-Rakic, P. S. (1993). Dissociation of object and spatial processing domains in primate prefrontal cortex. *Science, 260,* 1955–1958.

Yakovlev, P. I. (1948). Motility, behavior and the brain. *Journal of Nervous and Mental Disease, 107,* 313–335.

Yeterian, E. H., & Pandya, D. N. (1988). Corticothalamic connections of paralimbic regions in the rhesus monkey. *Journal of Comparative Neurology, 269*(1), 130–146.

Yonelinas, A. (2006). Unpacking explicit memory: Separating recollection and familiarity. In H. D. Zimmer, A. Mecklinger, & U. Lindenberger (Eds.), *Handbook of binding and memory: Perspectives from cognitive neuroscience*. Saarbrucken, NY: Oxford University Press.

Zola-Morgan, S., & Squire, L. R. (1993). Neuroanatomy of memory. *Annual Review of Neuroscience, 16,* 547–563.

Zola-Morgan, S., Squire, L. R., & Mishkin, M. (1982). The neuroanatomy of amnesia: Amygdala-hippocampus versus temporal stem. *Science, 218*(4579), 1337–1339.

## Chapter 3. Regulating Action

Aggleton, J. P., & Mishkin, M. (1986). The amygdala: Sensory gateway to the emotions. In R. Plutchik & H. Kellerman (Eds.), *Emotion: Theory, research and experience* (Vol. 3, pp. 281–299). New York: Academic Press.

Aggleton, J. P., & Brown, M. W. (1999). Episodic memory, amnesia, and the hippocampal-anterior thalamic axis. *Behavioral and Brain Sciences, 22*, 425–489.

Baddeley, A. (1986). *Working memory* (Vol. 11). Oxford: Clarendon Press.

Barbas, H., & Pandya, D. N. (1984). Topography of commissural fibers of the prefrontal cortex in the rhesus monkey. *Experimental Brain Research, 55*(1), 187–191.

Barbas, H., & Pandya, D. N. (1986). Architecture and frontal cortical connections of the premotor cortex (area 6) in the rhesus monkey. *Journal of Comparative Neurology, 256*, 211–228.

Barbas, H., & Pandya, D. N. (1989). Architecture and intrinsic connections of the prefrontal cortex in the rhesus monkey. *Journal of Comparative Neurology, 286*, 353–375.

Barbas, H., Zikopoulos, B., & Tiimbe, C. (2011). Sensory pathways and emotional context for action in primate prefrontal cortex. *Biological Psychiatry, 69*(12),1133–1139.

Cavanna, A. E., & Trimbel, M. R. (2006). The precuneus: a review of its functional anatomy and behavioural correlates. *Brain, 129*, 564–583.

Chao, L. L., & Knight, R. T. (1995). Human prefrontal lesions increase distractibility to irrelevant sensory inputs. *Neuroreport, 6*(12), 1605–1610.

Creem, S. H., & Proffitt, D. R. (2001). Defining the cortical visual systems: "what," "where," and "how." *Acta Psychologica, 107*, 43–68.

Crick, F. (1984). Function of the thalamic reticular complex: The searchlight hypothesis. *Proceedings of the National Academy of Sciences of the United States of America, 81*(14), 4586–4590.

Crick, F., & Koch, C. (1992). The problem of consciousness. *Scientific American, 267*(3), 152–159.

Damasio, A. R. (1996). The somatic marker hypothesis and the possible functions of the prefrontal cortex. *Philosophical Transactions of the Royal Society of London: B. Biological Sciences, 351*(1346), 1413–1420.

De Renzi, E., Liotti, M., & Nichelli, P. (1987). Semantic amnesia with preservation of autobiographic memory. A case report. *Cortex, 23*(4), 575–597.

Dum, R. P., & Strick, P. L. (1993). Cingulate motor areas. In B. A. Vogt & M. Gabriel (Eds.), *Neurobiology of the cingulate cortex and limbic thalamus* (pp. 415–441). Boston: Birkhauser.

Eidelberg, D., & Galaburda, A. M. (1984). Inferior parietal lobule: Divergent architectonic asymmetries in the human brain. *Archives of Neurology, 41*, 843–852(1).

Falk, D., Hildebolt, C., Smith, K., Morwood, M. J., Sutikna, T., Brown, P., et al. (2005). The brain of LB1, Homo floresiensis. *Science, 308*(5719), 242–245.

Fuster, J. M. (1985). The prefrontal cortex and temporal integration. In A. Peters & E. G. Jones (Eds.), *Cerebral cortex: Volume 4. Association and auditory cortices* (pp. 151–177). New York: Plenum Press.

Fuster, J. M. (1989). *The prefrontal cortex*. New York: Raven Press.

Futatsugi, Y., & Riviello, J. J. (1998). Mechanisms of generalized absence epilepsy. *Brain and Development, 20*, 75–79.

Gabriel, M., Kang, E., Poremba, A., Kubota, Y., Allen, M. T., Miller, D. P., et al. (1996). Neural substrates of discriminative avoidance learning and classical eyeblink conditioning in rabbits: A double dissociation. *Behavioural Brain Research, 82*(1), 23–30.

Gabriel, M., Lambert, R. W., Foster, K., Orona, E., Sparenborg, S., & Maiorca, R. R. (1983). Anterior thalamic lesions and neuronal activity in the cingulate and retrosplenial cortices during discriminative avoidance behavior in rabbits. *Behavioral Neuroscience, 97*(5), 675–696.

Gabriel, M., Vogt, B. A., Kubota, Y., Poremba, A., & Kang, E. (1991). Training-stage related neuronal plasticity in limbic thalamus and cingulate cortex during learning: A possible key to mnemonic retrieval. *Behavioural Brain Research, 46*(2), 175–185.

Gilbert, S. J., Spengler, S., Simons, J. S., Steele, J. D., Lawrie, S. M., Frith, C. D., et al. (2006). Functional specialization within rostral prefrontal cortex (area 10): A meta-analysis. *Journal of Cognitive Neuroscience, 18*(6), 932–948.

Gloor, P. (1978). Generalized epilepsy with bilateral synchronous spike and wave discharge: New findings concerning its physiological mechanisms. *Electroencephalography and Clinical Neurophysiology Supplement, 34*, 245–249.

Goldberg, G. (1985). Supplementary motor area structure and function: Review and hypotheses. *Behavioral and Brain Sciences, 8*, 567–616.

Goldman-Rakic, P. S. (1987). Circuitry of the primate prefrontal cortex and regulation of behavior by representational memory. In F. Plum (Ed.), *Handbook of physiology. Section 1: The nervous system. Volume V. Higher functions of the brain, Part 1.* (pp. 373–417). Bethesda, MD: American Physiological Society.

Goldman-Rakic, P. S. (1988). Changing concepts of cortical connectivity: Parallel distributed cortical networks. In P. Rakic & W. Singer (Eds.), *Neurobiology of neocortex* (pp. 177–202). New York: Wiley.

Goldman-Rakic, P. S., & Schwartz, M. L. (1982). Interdigitation of contralateral and ipsilateral columnar projections to frontal association cortex in primates. *Science, 216*, 755–757.

Goodale, M. A., & Milner, D. A. (1992). Seperate visual pathways for perception and action. *Trends in Neuroscience, 15*, 20–25.

Gurney, K., Prescott, T. J., & Redgrave, P. (2001). A computational model of action selection in the basal ganglia: II. Analysis and simulation of behaviour. *Biological Cybernetics, 84*(6), 411–423.

Haggard, P. (2008). Human volition: Towards a neuroscience of will. *Nature Reviews Neuroscience, 9*, 934–946.

Heims, S. J. (1991). *The cybernetics group*. Cambridge: MIT Press.

Helmich, R. C., Baumer, T., Siebner, H. R., Bloem, B. R., & Munchau, A. (2005). Hemispheric asymmetry and somatotopy of afferent inhibition in healthy humans. *Experimental Brain Research, 167*(2), 211–219.

Holmes, M. D., Brown, M., & Tucker, D. M. (2004). Are "generalized" seizures truly generalized? Evidence of localized mesial frontal and frontopolar discharges in absence. *Epilepsia, 45*(12), 1568–1579.

Hoshi, E., Sawamura, H., & Tanji, J. (2005). Neurons in the rostral cingulate motor area monitor multiple phases of visuomotor behavior with modest parametric selectivity. *Journal of Neurophysiology, 94,* 640–656.

Ingvar, D. H. (1985). "Memory for the future": An essay on the temporal organization of conscious awareness. *Human Neurobiology, 4,* 127–136.

Jackson, J. H. (1931). The evolution and dissolution of the nervous system. In *Selected writings of John Hughlings Jackson,* Vol II (pp. 45–75). London: Hodder and Stoughton.

Jeannerod, M. (1994). The representing brain: Neural correlates of motor intention and imagery. *Behavioral and Brain Sciences, 17,* 187–245.

Jeannerod, M., & Jacob, P. (2005). Visual cognition: A new look at the two-visual systems model. *Neuropsychologia, 43,* 301–312.

Jones, E. G. (2007). *The thalamus* (Vol. I). Cambridge, UK: Cambridge University Press.

Konow, A., & Pribram, K. H. (1970). Error recognition and utilization produced by injury to the frontal cortex in man. *Neuropsychologia, 8,* 489–491.

Lezak, M. D. (1983). *Neuropsychological Assessment.* New York: Oxford University Press.

Lhermitte, F., Pillon, B., & Serdaru, M. (1986). Human autonomy and the frontal lobes. Part I: Imitation and utilization behavior: A neuropsychological study of 75 patients. *Annals of Neurology, 19.*

Luria, A. R., & Homskaya, E. D. (1970). Frontal lobe and the regulation of arousal processes. In D. Mostofsky (Ed.), *Attention: Contemporary theory and research* (pp. 353–371). New York: Appleton, Century, Croft.

Luria, A. R., Pribram, K. H., & Homskaya, E. D. (1964). An experimental analysis of the behavioral disturbance produced by a left frontal arachnoidal endotheliomo (meningioma). *Neuropsychologia, 2,* 257–281.

Luu, P., & Tucker, D. M. (2004). Self-regulation by the medial frontal cortex: limbic representation of motive set-points. In M. Beauregard (Ed.), *Consciousness, emotional self-regulation and the brain* (pp. 123–161). Amsterdam: John Benjamin.

Luu, P., & Tucker, D. M. (2003a). Self-regulation and the executive functions: Electrophysiological clues. In A. Zani & A. M. Preverbio (Eds.), *The cognitive electrophysiology of mind and brain* (pp. 199–223). San Diego: Academic Press.

Luu, P., & Tucker, D. M. (2003b). Self-regulation by the medial frontal cortex: Limbic representation of motive set-points. In M. Beauregard (Ed.), *Consciousness, emotional self-regulation and the brain.* Amsterdam: John Benjamins.

Luu, P., Tucker, D. M., Derryberry, D., Reed, M., & Poulsen, C. (2003). Electrophysiological responses to errors and feedback in the process of action regulation. *Psychological Science, 14*(1), 47–53.

McEwen, B. S. (2000). The neurobiology of stress: from serendipity to clinical relevance. *Brain Research, 886,* 172–189.

McFarland, N. R., & Haber, S. N. (2002). Thalamic relay nuclei of the basal ganglia form both reciprocal and nonreciprocal cortical connections, linking mulitple frontal cortical areas. *Journal of Neuroscience, 22,* 8117–8132.

Miller, E. K., & Cohen, J. D. (2001). An integrative theory of prefrontal cortex function. *Annual Review of Neuroscience, 24,* 167–202.

Milner, B., & Pretrides, M. (1984). Behavioural effects of frontal-lobe lesions in man. *Trends in Neurosciences,* 403–407.

Mishkin, M. (1982). A memory system in the monkey. *Philosophical Transactions of the Royal Society of London: B. Biological Sciences, 298*(1089), 83–95.

Morrison, R. S., & Dempsey, E. W. (1942). A study of thalamocortical relations. *American Journal of Physiology, 135,* 281–292.

Morrison, R. S., & Dempsey, E. W. (1943). Mechanism of thalamocortical augmentation and repetition. *American Journal of Physiology, 138,* 297–308.

Murata, A., Fadiga, L., Fogassi, L., Gallese, V., Raos, V., & Rizzolatti, G. (1997). Object representation in the ventral premotor cortex (area F5) of the monkey. *Journal of Neurophysiology, 78,* 2226–2230.

Nauta, W. J. (1972). Neural associations of the frontal cortex. *Acta Neurobiologiae Experimentalis, 32*(2), 125–140.

Nauta, W. J. H. (1964). Some efferent connections of the prefrontal cortex in the monkey. In J. M. Warren & K. Akert (Eds.), *The frontal granular cortex and behavior* (pp. 397–409). New York: McGraw Hill.

Nauta, W. J. H. (1971). The problem of the frontal lobe: A reinterpretation. *Journal of Psychiatric Research, 8,* 167–187.

Neafsey, E. J. (1990). Prefrontal cortical control of the autonomic nervous system: Anatomical and physiological observations. In H. B. M. Uylings, C. G. Van Eden, J. P. C. De Bruin, M. A. Corner, & M. G. P. Feenstra (Eds.), *The prefrontal cortex: Its structure, function and pathology* (pp. 147–166). New York: Elsevier.

Neafsey, E. J., Terreberry, R. R., Hurley, K. M., Ruit, K. G., & Frysztak, R. J. (1993). Anterior cingulate cortex in rodents: Connections, visceral control functions, and implications for emotion. In B. A. Vogt & M. Gabriel (Eds.), *Neurobiology of the cingulate cortex and limbic thalamus* (pp. 206–223). Boston: Birkhauser.

Niedermeyer, E. (2000). Epileptic seizure disorders. In E. Niedermeyer & F. Lopes da Silva (Eds.), *Electroencephalography: Basic principles, clinical applications, and related fields.* Baltimore-Munich: Urban & Schwarzenberg.

O'Doherty, J., Kringelbach, M. L., Rolls, E. T., Hornak, J., & Andrews, C. (2001). Abstract reward and punishment representations in the human orbitofrontal cortex. *Nature Neuroscience, 4*(1), 95–102.

Pandya, D. N., Seltzer, B., & Barbas, H. (1988). Input-output organization of the primate cerebral cortex. *Comparative Primate Biology, 4,* 39–80.

Pandya, D. N., & Yeterian, E. H. (1990). Prefrontal cortex in relation to other cortical areas in rhesus monkey: Architecture and connections. In H. B. M. Uylings, C. G. Van Eden, J. P. C. De Bruin, M. A. Corner, & M. G. P. Feenstra (Eds.), *The prefrontal cortex: Its structure, function and pathology* (pp. 63–94). New York: Elsevier.

Pandya, D. N., & Yeterian, E. H. (1996). Comparison of prefrontal architecture and connections. *Philosophical Transactions of the Royal Society of London: B. Biological Sciences, 351*(1346), 1423–1432.

Passingham, R. E., Bengtsson, S. L., & Lau, H. C. (2010). Medial frontal cortex: from self-generated action to reflection on one's own performance. *Trends in Cognitive Sciences, 14*, 16–21.

Pisella, L., Sergio, L., Blangero, A., Torchin, H., Vighetto, A., & Rossetti, Y. (2009). Optic ataxia and the function of the dorsal stream: Contributions to perception and action. *Neuropsychologia, 47*, 3033–3044.

Postle, B. R. (2005). Delay-period activity in the prefrontal cortex: One function is sensory gating. *Journal of Cognitive Neuroscience, 17*(11), 1679–1690.

Postle, B. R. (2006). Working memory as an emergent property of the mind and brain. *Neuroscience, 139*(1), 23–38.

Pribram, K. H., & Tubbs, W. E. (1967). Short-term memory, parsing and the primate frontal cortex. *Science, 156*, 1765–1767.

Rizzolatti, G., & Matelli, M. (2003). Two different streams form the dorsal visual system: Anatomy and functions. *Experimental Brain Research, 153*, 146–157.

Rodin, E. (1999). Decomposition and mapping of generalized spike-wave complexes. *Clinical Neurophysiology, 110*(11), 1868–1875.

Rolls, E. T. (2000). Precis of *The brain and emotion. Behavioral and Brain Sciences, 23*(2), 177–191; discussion 192–233.

Rosenblueth, A., Weiner, N., & Bigelow, J. (1943). Behaviour, purpose and teleology. *Philosophy of Science, 10*, 18–24.

Schulkin, J., McEwen, B. S., & Gold, P. W. (1994). Allostasis, amygdala, and anticipatory angst. *Neuroscience and Biobehavioral Reviews, 18*, 385–396.

Semendeferi, K., Armstrong, E., Schleicher, A., Zilles, K., & Van Hoesen, G. W. (2001). Prefrontal cortex in humans and apes: A comparative study of area 10. *American Journal of Physical Anthropology, 114*(3), 224–241.

Shipp, S. (2005). The importance of being agranular: A comparative account of visual and motor cortex. *Philosophical Transactions of the Royal Society of London: B. Biological Sciences, 360*(1456), 797–814.

Stamm, J. S. (1987). The riddle of the monkey's delayed-response deficit has been solved. In E. Perceman (Ed.), *The frontal lobes revisited* (pp. 73–89). New York: IRBN Press.

Steriade, M. (2003). *Neuronal substrates of sleep and epilepsy*. New York: Cambridge University Press.

Steriade, M. (2004). Sleep and neuronal plasticity: Cellular mechanisms of corticothalamic oscillations. In P.-H. Luppi (Ed.), *Sleep: Circuits and functions* (pp. 1–24). Boca Raton: CRC Press.

Steriade, M., & Amzica, F. (2003). Sleep oscillations developing into seizures in corticothalamic systems. *Epilepsia, 44*(Suppl. 12), 9–20.

Teuber, H. L. (1964). The riddle of the frontal lobe function in man. In J. M. Warren & K. Akert (Eds.), *The frontal granular cortex and behavior* (pp. 410–444). New York: McGraw Hill.

Tucker, D. M., Brown, M., Luu, P., & Holmes, M. D. (2007). Discharges in ventromedial frontal cortex during absence spells. *Epilepsy and Behavior, 11*(4), 546–557.

Tucker, D. M., & Holmes, M. D. (2010). Fractures and bindings of consciousness. *American Scientist, 99*, 32–39.

Tucker, D. M., & Luu, P. (2007). Neurophysiology of motivated learning: Adaptive mechanisms of cognitive bias in depression. *Cognitive Therapy and Research, 31*, 189–209.

Tucker, D. M., Luu, P., & Poulsen, C. (January 1, 2009). Neural mechanisms of recursive processing in cognitive and linguistic complexity. *Typological Studies in Language, 85*, 461–490.

Umilta, M. A., Brochier, T., Spinks, R. L., & Lemon, R. N. (2007). Simultaneous recording of macaque premotor and primary motor cortex neuronal populations reveals different functional contributions to visuomotor grasp. *Journal of Neurophysiology, 98*(1), 488–501.

Ungerleider, L. G., & Mishkin, M. (1982). Two cortical visual systems. In D. J. Ingle, M. A. Goodale & R. J. W. Mansfield (Eds.), *Analysis of visual behavior* (pp. 549–586). Cambridge: MIT Press.

Vogt, B. A. (2005). Pain and emotion interactions in subregions of the cingulate gyrus. *Nature Reviews Neuroscience, 6*, 533–544.

Vogt, B. A., Vogt, L., & Laureys, S. (2006). Cytology and functionally correlated circuits of human posterior cingulate areas. *Neuroimage, 29*(2), 452–466.

Vorobiev, V., Govoni, P., Rizzolatti, G., Matelli, M., & Lippino, G. (1998). Parcellation of human mesial area 6: Cytoarchitectonic evidence for three separate areas. *European Journal of Neuroscience, 10*, 2199–2203.

Wang, P. (1987). Concept formation and the frontal lobe function: The search for a clinical frontal lobe test. In E. Perecman (Ed.), *The frontal lobes revisited* (pp. 189–205). New York: IRBN.

Werner, H. (1957). *The comparative psychology of mental development*. New York: Harper.

Winn, P. (1995). The lateral hypothalamus and motivated behavior: An old syndrome reassessed and a new perspective gained. *Current Directions in Psychological Science, 4*, 182–187.

Yakovlev, P. I. (1948). Motility, behavior and the brain. *Journal of Nervous and Mental Disease, 107*, 313–335.

Yamagata, T., Nakayama, Y., Tanji, J., & Hoshi, E. (2009). Processing of visual signals for direct specification of motor targets and for conceptual representation of action targets in the dorsal and ventral premotor cortex. *Journal of Neurophysiology, 102*, 3280–3294.

Yeterian, E. H., & Pandya, D. N. (1994). Laminar origin of striatal and thalamic projections of the prefrontal cortex in rhesus monkeys. *Experimental Brain Research, 99*(3), 383–398.

Yingling, C. D., & Skinner, J. E. (1976). Selective regulation of thalamic sensory relay nuclei by nucleus reticularis thalami. *Electroencephalography and Clinical Neurophysiology, 41*, 476–482.

Yingling, C. D., & Skinner, J. E. (1977). Gating of thalamic input to cerebral cortex by nucleus reticularis thalami. In J. E. Desmedt (Ed.), *Attention voluntary contraction and event-related cerebral potentials, progress in clinical neurophysiology* (Vol. 1, pp. 70–96). Basel: Karger.

Zikopoulos, B., & Barbas, H. (2006). Prefrontal projections to the thalamic reticular nucleus form a unique circuit for attentional mechanisms. *Journal of Neuroscience, 26*(28), 7348–7361.

Zola-Morgan, S., & Squire, L. R. (1993). Neuroanatomy of memory. *Annual Review of Neuroscience, 16,* 547–563.

## Chapter 4. Opponent Complementarity in Psychological Function

Achenbach, T. M. (1982). *Developmental psychopathology.* New York: Wiley.

Allport, F. H. (1924). *Social psychology.* Boston: Houghton Mifflin.

Andreasen, N. J. C., & Canter, A. (1974). The creative writer: Psychiatric symptoms and family history. *Comprehensive Psychiatry, 15,* 123–131.

Andreasen, N. J. C., & Powers, P. S. (1975). Creativity and psychosis: An examination of conceptual style. *Archives of General Psychiatry, 32,* 70–73.

Baldwin, D. A. (1991). Infants' contribution to the achievement of joint reference. *Child Development, 62*(5), 875–890.

Baldwin, D. A. (1993). Infants' ability to consult the speaker for clues to word reference. *Journal of Child Language, 20*(2), 395–418.

Baldwin, D. A., & Markman, E. M. (1989). Establishing word-object relations: A first step. *Child Development, 60*(2), 381–398.

Barbas, H., Zikopoulos, B., & Timbie, C. (2011). Sensory pathways and emotional context for action in primate prefrontal cortex. *Biol Psychiatry, 69*(12), 1133–1139.

Barbas, H., & Pandya, D. N. (1984). Topography of commissural fibers of the prefrontal cortex in the rhesus monkey. *Experimental Brain Research, 55*(1), 187–191.

Bear, M. F., & Singer, W. (1986). Modulation of visual cortical plasticity by acetylcholine and noradrenaline. *Nature, 320*(6058), 172–176.

Beeman, M. (1993). Semantic processing in the right hemisphere may contribute to drawing inferences from discourse. *Brain and Language, 44*(1), 80–120.

Beeman, M. J., Bowden, E. M., & Gernsbacher, M. A. (2000). Right and left hemisphere cooperation for drawing predictive and coherence inferences during normal story comprehension. *Brain and Language, 71*(2), 310–336.

Blumer, D., & Benson, D. F. (1975). Personality changes with frontal and temporal lobe lesions. In D. F. Benson & D. Blumer (Eds.), *Psychiatric aspects of neurologic disease* (pp. 151–170). New York: Gruen and Stratton.

Botvinick, M. W., Cohen, J. D., & Carter, C. S. (2004). Conflict monitoring and anterior cingulate cortex: and update. *Trends Cogn Sci, 8,* 539–546.

Bowlby, J. (1997). *Attachment.* London: Pimlico.

Bradley, M. M., Codispoti, M., Cuthbert, B. N., & Lang, P. J. (2001). Emotion and motivation: I. Defensive and appetitive reactions in picture processing. *Emotion, 1*(3), 276–298.

Bradley, M. M., Codispoti, M., Sabatinelli, D., & Lang, P. J. (2001). Emotion and motivation: II. Sex differences in picture processing. *Emotion, 1*(3), 300–319.

Brown, J. W. (1979). Language representation in the brain. In I. H. D. Steklis & M. J. Raleigh (Eds.), *Neurobiology of social communication in primates* (pp. 133–195). New York: Academic Press.

Brown, J. W. (1988). *The life of the mind: Selected papers.* Hillsdale, NJ: Erlbaum.

Bussey, T. J., Everitt, B. J., & Robbins, T. W. (1997). Dissociable effects of cingulate and medial frontal cortex lesions on stimulus-reward learning using a novel pavlovian autoshaping procedure for the rat: Implications for the neurobiology of emotion. *Behavioral Neuroscience, 5,* 908–919.

Bussey, T. J., Muir, J. L., Everitt, B. J., & Robbins, T. W. (1996). Dissociable effects of anterior and posterior cingulate cortex lesions on the acquisition of a conditional visual discrimination: Facilitation of early learning vs. impairment of late learning. *Behavioral Brain Research, 82,* 45–56.

Bussey, T. J., Wise, S. P., & Murray, E. A. (2001). The role of ventral and orbital prefrontal cortex in conditional visuomotor learning and strategy use in rhesus monkeys (macaca mulatta). *Behavioral Neuroscience, 115,* 971–982.

Bussey, T. J., Wise, S. P., & Murray, E. A. (2002). Interaction of ventral and orbital prefrontal cortex with inferotemporal cortex in conditional visuomotor learning. *Behavioral Neuroscience, 116,* 703–715.

Brass, M., Derrfuss, J., Forstmann, B., & von Cramon, D. Y. (2005). The role of the inferior frontal junction area in cognitive control. *Trends in Cognitive Sciences, 9,* 314–316.

Carlson, S. M., & Moses, L. J. (2001). Individual differences in inhibitory control and children's theory of mind. *Child Development, 72*(4), 1032–1053.

Catani, M., Jones, D. K., & ffytche, D. H. (2005). Perisylvian language networks of the human brain. *Annals of Neurology, 57*(1), 8–16.

Christakou, A., Brammer, M., Giampietro, V., & Rubia, K. (2009). Right ventromedial and dorsolateral prefrontal cortices mediate adaptive decisions under ambiguity by integrating choice utility and outcome evaluation. *Journal of Neurosciences, 29*(35), 11020–11028.

Cicchetti, D., & Tucker, D. M. (1994). Development of self-regulatory structures of the mind. *Development and Psychopathology, 6,* 533–549.

Chomsky, N. (1965). *Aspects of the Theory of Syntax.* MIT Press.

Corbetta, M., Patel, G., & Shulman, G. L. (2008). The reorienting system of the human brain: from environment to theory of mind. *Neuron, 58*(3), 306–324.

Crick, F., & Koch, C. (1990). Towards a neurobiological theory of consciousness. *Seminars in the Neurosciences, 2,* 263–275.

Cuthbert, B. N., Bradley, M. M., & Lang, P. J. (1996). Probing picture perception: Activation and emotion. *Psychophysiology, 33*(2), 103–111.

Damasio, H., Grabowski, T., Frank, R., Galaburda, A. M., & Damasio, A. R. (1994). The return of Phineas Gage: Clues about the brain from the skull of a famous patient. *Science, 264*(5162), 1102–1105.

Dehaene, S., Posner, M. I., & Tucker, D. M. (1994). Localization of a neural system for error detection and compensation. *Psychological Science, 5,* 303–305.

Delgado, M. R., Frank, R. H., & Phelps, E. A. (2005). Perceptions of moral character modulate the neural systems of reward during the trust game. *Nature Neuroscience, 8*(11), 1611–1618.

DellaSala, S., Marchetti, C., & Spinnler, H. (1991). Right-sided anarchic (alien) hand: A longitudinal study. *Neuropsychologia, 29,* 1113–1127.

De Martino, B., Kumaran, D., Seymour, B., & Dolan, R. J. (2006). Frames, biases, and rational decision-making in the human brain. *Science, 313*(5787), 684–687.

den Ouden, H. E., Frith, U., Frith, C., & Blakemore, S. J. (2005). Thinking about intentions. *Neuroimage, 28*(4), 787–796.

Derryberry, D., & Reed, M. A. (1994). Temperament and attention: Orienting toward and away from positive and negative signals. *Journal of Personality and Social Psychology, 66*(6), 1128–1139.

Derryberry, D., & Reed, M. A. (1998). Anxiety and attentional focusing: Trait, state, and hemispheric influences. *Personality and Individual Differences, 25,* 745–761.

Derryberry, D., & Rothbart, M. K. (1987). *Temperamental functioning of arousal affect and attention.* Unpublished manuscript, University of Oregon.

Derryberry, D., & Rothbart, M. K. (1988). Affect, arousal, and attention as components of temperament. *Journal of Personality and Social Psychology, 55,* 958–966.

Derryberry, D., & Rothbart, M. K. (1997). Reactive and effortful processes in the organization of temperament. *Development and Psychopathology, 9*(4), 633–652.

Derryberry, D., & Tucker, D. M. (1990). The adaptive base of the neural hierarchy: Elementary motivational controls on network function. *Nebraska Symposium on Motivation, 38,* 289–342.

Derryberry, D., & Tucker, D. M. (1992). Neural mechanisms of emotion. *Journal of Consulting and Clinical Psychology, 60*(3), 329–338.

Derryberry, D., & Tucker, D. M. (1994). Motivating the focus of attention. In P. Niedenthal & S. Kitayama (Eds.), *The heart's eye: Emotional influences in perception and attention* (pp. 166–196): Academic Press.

Derryberry, D., & Tucker, D. M. (2006). Motivation, self-regulation, and self-organization. In D. J. Cohen & D. Cicchetti (Eds.), *Handbook of developmental psychopathology: Vol. 2. Developmental neuroscience* (pp. 502–532). New York: Wiley.

Dobzhansky, T. (1964). Biology, molecular and organismic. *American Zoologist, 4,* 443–452.

Dosenbach, N. U., Fair, D. A., Miezin, F. M., Cohen, A. L., Wenger, K. K., Dosenbach, R. A., et al. (2007). Distinct brain networks for adaptive and stable task control in humans. *Proceedings of the National Academy of Sciences of the United States of America, 104*(26), 11073–11078.

Dosenbach, N. U., Visscher, K. M., Palmer, E. D., Miezin, F. M., Wenger, K. K., Kang, H. C., et al. (2006). A core system for the implementation of task sets. *Neuron, 50*(5), 799–812.

Eickhoff, S. B., Laird, A. R., Grefkes, C., Wang, L. E., Zilles, K., & Fox, P. T. (2009). Coordinate-based activation likelihood estimation meta-analysis of neuroimaging data: A random-effects approach based on empirical estimates of spatial uncertainty. *Human Brain Mapping, 30,* 2970–2926.

Eidelberg, D., & Galaburda, A. M. (1984). Inferior parietal lobule: Divergent architectonic asymmetries in the human brain. *Archives of Neurology, 41, 843–852*(1).

Eisenberg, N., Spinrad, T. L., & Eggum, N. D. (2010). Emotion-related self-regulation and its relation to children's maladjustment. *Annual Review of Clinical Psychology, 6,* 495–525.

Eisenberg, N., Valiente, C., Spinrad, T. L., Liew, J., Zhou, Q., Losoya, S. H., et al. (2009). Longitudinal relations of children's effortful control, impulsivity, and negative emotionality to their externalizing, internalizing, and co-occurring behavior problems. *Developmental Psychology, 45*(4), 988–1008.

Ellinwood, E. H. (1967). Amphetamine psychosis: I. Description of the individuals and process. *Journal of Nervous and Mental Disease, 144*(0), 273–283.

Elliott, R., & Dolan, R. J. (1998). Activation of different anterior cingulate foci in association with hypothesis testing and response selection. *Neuroimage, 8,* 17–29.

Eysenck, H. J. (1967). *The biological basis of personality.* Springfield, IL: C. C. Thomas.

Eysenck, M. W. (1976). Arousal, learning, and memory. *Psychological Bulletin, 83,* 359–404.

Fair, D. A., Dosenbach, N. U., Church, J. A., Cohen, A. L., Brahmbhatt, S., Miezin, F. M., et al. (2007). Development of distinct control networks through segregation and integration. *Proceedings of the National Academy of Sciences of the United States of America, 104*(33), 13507–13512.

Falkenstein, M., Hohnsbein, J., Hoormann, J., & Blanke, L. (1991). Effects of cross-modal divided attention on late ERP components. II. Error processing in choice reaction tasks. *Electroencephalography and Clinical Neurophysiology, 78,* 447–455.

Farrer, C., Franck, N., Frith, C. D., Decety, J., Georgieff, N., d'Amato, T., et al. (2004). Neural correlates of action attribution in schizophrenia. *Psychiatry Research, 131*(1), 31–44.

Farrer, C., & Frith, C. D. (2002). Experiencing oneself vs another person as being the cause of an action: The neural correlates of the experience of agency. *Neuroimage, 15*(3), 596–603.

Foote, S. L., & Morrison, J. H. (1987). Extrathalamic modulation of cortical function. *Annual Review of Neuroscience, 10,* 67–95.

Fox, M. D., Corbetta, M., Snyder, A. Z., Vincent, J. L., & Raichle, M. E. (2006). Spontaneous neuronal activity distinguishes human dorsal and ventral attention systems. *Proceedings of the National Academy of Sciences of the United States of America, 103*(26), 10046–10051.

Fox, M. D., Snyder, A. Z., Vincent, J. L., Corbetta, M., Van Essen, D. C., & Raichle, M. E. (2005). The human brain is intrinsically organized into dynamic, anticorrelated functional networks. *Proceedings of the National Academy of Sciences of the United States of America, 102*(27), 9673–9678.

Fuster, J. (1989). *The prefrontal cortex.* New York: Raven Press.

Gabriel, M., Burhans, L., Talk, A., & Scalf, P. (2002). Cingulate cortex. In V. S. Ramachandran (Ed.), *Encyclopedia of the human brain* (pp. 775–791). Elsevier Science.

Gabriel, M., Kang, E., Poremba, A., Kubota, Y., Allen, M. T., Miller, D. P., et al. (1996). Neural substrates of discriminative avoidance learning and classical eyeblink conditioning in rabbits: A double dissociation. *Behavioural Brain Research, 82*(1), 23–30.

Gabriel, M., Vogt, B. A., Kubota, Y., Poremba, A., & Kang, E. (1991). Training-stage related neuronal plasticity in limbic thalamus and cingulate cortex during learning: A possible key to mnemonic retrieval. *Behavioural Brain Research, 46*(2), 175–185.

Galaburda, A. M. (1984). The anatomy of language: Lessons from comparative anatomy. In D. Caplan, A. R. Lacours, & A. Smith (Eds.), *Biological perspectives on language.* Cambridge, MA: MIT Press.

Galaburda, A. M., & Pandya, D. N. (1982). Role of architectonics and connections in the study of primate brain evolution. In E. R. Armstrong & D. Falk (Eds.), *Primate brain and evolution: Methods and concepts.* New York: Plenum Press.

Gehring, W. J., Goss, B., Coles, M. G. H., Meyer, D. E., & Donchin, E. (1993). A neural system for error detection and compensation. *Psychological Science, 4,* 385–390.

Gehring, W. J., Himle, J., & Nisenson, L. G. (2000). Action monitoring dysfunction in obsessive-compulsive disorder. *Psychological Science, 11,* 1–6.

Givon, T. (2009). *The genesis of syntactic complexity: Diachrony, ontogeny, neuro-cognition, evolution.* Philadelphia: John Benjamins.

Goldberg, E., & Costa, L. D. (1981). Hemisphere differences in the acquisition and use of descriptive systems. *Brain and Language, 14,* 144–173.

Goldberg, G. (1985). Supplementary motor area structure and function: Review and hypotheses. *Behavioral and Brain Sciences, 8,* 567–616.

Goldberg, G., Mayer, N. H., & Toglia, J. U. (1981). Medial frontal cortex infarction and the alien hand sign. *Archives of Neurology, 38*(11), 683–686.

Goldberg, L. R., & Rosolack, T. K. (1994). The Big Five factor structure as an integrative framework: An empirical comparison with Esynck's P-E-N model. In C. F. Halverson, G. A. Kohnstamm, & R. P. Martin (Eds.), *The developing structure of temperament and personality from infancy to adulthood* (pp. 7–35). New York: Erlbaum.

Goldman-Rakic, P. S. (1987). Circuitry of the prefrontal cortex and the regulation of behavior by representational memory. In *Handbook of physiology* (Vol. 5, pp. 373–417). New York: American Physiological Society.

Goldman-Rakic, P. S., & Schwartz, M. L. (1982). Interdigitation of contralateral and ipsilateral columnar projections to frontal association cortex in primates. *Science, 216,* 755–757.

Goodglass, H. (1993). *Understanding aphasia.* New York: Academic Press.

Gray, J. A. (1990). Brain systems that mediate both emotion and cognition. *Cognition and Emotion, 4,* 269–288.

Grice, P. (1957). Meaning. *Philosophical Review, 66,* 377–388.

Grossberg, S. (2000). The imbalanced brain: From normal behavior to schizophrenia. *Biological Psychiatry, 48*(2), 81–98.

Haggard, P. (2008). Human volition: Towards a neuroscience of will. *Nature Reviews Neuroscience, 9,* 934–946.

Iversen, S. D. (1977). Brain dopamine systems and behavior. In L. L. Iversen, S. D. Iversen, & S. H. Snyder (Eds.), *Handbook of psychopharmacology: Vol. 8. Drugs, neurotransmitters and behavior* (pp. 333–384). New York: Plenum Press.

Jeannerod, M. (1994). The representing brain: Neural correlates of motor intention and imagery. *Behavioral and Brain Sciences, 17*, 187–245.

Jeannerod, M., & Jacob, P. (2005). Visual cognition: A new look at the two-visual systems model. *Neuropsychologia, 43*, 301–312.

Johannes, S., Wieringa, B. M., Nager, W., Dengler, R., & Münte, T. F. (2001). Oxazepam alters action monitoring. *Psychopharmacology, 155*, 100–106.

Jones, E. G. (2007). *The thalamus* (Vol. I). Cambridge: Cambridge University Press.

Jürgens, U., & Ploog, D. (1970). Cerebral representation of vocalization in the squirrel monkey. *Experimental Brain Research, 10*, 532–554.

Koechlin, E., Ody, C., & Kouneiher, F. (2003). The architecture of cognitive control in the human prefrontal cortex. *Science, 302*(5648), 1181–1185.

Kohut, H. (1978). *The search for the self.* New York: International Universities Press.

Kokkinidis, L., & Anisman, H. (1980). Amphetamine models of paranoid schizophrenia: An overview and elaboration of animal experimentation. *Psychological Bulletin, 88*, 551–578.

Kotz, S. A., & Schwartze, M. (2010). Cortical speech processing unplugged: A timely subcortico-cortical framework. *Trends in Cognitive Sciences, 14*(9), 392–399.

Krechevsky, I. (1932). "Hypotheses" in rats. *Psychological Review, 39*, 516–532.

Lang, P. J. (1995). The emotion probe. *American Psychologist, 50*, 372–385.

Lang, P. J., & Bradley, M. M. (2009). Emotion and the motivational brain. *Biological Psychology, 84*, 437–450.

Lang, P. J., Bradley, M. M., & Cuthbert, B. N. (1998a). Emotion and motivation: Measuring affective perception. *Journal of Clinical Neurophysiology, 15*(5), 397–408.

Lang, P. J., Bradley, M. M., & Cuthbert, B. N. (1998b). Emotion, motivation, and anxiety: Brain mechanisms and psychophysiology. *Biological Psychiatry, 44*(12), 1248–1263.

Laurens, K. R., Ngan, E. T. C., Bates, A. T., Kiehl, K. A., & Liddle, P. F. (2003). Rostral anterior cingulate cortex dysfunction during error processing in schizophrenia. *Brain, 126*, 610–622.

Lewis, M. D. (2005). Bridging emotion theory and neurobiology through dynamic systems modeling. *Behavioral and Brain Sciences, 28*(2), 169–194; discussion 194–245.

Lieberman, P. (1984). *The biology and evolution of language.* Cambridge: Harvard University Press.

Liotti, M., & Tucker, D. M. (1994). Emotion in asymmetric corticolimbic networks. In R. J. Davidson & K. Hugdahl (Eds.), *Human brain laterality* (pp. 389–424). New York: Oxford.

Luu, P., Flaisch, T., & Tucker, D. M. (2000). Medial frontal cortex in action monitoring. *Journal of Neuroscience, 20*, 464–469.

Luu, P., Shane, M., Pratt, N. L., & Tucker, D. M. (2009). Corticolimbic mechanisms in the control of trial and error learning. *Brain Research, 1247*, 100–113.

Luu, P., & Tucker, D. M. (2003). Self-regulation by the medial frontal cortex: Limbic representation of motive set-points. In M. Beauregard (Ed.), *Consciousness, emotional self-regulation and the brain.* Amsterdam: John Benjamins.

Luu, P., Tucker, D. M., & Derryberry, D. (1998). Anxiety and the motivational basis of working memory. *Cognitive Therapy and Research, 22*, 577–594.

Luu, P., Tucker, D. M., Derryberry, D., Reed, M., & Poulsen, C. (2003). Electrophysiological responses to errors and feedback in the process of action regulation. *Psychological Science, 14*, 47–53.

Luu, P., Tucker, D. M., & Stripling, R. (2007). Neural mechanisms for learning actions in context. *Brain Research, 179*, 89–105.

Mahler, M. S. (1968). *On human symbiosis and the vicissitudes of individuation*. New York: International Universities Press.

McCrae, R. R., & Costa, P. T. (1985). Updating Norman's "adequate taxonomy": Intelligence and personality dimensions in natural language and in questionnaires. *Journal of Personality and Social Psychology, 49*(3), 710–721.

Menon, V., Adleman, N. E., White, C. D., Glover, G. H., & Reiss, A. L. (2001). Error-related brain activation during a go/nogo response inhibition task. *Human Brain Mapping, 12*, 131–143.

Miller, E. K., & Cohen, J. D. (2001). An integrative theory of prefrontal cortex function. *Annual Review of Neuroscience, 24*, 167–202.

Mishkin, M. (1982). A memory system in the monkey. *Philosophical Transactions of the Royal Society of London: B. Biological Sciences, 298*(1089), 83–95.

Monrad-Krohn, G. H. (1924). On the dissociation of voluntary and emotional innervation in facial paresis of central origin. *Brain, 47*, 22–35.

Morrison, J. H., & Foote, S. L. (1986). Noradrenergic and serotonergic innervation of cortical, thalamic, and tectal visual structures in old and new world monkey. *Journal of Comparative Neurology, 243*, 117–138.

Moses, L. J. (2001). Executive accounts of theory-of-mind development. *Child Development, 72*(3), 688–690.

Murata, A., Fadiga, L., Fogassi, L., Gallese, V., Raos, V., & Rizzolatti, G. (1997). Object representation in the ventral premotor cortex (area F5) of the monkey. *Journal of Neurophysiology, 78*, 2226–2230.

Neafsey, E. J. (1990). Prefrontal cortical control of the autonomic nervous system: Anatomical and physiological observations. In H. B. M. Uylings, C. G. Van Eden, J. P. C. De Bruin, M. A. Corner, & M. G. P. Feenstra (Eds.), *The prefrontal cortex: Its structure, function and pathology* (pp. 147–166). New York: Elsevier.

Neafsey, E. J., Terreberry, R. R., Hurley, K. M., Ruit, K. G., & Frysztak, R. J. (1993). Anterior cingulate cortex in rodents: Connections, visceral control functions, and implications for emotion. In B. A. Vogt & M. Gabriel (Eds.), *Neurobiology of the cingulate cortex and limbic thalamus* (pp. 206–223). Boston: Birkhauser.

Norman, D. A., & Shallice, T. (1986). Attention to action: Willed and automatic control of behavior. In R. J. Davidson, G. E. Schwartz, & D. Shapiro (Eds.), *Consciousness and self-regulation* (pp. 1–18). New York: Plenum.

Pandya, D. N., & Seltzer, B. (1982). Association areas of the cerebral cortex. *Trends in Neurosciences, 5*, 386–390.

Pandya, D. N., & Yeterian, E. H. (1984). Proposed neural circuitry for spatial memory in the primate brain. *Neuropsychologia, 22*(2), 109–122.

Papini, M. R. (2003). Comparative psychology of surprising nonreward. *Brain, Behavior and Evolution, 62*(2), 83–95.

Papini, M. R., & Bitterman, M. E. (1990). The role of contingency in classical conditioning. *Psychological Review, 97*(3), 396–403.

Poremba, A., & Gabriel, M. (1997). Amygdalar lesions block discriminative avoidance learning and cingulothalamic training-induced neuronal plasticity in rabbits. *Journal of Neuroscience, 17*, 5237–5244.

Posner, M. I. (1978). *Chronometric explorations of mind*. Hillsdale, NJ: Erlbaum.

Posner, M. I., & Dehaene, S. (1994). Attentional networks. *Trends Neurosci, 17*, 75–79.

Posner, M. I., & Rothbart, M. K. (1998). Attention, self regulation and consciousness. *Philosophical Transactions of the Royal Society of London: Series B, 353*, 1–13.

Raichle, M. E., & Gusnard, D. A. (2005). Intrinsic brain activity sets the stage for expression of motivated behavior. *Journal of Comparative Neurology, 493*(1), 167–176.

Redgrave, P., Prescott, T. J., & Gurney, K. (1999). The basal ganglia: A vertebrate solution to the selection problem? *Neuroscience, 89*(4), 1009–1023.

Rescorla, R. A., & Wagner, A. R. (1972). A theory of Pavlovian conditioning: Variations in the effectiveness of reinforcement and nonreinforcement. In A. H. Black & W. F. Prokasy (Eds.), *Classical conditioning: II. Current research and theory* (pp. 65–99). New York: Appleton-Century-Crofts.

Ridderinkhof, K. R., Ullsperger, M., Crone, E. A., & Nieuwenhuis, S. (2004). The role of the medial frontal cortex in cognitive control. *Science, 306*, 443–447.

Rilling, J. K., Glasser, M. F., Preuss, T. M., Ma, X., Zhao, T., Hu, X., et al. (2008). The evolution of the arcuate fasciculus revealed with comparative DTI. *Nature Neuroscience, 11*(4), 426–428.

Rizzolatti, G., Fadiga, L., Fogassi, L., & Gallese, V. (1999). Resonance behaviors and mirror neurons. *Archives Italiennes de Biologie, 137*(2–3), 85–100.

Rushworth, M. F. S., Walton, M. E., Kennerley, S. W., & Bannerman, D. M. (2004). Action sets and decisions in the medial frontal cortex. *Trends in Cognitive Science, 8*, 410–417.

Saucier, G., & Goldberg, L. R. (1998). What is beyond the big five? *Journal of Personality, 66*(4), 495–524.

Saucier, G. (2009). Recurrent personality dimensions in inclusive lexical studies: Indications for a big six structure. *Journal of Personality, 77*(5), 1577–1614.

Semendeferi, K., Armstrong, E., Schleicher, A., Zilles, K., & Van Hoesen, G. W. (2001). Prefrontal cortex in humans and apes: A comparative study of area 10. *American Journal of Physical Anthropology, 114*(3), 224–241.

Shapiro, D. (1965). *Neurotic styles*. New York: Basic Books.

Shapiro, D. (1981). *Autonomy and the rigid character*. New York: Basic Books.

Shaw, E. D., Mann, J. J., Stokes, P. E., & Manevitz, Z. A. (1986). Effects of lithium carbonate on associative productivity and idiosyncracy in bipolar outpatients. *American Journal of Psychiatry, 143*, 1166–1169.

Shepard, R. N. (1984). Ecological constraints on internal representation: Resonant kinematics of perceiving, imagining, thinking, and dreaming. *Psychological Review, 91*(4), 417–447.

Shima, K., & Tanji, J. (1998). Role for cingulate motor area cells in voluntary movement selection based on reward. *Science, 282*, 1335–1338.

Smith, D. M., Freeman, J. H., Jr., Nicholson, D., & Gabriel, M. (2002). Limbic thalamic lesions, appetitively motivated discrimination learning, and training-induced neuronal activity in rabbits. *Journal of Neuroscience, 22*(18), 8212–8221.

Spitz, R. A. (1965). *The first year of life: A psychoanalytic study of normal and deviant development of object relations*. New York: International Universities Press.

Tachibana, K., Suzuki, K., Mori, E., Miura, N., Kawashima, R., Horie, K., et al. (2009). Neural activity in the human brain signals logical rule identification. *Journal of Neurophysiology, 102*(3), 1526–1537.

Tellegen, A. (1985). Structures of mood and personality and their relevance to assessing anxiety, with an emphasis on self-report. In A. H. Tuma & J. D. Maser (Eds.), *Anxiety and the anxiety related disorders* (pp. 681–706). Hillsdale, NJ: Erlbaum.

Tellegen, A. (1988). The analysis of consistency in personality assessment. *Journal of Personality, 56*, 621–663.

Tellegen, A. (1993). Folk concepts and psychological concepts of personality and personality disorders. *Psychological Inquiry, 4*, 122–130.

Thalmayer, A. G., Saucier, G., & Eigenhuis, A. (2011). Comparative validity of brief to medium-length Big Five and Big Six Personality Questionnaires. *Psychological Assessment, 23*(4), 995–1009.

Thayer, R. E. (1978). Toward a psychological theory of multidimensional activation. *Motivation and Emotion, 2*, 1–34.

Thayer, R. E. (1989). *The biopsychology of mood and arousal*. New York: Oxford University Press.

Tomlin, D., Kayali, M. A., King-Casas, B., Anen, C., Camerer, C. F., Quartz, S. R., et al. (2006). Agent-specific responses in the cingulate cortex during economic exchanges. *Science, 312*(5776), 1047–1050.

Trevarthen, C. (1985). Neuroembryology and the development of perceptual mechanisms. In F. Falkner & J. M. Tanner (Eds.), *Human growth, a comprehensive treatise: Volume 2. Postnatal growth, neurobiology*. New York: Plenum Press.

Tucker, D. M. (1989). Neural and psychological maturation in a social context. In D. Cicchetti (Ed.), *The emergence of a discipline: Rochester symposium on developmental psychopathology* (pp. 69–88). Hillsdale, NJ: Erlbaum.

Tucker, D. M. (1992). Developing emotions and cortical networks. In M. Gunnar & C. Nelson (Eds.), *Developmental behavioral neuroscience: Minnesota symposium on child psychology* (Vol. 24, pp. 75–127). Hillsdale, NJ: Erlbaum.

Tucker, D. M. (2001). Motivated anatomy: A core-and-shell model of corticolimbic architecture. In G. Gainotti (Ed.), *Handbook of neuropsychology, 2nd edition: Volume 5. Emotional behavior and its disorders* (pp. 125–160). Amsterdam: Elsevier.

Tucker, D. M. (2002). Embodied meaning: An evolutionary-developmental analysis of adaptive semantics. In T. Givon & B. Malle (Eds.), *The evolution of language out of pre-language*. Amsterdam: John Benjamins.

Tucker, D. M. (2007). *Mind from body: Experience from neural structure*. New York: Oxford University Press.

Tucker, D. M. (2008). Self-organizing ontogenesis on the phyletic frame In M. Pachalska & M. Weber (Eds.), *Philosophy of mind in process: Essays in honor of Jason Brown* (pp. 371–400). Ontos: Verlag.

Tucker, D. M., Brown, M., Luu, P., & Holmes, M. D. (2007). Discharges in ventromedial frontal cortex during absence spells. *Epilepsy and Behavior, 11*, 546–557.

Tucker, D. M., & Derryberry, D. (1992). Motivated attention: Anxiety and the frontal executive functions. *Neuropsychiatry, Neuropsychology, and Behavioral Neurology, 5*, 233–252.

Tucker, D. M., Derryberry, D., & Luu, P. (2000). Anatomy and physiology of human emotion: Vertical integration of brainstem, limbic, and cortical systems. In J. Borod (Ed.), *Handbook of the neuropsychology of emotion* (pp. 56–79). New York: Oxford.

Tucker, D. M., Frishkoff, G. A., & Luu, P. (2008). Microgenesis of language: Vertical integration of neurolinguistic mechanisms across the neuraxis. In B. Stemmer & H. A. Whitaker (Eds.), *Handbook of the neuroscience of language*. New York: Oxford.

Tucker, D. M., & Luu, P. (2006). Adaptive binding. In H. Zimmer, A. Mecklinger, & U. Lindenberger (Eds.), *Binding in human memory: A neurocognitive approach* (pp. 85–108). New York: Oxford University Press.

Tucker, D. M., & Luu, P. (2007). Neurophysiology of motivated learning: Adaptive mechanisms of cognitive bias in depression. *Cognitive Therapy and Research, 31*, 189–209.

Tucker, D. M., Luu, P., & Poulsen, C. (2009). Neural mechanisms of recursive processing in cognitive and linguistic complexity. In T. Givón & M. Shibatani (Eds.), *Syntactic complexity: Diachrony, acquisition, neuro-cognition, evolution* (pp. 461–489). Philadelphia: Benjamins.

Tucker, D. M., & Williamson, P. A. (1984). Asymmetric neural control systems in human self-regulation. *Psychological Review, 91*(2), 185–215.

Umilta, M. A., Brochier, T., Spinks, R. L., & Lemon, R. N. (2007). Simultaneous recording of macaque premotor and primary motor cortex neuronal populations reveals different functional contributions to visuomotor grasp. *Journal of Neurophysiology, 98*(1), 488–501.

Ungerleider, L. G., & Mishkin, M. (1982). Two cortical visual systems. In D. J. Ingle, R. J. W. Mansfield, & M. A. Goodale (Eds.), *The analysis of visual behavior* (pp. 549–586). Cambridge, MA: MIT Press.

Watson, D., Clark, L. A., & Tellegen, A. (1988). Development and validation of brief measures of positive and negative affect: The PANAS scales. *Journal of Personality and Social Psychology, 54*, 1063–1070.

Watson, D., & Tellegen, A. (1985). Toward a consensual structure of mood. *Psychological Bulletin, 98*(2), 219–235.

Williams, Z. M., Bush, G., Rauch, S. L., Cosgrove, G. R., & Eskander, E. N. (2004). Human anterior cingulate neurons and the integration of monetary reward with motor responses. *Nature Neuroscience, 7*, 1370–1375.

Yamagata, T., Nakayama, Y., Tanji, J., & Hoshi, E. (2009). Processing of visual signals for direct specification of motor targets and for conceptual representation of action targets in the dorsal and ventral premotor cortex. *Journal of Neurophysiology, 102*, 3280–3294.

Zikopoulos, B., & Barbas, H. (2006). Prefrontal projections to the thalamic reticular nucleus form a unique circuit for attentional mechanisms. *Journal of Neuroscience, 26*(28), 7348–7361.

## Chapter 5. Structural Clues to Dorsal-Ventral Specialization

Aggleton, J. R., & Brown, M. W. (1999). Episodic memory, amnesia, and the hippocampal-anterior thalamic axis. *Behavioral and Brain Sciences, 22*, 425–489.

Arnsten, A. F. T., & Goldman-Rakic, P. S. (1984). Selective prefrontal cortical projections to the region of the locus coeruleus and raphe nuclei in the rhesus monkey. *Brain Research, 306*, 9–18.

Aston-Jones, G., & Cohen, J. D. (2005). Adaptive gain and the role of the locus coeruleus-norepinephrine system in optimal performance. *Journal of Comparative Neurology, 493*(1), 99–110.

Barbas, H., & Pandya, D. N. (1989). Architecture and intrinsic connections of the prefrontal cortex in the rhesus monkey. *Journal of Comparative Neurology, 286*, 353–375.

Derryberry, D., & Rothbart, M. K. (1997). Reactive and effortful processes in the organization of temperament. *Development and Psychopathology, 9*(4), 633–652.

Derryberry, D., & Tucker, D. M. (2006). Motivation, self-regulation, and self-organization. In D. J. Cohen & D. Cicchetti (Eds.), *Handbook of developmental psychopathology: Vol. 2. Developmental neuroscience* (pp. 502–532). New York: Wiley.

Eidelberg, D., & Galaburda, A. M. (1984). Inferior parietal lobule: Divergent architectonic asymmetries in the human brain. *Archives of Neurology, 41*, 843–852.

Foote, S. L., & Morrison, J. H. (1987). Extrathalamic modulation of cortical function. *Annual Review of Neuroscience, 10*, 67–95.

Galaburda, A. M. (1984). The anatomy of language: Lessons from comparative anatomy. In D. Caplan, A. R. Lacours, & A. Smith (Eds.), *Biological perspectives on language*. MIT Press, Cambridge, MA.

Goldberg, E., & Costa, L. D. (1981). Hemisphere differences in the acquisition and use of descriptive systems. *Brain and Language, 14*, 144–173.

Groenewegen, H. J., Berendse, H. W., Wolters, J. G., & Lohman, A. H. (1990). The anatomical relationship of the prefrontal cortex with the striatopallidal system, the thalamus and the amygdala: Evidence for a parallel organization. *Progress in Brain Research, 85*, 95–116; discussion 116–118.

Helmich, R. C., Baumer, T., Siebner, H. R., Bloem, B. R., & Munchau, A. (2005). Hemispheric asymmetry and somatotopy of afferent inhibition in healthy humans. *Experimental Brain Research, 167*(2), 211–219.

Herrick, C. J. (1948). *The brain of the tiger salamander*. Chicago: University of Chicago Press.

Johnson, L. R., Ledoux, J. E., & Doyere, V. (2009). Hebbian reverberations in emotional memory micro circuits. *Front Neurosci, 3*(2), 198–205.

Jones, E. G. (2007). *The thalamus* (Vol. 1). Cambridge, UK: Cambridge University Press.

Liotti, M., & Tucker, D. M. (1994). Emotion in asymmetric corticolimbic networks. In R. J. Davidson & K. Hugdahl (Eds.), *Human brain laterality* (pp. 389–424). New York: Oxford.

Massimini, M., Ferrarelli, F., Huber, R., Esser, S. K., Singh, H., & Tononi, G. (2005). Breakdown of cortical effective connectivity during sleep. *Science, 309*(5744), 2228–2232.

Mesulam, M. M., & Mufson, E. J. (1984). Neural inputs into the nucleus basalis of the substantia innominata (Ch4) in the rhesus monkey. *Brain, 107,* 253–274.

Mesulam, M. M., Mufson, E. J., Levey, A. I., & Wainer, B. H. (1983). Cholinergic innervation of cortex by the basal forebrain: Cytochemistry and cortical connections of the septal area, diagonal band nuclei, nucleus basalis (substantia innominata), and hypothalamus in the rhesus monkey. *Journal of Comparative Neurology, 214,* 170–197.

Mishkin, M. (1982). A memory system in the monkey. *Philosophical Transactions of the Royal Society of London: B. Biological Sciences, 298*(1089), 83–95.

Neafsey, E. J. (1990). Prefrontal cortical control of the autonomic nervous system: Anatomical and physiological observations. In H. B. M. Uylings, C. G. Van Eden, J. P. C. De Bruin, M. A. Corner, & M. G. P. Feenstra (Eds.), *The prefrontal cortex: Its structure, function and pathology* (pp. 147–166). New York: Elsevier.

Neafsey, E. J., Terreberry, R. R., Hurley, K. M., Ruit, K. G., & Frysztak, R. J. (1993). Anterior cingulate cortex in rodents: Connections, visceral control functions, and implications for emotion. In B. A. Vogt & M. Gabriel (Eds.), *Neurobiology of the cingulate cortex and limbic thalamus* (pp. 206–223). Boston: Birkhauser.

Papez, J. W. (1937). A proposed mechanism of emotion. *Archives of Neurology and Psychiatry, 38,* 725–743.

Schamahmann, J. D., & Pandya, D. N. (2006). *Fiber pathways of the brain.* New York: Oxford.

Semmes, J. (1968). Hemispheric specialization: A possible clue to mechanism. *Neuropsychologia, 6,* 11–26.

Shin, H. W., Sohn, Y. H., & Hallett, M. (2009). Hemispheric asymmetry of surround inhibition in the human motor system. *Clinical Neurophysiology, 120*(4), 816–819.

Tucker, D. M., Brown, M., Luu, P., & Holmes, M. D. (2007). Discharges in ventromedial frontal cortex during absence spells. *Epilepsy and Behavior, 11*(4), 546–557.

Tucker, D. M., Frishkoff, G. A., & Luu, P. (2008). Microgenesis of language: Vertical integration of neurolinguistic mechanisms across the neuraxis. In B. Stemmer & H. A. Whitaker (Eds.), *Handbook of the neuroscience of language.* New York: Oxford.

Tucker, D. M., & Holmes, M. D. (2010). Fractures and bindings of consciousness. *American Scientist, 99,* 32–39.

Tucker, D. M., & Luu, P. (2006). Adaptive binding. In H. Zimmer, A. Mecklinger, & U. Lindenberger (Eds.), *Binding in human memory: A neurocognitive approach* (pp. 85–108). New York: Oxford University Press.

Tucker, D. M., & Luu, P. (2007). Neurophysiology of motivated learning: Adaptive mechanisms of cognitive bias in depression. *Cognitive Therapy and Research, 31,* 189–209.

Tucker, D. M., Luu, P., & Derryberry, D. (2005). Love hurts: The evolution of empathic concern through the encephalization of nociceptive capacity. *Development and Psychopathology, 17*(3), 699–713.

Tucker, D. M., Luu, P., & Poulsen, C. (2009). Neural mechanisms of recursive processing in cognitive and linguistic complexity. In T. Givon & M. Shibatani (Eds.), *Syntactic complexity: Diachrony, acquisition, neuro-cognition, evolution*. Amsterdam: John Benjamins.

Tucker, D. M., & Moller, L. (2007). The metamorphosis: Individuation of the adolescent brain. In D. Romer & E. F. Walker (Eds.), *Adolescent psychopathology and the developing brain: Integrating brain and prevention science*. New York: Oxford.

Tucker, D. M., & Williamson, P. A. (1984). Asymmetric neural control systems in human self-regulation. *Psychol Rev, 91*(2), 185–215.

Vogt, B. A. (1993). Structural organization of cingulate cortex: Areas, neurons, an som atodendritic transmitter receptors. In B. A. Vogt & M. Gabriel (Eds.), *Neurobiology of cingulate cortex and limbic thalamus* (pp. 19–70). Boston: Birkhäuser.

Vogt, B. A., Vogt, L., & Laureys, S. (2006). Cytology and functionally correlated circuits of human posterior cingulate areas. *Neuroimage, 29*(2), 452–466.

Williams, S., & Goldman-Rakic, P. (1993). Characterization of the dopaminergic innervation of the primate frontal cortex using a dopamine-specific antibody. *Cerebral Cortex, 3*(3), 199.

Zikopoulos, B., & Barbas, H. (2006). Prefrontal projections to the thalamic reticular nucleus form a unique circuit for attentional mechanisms. *Journal of Neuroscience, 26*(28), 7348–7361.

# Chapter 6. The Evolved Structure of Mammalian Memory

Abbie, A. A. (1940). Cortical lamination in the monotremata. *Journal of Comparative Neurology, 72,* 428–467.

Abbie, A. A. (1942). Cortical lamination in the polyprotodont marsupial, Perameles nasuta. *Journal of Comparative Neurology, 76,* 509–536.

Aboitiz, F., Morales, D., & Montiel, J. (2003). The evolutionary origin of the mammalian isocortex: towards an integrated developmental and functional approach. *Behavioral and Brain Sciences, 26,* 535–586.

Arnsten, A. F. T., & Goldman-Rakic, P. S. (1984). Selective prefrontal cortical projections to the region of the locus coeruleus and raphe nuclei in the rhesus monkey. *Brain Research, 306,* 9–18.

Bagley, C., & Richter, C. P. (1924). Electrically excitable region of the forebrain of the alligator. *Archives of Neurology and Psychiatry, 11,* 257–263.

Barbas, H., & Pandya, D. N. (1986). Architecture and frontal cortical connections of the premotor cortex (area 6) in the rhesus monkey. *Journal of Comparative Neurology, 256,* 211–228.

Barbas, H., & Pandya, D. N. (1989). Architecture and intrinsic connections of the prefrontal cortex in the rhesus monkey. *Journal of Comparative Neurology, 286,* 353–375.

Brown, J. (2011). Theoretic note: The relation of embryology to linguistic and cognitive process. *Journal of Psycholinguistic Research, 40,* 189–194.

Butler, A. B., & Hodos, W. (2005). *Comparative vertebrate neuroanatomy: evolution and adaptation* (Second ed.). Hoboken: John Wiley & Sons.

Butler, A. B., & Molnar, Z. (2002). Development and evolution of the collopallium in amniotes: A new hypothesis of field homology. *Brain Research Bulletin, 57*, 475–479.

Dart, R. A. (1934). The dual structure of the neopallium: Its history and significance. *Journal of Anatomy, 69*, 3–19.

De Carlos, J. A., Lopez-Mascaraque, L., & Valverde, F. (1996). Dynamics of cell migration from the lateral ganglionic eminence in the rat. *Journal of Neuroscience, 16*, 6146–6156.

Denny-Brown, D. (1966). *The cerebral control of movement.* Springfield, IL: Charles C. Thomas.

Foote, S. L., & Morrison, J. H. (1987). Extrathalamic modulation of cortical function. *Annual Review of Neuroscience, 10*, 67–95.

Goldberg, G. (1985). Supplementary motor area structure and function: Review and hypotheses. *Behavioral and Brain Sciences, 8*, 567–615.

Gould, S. J. (1977). *Ontogeny and phylogeny.* Cambridge, MA: Harvard University Press.

Gurney, K., Prescott, T. J., & Redgrave, P. (2001). A computational model of action selection in the basal ganglia: II. Analysis and simulation of behaviour. *Biological Cybernetics, 84*(6), 411–423.

Hagmann, P., Cammoun, L., Gigandet, X., Meuli, R., Honey, C. J., Wedeen, V. J., et al. (2008). Mapping the structural core of human cerebral cortex. *PLoS Biology, 6*(7), e159.

Herrick, C. J. (1948). *The brain of the tiger salamander.* Chicago: University of Chicago Press.

Honey, C. J., Sporns, O., Cammoun, L., Gigandet, X., Thiran, J. P., Meuli, R., et al. (2009). Predicting human resting-state functional connectivity from structural connectivity. *Proceedings of the National Academy of Sciences of the United States of America, 106*(6), 2035–2040.

Isaacson, R. L. (1982). *The limbic system* (2nd ed.). New York: Plenum.

Jones, E. G. (2007a). *The thalamus* (Vol. 1). Cambridge, UK: Cambridge University Press.

Jones, E. G. (2007b). *The thalamus* (Vol. 2). Cambridge, UK: Cambridge University Press.

Jones, E. G. (2009). Synchrony in the interconnected circuitry of the thalamus and cerebral cortex. *Annals of the New York Academy of Sciences, 1157*, 10–23.

Kaas, J. H. (1988). Development of cortical sensory maps.

Kaas, J. H. (1989). Why does the brain have so many visual areas? *Journal of Cognitive Neuroscience, 1*(2), 121–135.

Kubota, Y., & Gabriel, M. (1995). Studies of the limbic comparator: Limbic circuit training-induced unit activity and avoidance behavior in rabbits with anterior dorsal thalamic lesions. *Behavioral Neuroscience, 109*(2), 258–277.

Luu, P., & Tucker, D. M. (2003a). Self-regulation and the executive functions: Electrophysiological clues. In A. Zani & A. M. Preverbio (Eds.), *The cognitive electrophysiology of mind and brain* (pp. 199–223). San Diego: Academic Press.

Luu, P., & Tucker, D. M. (2003b). Self-regulation by the medial frontal cortex: Limbic representation of motive set-points. In M. Beauregard (Ed.), *Consciousness, emotional self-regulation and the brain.* Amsterdam: John Benjamins.

MacLean, P. D. (1949). Psychosomatic disease and the "visceral brain." *Psychosomatic Medicine, 11,* 338–353.

MacLean, P. D. (1986). Culminating developments in the evolution of the limbic system: The thalamocingulate division. In D. K. Benjamin & K. E. Livingston (Eds.), *The limbic system: Functional organization and clinical disorders* (pp. 1–28). New York: Raven Press.

Marin, O., & Rubenstein, J. L. (2001). A long, remarkable journey: Tangential migration in the telencephalon. *Nature Reviews Neuroscience, 2*(11), 780–790.

Marin-Padilla, M. (1998). Cajal-Retzius cells and the development of the neocortex. *Trends in Neurosciences, 21*(2), 64–71.

Mesulam, M. M. (2000). Behavioral neuroanatomy: Large-scale networks, association, cortex, frontal syndromes, the limbic system, and hemispheric specializations. In M. M. Mesulam (Ed.), *Principles of behavioral and cognitive neurology* (pp. 1–120). Oxford: Oxford University Press.

Mesulam, M. M., & Mufson, E. J. (1984). Neural inputs into the nucleus basalis of the substantia innominata (Ch4) in the rhesus monkey. *Brain, 107,* 253–274.

Mesulam, M. M., Mufson, E. J., Levey, A. I., & Wainer, B. H. (1983). Cholinergic innervation of cortex by the basal forebrain: Cytochemistry and cortical connections of the septal area, diagonal band nuclei, nucleus basalis (substantia innominata), and hypothalamus in the rhesus monkey. *Journal of Comparative Neurology, 214,* 170–197.

Morrison, J. H., & Foote, S. L. (1986). Noradrenergic and serotonergic innervation of cortical, thalamic, and tectal visual structures in old and new world monkey. *Journal of Comparative Neurology, 243,* 117–138.

Nauta, W. J. H., & Karten, H. J. (1970). A general profile of the vertebrate brain, with sidelights on the ancestry of cerebral cortex. In G. C. Quarton, T. Melnechuck, & G. Adelman (Eds.), *The neurosciences* (pp. 7–26). New York: Rockefeller University Press.

Nieuwenhuys, R., Ten Donkelaar, H. J., & Nicholson, C. (1998). *The central nervous system of vertebrates* (Vols. 1–3). Berlin: Springer Verlag.

Northcutt, G. R., & Kaas, J. H. (1995). The emergence and evolution of the mammalian neocortex. *Trends in Neurosciences, 18,* 373–379.

Nowakowski, R. S., & Rakic, P. (1981). The site of origin and route and rate of migration of neurons to the hippocampal region of the rhesus monkey. *Journal of Comparative Neurology, 196,* 129–154.

Pandya, D. N., & Barnes, C. L. (1987). Architecture and connections of the frontal lobe. In E. Perecman (Ed.), *The frontal lobes revisited* (pp. 41–72). New York: IRBN.

Pandya, D. N., & Seltzer, B. (1982). Association areas of the cerebral cortex. *Trends in Neurosciences, 5,* 386–390.

Pandya, D. N., & Yeterian, E. H. (1984). Proposed neural circuitry for spatial memory in the primate brain. *Neuropsychologia, 22*(2), 109–122.

Pandya, D. N., & Yeterian, E. H. (1985). Architecture and connections of cortical association areas. In A. Peters & E. G. Jones (Eds.), *Cerebral cortex: Volume 4. Association and auditory cortices* (pp. 3–61). New York: Plenum Press.

Pribram, K. H., & MacLean, P. D. (1953). Neuronographic analysis of medial and basal cerebral cortex: II. Monkey. *Journal of Neurophysiology, 16,* 324–340.

Rakic, P. (2009a). Evolution of the neocortex: A perspective from developmental biology. *Nature Reviews: Neuroscience, 10*(10), 724–735.

Rakic, P., Ayoub, A. E., Breunig, J. J., & Dominguez, M. H. (2009b). Decision by division: making cortical maps. *Trends in Neurosciences, 32*(5), 291–301.

Redgrave, P., Prescott, T. J., & Gurney, K. (1999). The basal ganglia: A vertebrate solution to the selection problem? *Neuroscience, 89*(4), 1009–1023.

Sanides, F. (1970). Functional architecture of motor and sensory cortices in primates in the light of a new concept of neocortex evolution. In C. R. Noback & W. Montagna (Eds.), *The primate brain: Advances in primatology* (Vol. 1, pp. 137–208). New York: Appleton-Century-Crofts.

Schafe, G. E., Nader, K., Blair, H. T., & LeDoux, J. E. (2001). Memory consolidation of Pavlovian fear conditioning: A cellular and molecular perspective. *Trends in Neurosciences, 24*(9), 540–546.

Semmes, J. (1968). Hemispheric specialization: A possible clue to mechanism. *Neuropsychologia, 6,* 11–26.

Sherman, S. M., & Guillery, R. W. (2006). *Exploring the thalamus and its role in cortical function.* Cambridge, MA: MIT Press.

Shipp, S. (2005). The importance of being agranular: A comparative account of visual and motor cortex. *Philosophical Transactions of the Royal Society of London: B. Biological Sciences, 360*(1456), 797–814.

Shipp, S. (2007). Structure and function of the cerebral cortex. *Current Biology, 17*(12), R443–449.

Singer, W. (1987). Activity-dependent self-organization of synaptic connections as a substrate of learning. In J. P. Changeux & M. Konishi (Eds.), *The neural and molecular basis of learning* (pp. 301–336). New York: Wiley.

Sporns, O., & Honey, C. J. (2006). Small worlds inside big brains. *Proceedings of the National Academy of Sciences of the United States of America, 103*(51), 19219–19220.

Striedter, G. F., Marchant, T. A., & Beydler, S. (1998). The "neostriatum" develops as part of the lateral pallium in birds. *Journal of Neuroscience, 18,* 5839–5849.

Trevarthen, C. (1985). Neuroembryology and the development of perceptual mechanisms. In F. Falkner & J. M. Tanner (Eds.), *Human growth, a comprehensive treatise: Volume 2. Postnatal growth, neurobiology.* New York: Plenum.

Tucker, D. M. (1992). Developing emotions and cortical networks. In M. Gunnar & C. Nelson (Eds.), *Developmental behavioral neuroscience: Minnesota symposium on child psychology* (Vol. 24, pp. 75–127). Hillsdale: Erlbaum.

Tucker, D. M. (2001). Motivated Anatomy: A core-and-shell model of corticolimbic architecture. In G. Gainotti (Ed.), *Handbook of neuropsychology, 2nd edition: Volume 5. Emotional behavior and its disorders* (pp. 125–160). Amsterdam: Elsevier.

Tucker, D. M. (2007). *Mind from body: Experience from neural structure.* New York: Oxford University Press.

Tucker, D. M., Brown, M., Luu, P., & Holmes, M. D. (2007). Discharges in ventromedial frontal cortex during absence spells. *Epilepsy and Behavior, 11,* 546–557.

Tucker, D. M., Frishkoff, G. A., & Luu, P. (2008). Microgenesis of language: Vertical integration of neurolinguistic mechanisms across the neuraxis. In B. Stemmer & H. A. Whitaker (Eds.), *Handbook of the neuroscience of language*. New York: Oxford.

Tucker, D. M., & Holmes, M. D. (2010). Fractures and bindings of consciousness. *American Scientist, 99*, 32–39.

Tucker, D. M., & Luu, P. (2006). Adaptive binding. In H. Zimmer, A. Mecklinger & U. Lindenberger (Eds.), *Binding in human memory: A neurocognitive approach* (pp. 85–108). New York: Oxford University Press.

Tucker, D. M., & Luu, P. (2007). Neurophysiology of motivated learning: Adaptive mechanisms of cognitive bias in depression. *Cognitive Therapy and Research, 31*, 189–209.

Tucker, D. M., Luu, P., & Poulsen, C. (2009). Neural mechanisms of recursive processing in cognitive and linguistic complexity. In T. Givon & M. Shibatani (Eds.), *Syntactic complexity: Diachrony, acquisition, neuro-cognition, evolution* (p. 461). Amsterdam: Benjamins.

Tucker, D. M., & Williamson, P. A. (1984). Asymmetric neural control systems in human self-regulation. *Psychological Review, 91*(2), 185–215.

Yakovlev, P. I. (1948). Motility, behavior and the brain. *Journal of Nervous and Mental Disease, 107*, 313–335.

Yingling, C. D., & Skinner, J. E. (1976). Selective regulation of thalamic sensory relay nuclei by nucleus reticularis thalami. *Electroencephalography and Clinical Neurophysiology, 41*, 476–482.

Zikopoulos, B., & Barbas, H. (2006). Prefrontal projections to the thalamic reticular nucleus form a unique circuit for attentional mechanisms. *Journal of Neuroscience, 26*(28), 7348–7361.

## Chapter 7. Self-Organizing Ontogenesis on the Phyletic Frame

Aboitiz, F., Morales, D., & Montiel, J. (2003). The evolutionary origin of the mammalian isocortex: towards an integrated developmental and functional approach. *Behavioral and Brain Sciences, 26*, 535–586.

Allman, J. M., Watson, K. K., Tetreault, N. A., & Hakeem, A. Y. (2005). Intuition and autism: A possible role for Von Economo neurons. *Trends in Cognitive Sciences, 9*(8), 367–373.

Barbas, H. (1995). Anatomic basis of cognitive-emotional interactions in the primate prefrontal cortex. *Neuroscience and Biobehavioral Reviews, 19*(3), 499–510.

Beeman, M. (1993). Semantic processing in the right hemisphere may contribute to drawing inferences from discourse. *Brain and Language, 44*(1), 80–120.

Birch, H. G., Belmont, I., & Karp, E. (1967). Delayed information processing and extinction following cerebral damage. *Brain, 90*, 113–130.

Blumer, D., & Benson, D. F. (1975). Personality changes with frontal and temporal lobe lesions. In D. F. Benson & D. Blumer (Eds.), *Psychiatric aspects of neurologic disease* (pp. 151–170.). New York: Gruen and Stratton.

Bogen, J. E., & Bogen, G. M. (1969). The other side of the brain: III. The corpus callosum and creativity. *Bulletin of the Los Angeles Neurological Society, 34,* 191–220.

Borod, J. C. (1992). Interhemispheric and intrahemispheric control of emotion: A focus on unilateral brain damage. *Journal of Consulting and Clinical Psychology, 60,* 339–348.

Borod, J. C. (2000). *The neuropsychology of emotion.* New York: Oxford.

Bowlby, J. (1997). *Attachment.* London: Pimlico.

Brown, J. W. (1977). *Mind, brain, and consciousness: The neuropsychology of cognition.* New York: Academic Press.

Brown, J. W. (1987). The microstructure of action. In E. Perceman (Ed.), *The frontal lobes revisited* (pp. 251–272). New York: IRBN.

Brown, J. W. (1988). *The life of the mind: Selected papers.* Hillsdale, NJ: Erlbaum.

Buzsaki, G. (1996). The hippocampal-neocortical dialogue. *Cerebral Cortex, 6*(81–92).

Buzsaki, G. (2006). *Rhythms of the brain.* New York: Oxford.

Buzsaki, G., Kaila, K., & Raichle, M. (2007). Inhibition and brain work. *Neuron, 56*(5), 771–783.

Cicchetti, D., & Tucker, D. M. (1994). Development of self-regulatory structures of the mind. *Development and Psychopathology, 6,* 533–549.

Deacon, T. W. (1997). *The symbolic species: The co-evolution of language and the brain.* New York: W. W. Norton.

Denny-Brown, D. (1966). *The cerebral control of movement.* Springfield, IL: Charles C. Thomas.

Dosenbach, N. U., Nardos, B., Cohen, A. L., Fair, D. A., Power, J. D., Church, J. A., et al. (2010). Prediction of individual brain maturity using fMRI. *Science, 329*(5997), 1358–1361.

Fleming, S. M., Weil, R. S., Nagy, Z., Dolan, R. J., & Rees, G. (2010). Relating introspective accuracy to individual differences in brain structure. *Science, 329*(5998), 1541–1543.

Freud, S. (1895). Project for a scientific psychology. In J. Strachey (Ed.), *The standard edition of the complete psychological works of Sigmund Freud* (Vol. 1, pp. 295–344). London: Hogarth Press.

Freud, S. (1953). *The interpretation of dreams.* London: Hogarth Press. (First German Edition, 1900).

Galaburda, A. M. (1984). The anatomy of language: Lessons from comparative anatomy. In D. Caplan, A. R. Lacours, & A. Smith (Eds.), *Biological perspectives on language.* Cambridge, MA: MIT Press.

Galaburda, A. M., & Geschwind, N. (1981). Anatomical asymmetries in the adult and developing brain and their implications for function. *Advances in Pediatrics, 28,* 271–292.

Galaburda, A. M., LeMay, M., Kemper, T. L., & Geschwind, N. (1978). Right-left asymmetrics in the brain. *Science, 199*(4331), 852–856.

Geschwind, N., & Levitsky, W. (1968). Human brain: Left-right asymmetries in temporal speech regions. *Science, 161,* 186–187.

Gilbert, S. J., Spengler, S., Simons, J. S., Steele, J. D., Lawrie, S. M., Frith, C. D., et al. (2006). Functional specialization within rostral prefrontal cortex (area 10): A meta-analysis. *Journal of Cognitive Neuroscience, 18*(6), 932–948.

Goldberg, E., & Costa, L. D. (1981). Hemisphere differences in the acquisition and use of descriptive systems. *Brain and Language, 14,* 144–173.

Goldberg, G. (1985). Supplementary motor area structure and function: Review and hypotheses. *Behavioral and Brain Sciences, 8,* 567–616.

Goodglass, H. (1993). *Understanding aphasia.* New York: Academic Press.

Greicius, M., Supekar, K., Menon, V., & Dougherty, R. (2008). Resting-state functional connectivity reflects structural connectivity in the default mode network. *Cerebral Cortex, 19,* 72–78.

Grossberg, S. (1980). How does a brain build a cognitive code? *Psychological Review, 87*(1), 1–51.

Grossberg, S., & Versace, M. (2008). Spikes, synchrony, and attentive learning by laminar thalamocortical circuits. *Brain Research, 1218,* 278–312.

Hagmann, P., Cammoun, L., Gigandet, X., Meuli, R., Honey, C. J., Wedeen, V. J., et al. (2008). Mapping the structural core of human cerebral cortex. *PLoS Biology, 6*(7), e159.

Harvey, O. J., Hunt, D. E., & Schroder, H. M. (1961). *Conceptual systems and personality organization.* New York: Wiley.

Hecaen, H. (1962). Clinical symptomatology in right and left hemisphere lesions. In V. B. Mountcastle (Ed.) *Interhemispheric relations and cerebral dominance.* Baltimore, MD: Johns Hopkins Press.

Heilman, K. M. (1979). Neglect and related disorders. In K. M. Heilman & E. Valenstein (Eds.), *Clinical neuropsychology.* New York: Oxford University Press.

Heilman, K. M., & Van Den Able, T. (1979). Right hemisphere dominance for mediating cerebral activation. *Neuropsychologia, 17,* 315–321.

Honey, C. J., Sporns, O., Cammoun, L., Gigandet, X., Thiran, J. P., Meuli, R., et al. (2009). Predicting human resting-state functional connectivity from structural connectivity. *Proceedings of the National Academy of Sciences of the United States of America, 106*(6), 2035–2040.

Jackson, J. H. (1879). On affections of speech from diseases of the brain. *Brain, 2,* 203–222.

Jeannerod, M. (1994). The representing brain: Neural correlates of motor intention and imagery. *Behavioral and Brain Sciences, 17,* 187–245.

Jones, E. G. (2007a). *The thalamus* (Vol. 1). Cambridge, UK: Cambridge University Press.

Jones, E. G. (2007b). *The thalamus* (Vol. 2). Cambridge, UK: Cambridge University Press.

Jones, E. G. (2009). Synchrony in the interconnected circuitry of the thalamus and cerebral cortex. *Annals of the New York Academy of Sciences, 1157,* 10–23.

Kohonen, T., & Honkela, T. (2007). Kohonen network. *Scholarpedia.*

Kohut, H. (1978). *The search for the self.* New York: International Universities Press.

Lacruz, M., Garcia Seoane, J., Valentin, A., Selway, R., & Alarcon, G. (2007). Frontal and temporal functional connections of the living human brain. *European Journal of Neuroscience, 26*(5), 1357–1370.

Liotti, M., & Tucker, D. M. (1994). Emotion in asymmetric corticolimbic networks. In R. J. Davidson & K. Hugdahl (Eds.), *Human brain laterality* (pp. 389–424). New York: Oxford.

Luu, P., Tucker, D. M., & Makeig, S. (2004). Frontal midline theta and the error-related negativity: Neurophysiological mechanisms of action regulation. *Clinical Neurophysiology, 115*(8), 1821–1835.

MacNeilage, P. (1986). *On the evolution of handedness.* Paper presented at the symposium, "The Dual Brain: Unified functioning and specialization of the cerebral hemispheres." Stockholm, May.

Mahler, M. S. (1968). *On human symbiosis and the vicissitudes of individuation.* New York: International Universities Press.

Makeig, S., Luu, P., Briggman, K., Visser, E., Sejnowski, T. J., & Tucker, D. M. (in preparation). Error-related dynamics in distributed brain networks. Manuscript in preparation.

Marin-Padilla, M. (1998). Cajal-Retzius cells and the development of the neocortex. *Trends in Neurosciences, 21*(2), 64–71.

Maslow, A. (1968). *Toward a psychology of being.* New York: Wiley.

Nimchinsky, E. A., Gilissen, E., Allman, J. M., Perl, D. P., Erwin, J. M., & Hof, P. R. (1999). A neuronal morphologic type unique to humans and great apes. *Proceedings of the National Academy of Sciences of the United States of America, 96*(9), 5268–5273.

Piaget, J. (1936/1992). *The origins of intelligence in children.* New York: International Universities Press.

Pisella, L., Sergio, L., Blangero, A., Torchin, H., Vighetto, A., & Rossetti, Y. (2009). Optic ataxia and the function of the dorsal stream: contributions to perception and action. *Neuropsychologia, 47*, 3033–3044.

Posner, M. I., Walker, J. A., Friedrich, F. A., & Rafal, R. D. (1987). How do the parietal lobes direct covert attention? *Neuropsychologia, 25*(1A), 135–145.

Pribram, K. H. (1971). *Languages of the brain: Experimental paradoxes and principles in neuropsychology.* Monterey: Brooks/Cole.

Pribram, K. H., & McGuinness, D. (1975). Arousal, activation, and effort in the control of attention. *Psychological Review, 82*, 6–149.

Protzner, A., & McIntosh, A. (2008). Modulation of ventral prefrontal cortex functional connections reflects the interplay of cognitive processes and stimulus characteristics. *Cerebral Cortex, 19*, 1042–1054.

Rakic, P. (2009). Evolution of the neocortex: A perspective from developmental biology. *Nature Reviews: Neuroscience, 10*(10), 724–735.

Redgrave, P., Prescott, T. J., & Gurney, K. (1999). The basal ganglia: A vertebrate solution to the selection problem? *Neuroscience, 89*(4), 1009–1023.

Rizzolatti, G., Fadiga, L., Fogassi, L., & Gallese, V. (1999). Resonance behaviors and mirror neurons. *Archives Italiennes de Biologie, 137*(2–3), 85–100.

Robertson, L. C., & Lamb, M. R. (1991). Neuropsychological contributions to theories of part/whole organization. *Cognitive Psychology, 23*, 299–330.

Schmahmann, J., & Pandya, D. N. (2006). *Fiber pathways of the brain.* New York: Oxford.

Semendeferi, K., Armstrong, E., Schleicher, A., Zilles, K., & Van Hoesen, G. W. (2001). Prefrontal cortex in humans and apes: A comparative study of area 10. *American Journal of Physical Anthropology, 114*(3), 224–241.

Semmes, J. (1968). Hemispheric specialization: A possible clue to mechanism. *Neuropsychologia, 6*, 11–26.

Shapiro, D. (1965). *Neurotic styles*. New York: Basic Books.
Shin, H. W., Sohn, Y. H., & Hallett, M. (2009). Hemispheric asymmetry of surround inhibition in the human motor system. *Clinical Neurophysiology, 120*(4), 816–819.
Thatcher, R. W., Krause, P. J., & Rhybyk, M. (1986). Cortico-cortical associations and EEG coherence: A two-compartmental model. *Electroencephalography and Clinical Neurophysiology, 64*, 123–143.
Tucker, D. M. (1981). Lateral brain function, emotion, and conceptualization. *Psychological Bulletin, 89*(1), 19–46.
Tucker, D. M. (1989). Neural and psychological maturation in a social context. In D. Cicchetti (Ed.), *The emergence of a discipline: Rochester symposium on developmental psychopathology* (pp. 69–88). Hillsdale, NJ: Erlbaum.
Tucker, D. M. (1992). Developing emotions and cortical networks. In M. Gunnar & C. Nelson (Eds.), *Developmental behavioral neuroscience: Minnesota symposium on child psychology* (Vol. 24, pp. 75–127). Hillsdale: Erlbaum.
Tucker, D. M., Brown, M., Luu, P., & Holmes, M. D. (2007). Discharges in ventromedial frontal cortex during absence spells. *Epilepsy and Behavior, 11*(4), 546–557.
Tucker, D. M., & Derryberry, D. (1992). Motivated attention: Anxiety and the frontal executive functions. *Neuropsychiatry, Neuropsychology, and Behavioral Neurology, 5*, 233–252.
Tucker, D. M., Frishkoff, G. A., & Luu, P. (2008). Microgenesis of language: Vertical integration of neurolinguistic mechanisms across the neuraxis. In B. Stemmer & H. A. Whitaker (Eds.), *Handbook of the neuroscience of language*. New York: Oxford.
Tucker, D. M., & Holmes, M. D. (2010). Fractures and bindings of consciousness. *American Scientist, 99*, 32–39.
Tucker, D. M., & Moller, L. (2007). The metamorphosis: Individuation of the adolescent brain. In D. Romer & E. F. Walker (Eds.), *Adolescent psychopathology and the developing brain: Integrating brain and prevention science*. New York: Oxford.
Tucker, D. M., Roth, D. L., & Bair, T. B. (1986). Functional connections among cortical regions: Topography of EEG coherence. *Electroencephalography and Clinical Neurophysiology Supplement, 63*(3), 242–250.
Tucker, D. M., & Williamson, P. A. (1984). Asymmetric neural control systems in human self-regulation. *Psychological Review, 91*(2), 185–215.
Tyler, S. K., & Tucker, D. M. (1982). Anxiety and perceptual structure: Individual differences in neuropsychological function. *Journal of Abnormal Psychology, 91*(3), 210–220.
Wedeen, V. J., Wang, R. P., Schmahmann, J. D., Benner, T., Tseng, W. Y., Dai, G., et al. (2008). Diffusion spectrum magnetic resonance imaging (DSI) tractography of crossing fibers. *Neuroimage, 41*(4), 1267–1277.
Werner, H. (1957). *The comparative psychology of mental development*. New York: Harper.
Wynn, T., & Coolidge, F. (2008). Did a small but significant enhancement in working memory capacity power the evolution of modern thinking? In P. Mellars (Ed.), *Rethinking the Human Revolution* (pp. 79–90). Cambridge: McDonald Institute for Archaeological Research.
Zikopoulos, B., & Barbas, H. (2006). Prefrontal projections to the thalamic reticular nucleus form a unique circuit for attentional mechanisms. *Journal of Neuroscience, 26*(28), 7348–7361.

# Index

absence discharges
 spike-wave, 81
absence seizures, 81
 frontopolar control, 83
 frontothalamic circuit, 82
absence spells, 68, 82
academic psychology, 7
accommodation
 artus, 210
 cognition, 209
 development, 8
 mammalian self-regulation, 214
 neurodevelopment algorithm, 8
 Piagetian concept, 210
 redundancy bias, 216
acetylcholine
 activation control, 212
 neuromodulator, 147, 182–83, 188
 neurotransmitter of basal ganglia, 52
 ventral corticolimbic neuromodulation, 148–49
action
 elementary organization of, 90–91
 frames for, 72–74
 frames for meaning, 74–75
 internal vs. external generation, 98–101
 projectional and reactive vectors of, 65–66
 receptive brain and, 70–78
 regulating, 59–60
 thalamic regulation, 80–82
 vectoral cybernetics, 34–38
action control
 constraints by reality, 101–2
action regulation
 child's language, 136
 complementarity of pragmatic and semantic modes, 75–78
 control and representation, 88
 corticolimbic analysis, 119
 corticolimbic systems of, 58
 cybernetics, 63–64
 dorsal and ventral corticolimbic systems, 116
 evaluation, 109
 extension of, 84
 language, 135
 limbifugal dominance, 185
 model, 36
 motive cybernetics and executive functions, 68–70
 neuropsychological theory, 89
 posterior brain, 70–72
 primitive cybernetics, 86
 requirements, 60
action-relevant
 information, 72
activation
 neurocybernetics, 114–16
 psychological theory, 105–7
 self-regulation, 86–88
 self-regulation of, and arousal, 15–21
adaptive learning
 mammals, 5
adaptive resonance
 dorsal and ventral divisions, 181
 theory, 41
adaptive resonance model
 cortical learning, 202
affect
 psychological theory, 105–7
agreeableness
 personality, 107
alien hand sign, 124
alligator
 electrical stimulation of brain, 164
allocentric orientations, 73
allocentric reference frame
 meaning, 75
allocentric shift
 cognition, 119
allostasis
 concept, 70
allostatic
 self-regulation, 49

American Psychiatric Association Diagnostic and Statistical Manual (DSM), 110
amnesia
 cognition, 196
 retrograde, 25
amphetamine
 paranoia, 128
amphibians, 160
 evolution progression, 170
 pallium, 159
amygdala
 feedback guidance, 12
 learning, 10
 memory circuit, 55
amygdala-centric system
 avoidance learning, 12
anatomical asymmetries
 brain, 199
animal behavior
 learning, 5
animal learning
 control and communication, 6–8
 expectancy confirmation vs. discrepancy, 41
 signal for new learning, 13
 ventral mechanisms, 102
anlagen
 pallial, of dorsal hemisphere, 183–86
 subpallial and pallial, 154–55
 subpallial, of ventral hemisphere, 186–89
 vertebrate neuraxis, 171
anterior cingulate cortex (ACC)
 emotional evaluation, 80
 executive control, 92–94, 94–96
 functions, 75
 lesions and learning, 103
 regulating visceromotor functions, 49
anterior dorsal ventricular ridge (ADVR), 168
 external striatum, 156
anterior insular cortex (AIC)
 executive control, 92–94
anterior networks
 dorsal and ventral, 113
anterior ventral (AV) nuclei
 learning, 10
anterior-posterior complementarity, 38
anxiety, 108, 127
 experience, 106
 self-regulation, 111
apes, 61
approach discrimination learning, 12–13
approach learning
 cognition, 90
apraxia
 ideational and ideomotor, 74
archicortex
 dorsal neocortex, 166
 dorsal system, 140
 limbic base of, 175
 neocortex evolution, 152
archicortical base
 learning and consolidation, 43

architecture
 corticolimbic pathways, 34
 fused or hybrid, 186
 parallel, for attention and intention, 78–83
arcuate premotor area (APA), 93
 action coordination, 69
 cytoarchitectonics, 66
 motor control, 101
 ventrolateral front lobe, 65
arousal, 106
 attention and cognition, 16
 concept, 15, 147
 drug manipulations, 127–28
 neural mechanism, 109
 neurocybernetics, 114–16
 psychological theory, 105–7
 self-regulation, 86–88
 self-regulation through activation and, 15–21
 valenced, 109
arousal control
 pallidal-thalamic projections, 53
articulation
 language, 136
 skills in social context, 137
 speech and language, 133–37
artificial neural network
 simulations, 7
artus
 accommodation, 210
 anxiety, 119, 129
 cognition, 206
 constraint, viii
 control mode, 129
 external control, 214
 mechanisms, ix
 selective attention, 133
 ventral reactive mode, 91
assimilation
 cognition, 209
 development, 8
 mammalian self-regulation, 214
 neurodevelopment algorithm, 8
 Piagetian concept, 210
association cortex
 concept components, 134
associative attention
 cognitive control, 94–96
asymmetric structure
 memory, 38–42
attachment
 object relation, 111
attention
 cognitive control and, 94–96
 differentiated, vs. integrative intention, 82–83
 frontal networks, 78
 hemispheric asymmetries, 119
 language and focused, 135
 parallel architecture for, and intention, 78–83
 regulation in mammalian brain, 216

shaping connections through reactive, 132–33
attentional control, 96
  consciousness, 204
  dual model, 96–97
  goal- vs. stimulus-driven, 95
  learning and memory, 96
attentional mode
  self-regulation, 126
auditory cortex
  speech comprehension, 134
augmenting
  limbic regions, 40
autism, 124
  interpersonal attachment, 122
  intersubjectivity, 123
aversive/discrepant expectancies
  primitive cognition, 85–86
avoidance discrimination learning, 12–13

back projections
  term, 33
back-projections, 35
basal ganglia, 51
  action readiness and inhibitory specification, 52–53
  circuitry, 50
  embryonic migration, 171
  loops, 50
  motivational control, 57
  neocortex origins, 166–67
  subcortical system, 60
Big Five
  dimensions of personality, 107
biology
  development, v
bipolar disorder, 110
blocking
  active cognition of mammals, 6
  learning and new cues, 6
brain
  activation mechanisms, 16
  connectivity, 191–93
  controlled experiments, vi
  evolutionary analysis, 117
  evolutionary order, 54
  hemispheric specialization, 198–201
  learning, ix
  mind, v
  neurodevelopmental theory, 193–95
  receptive, and action, 70–78
brainstem, 16
  controls, 49
  dopamine and acetylcholine, 148–49
  neuromodulators, 147–49, 181–83
  structures, 42
  subcortical system, 60
  telencephalon, 8
brainstem activation
  neurotransmitter release, 15
brainstem neuromodulator systems
  mechanism, 16–18

brainstem neurotransmitter systems
  sleep stages, 50
Broca's
  speech, 135
Broca's aphasia, 196

Cajal-Retzius (CR) cells, 158, 170, 177, 181
calbindin matrix projections
  cortical differentiation, 180
calm-anxiety dimension
  tonic activation, 106
calretinin matrix cells
  embryogenesis, 179
catastrophic interference
  learning, 26
catecholamine hypothesis
  affective disorders, 19
categorical perception
  speech, 135
cats
  orbital frontal lobe, 81
caudodorsomedial region
  human brain, 175
cell assembly
  Hebb's model, 4, 5
child development
  hemispheric patterns, 201
  nested frames of evolution and, 207–8
childhood
  self-awareness and consciousness, 217
cholinergic nuclei
  controlling input, 47–48
cingulate motor area (CMA)
  behavior adjustments, 103
cingulothalamic networks
  learning, 10, 11
classically conditioned
  memory consolidation, 40
cognition
  adaptive control of, 49
  de novo operation, 208
  developing language, 133–37
  dorsal capacity for organization, 184
  feedforward and feedback, 102–5
  integral mechanisms of primitive, 85–86
  inverting locus of control, 215–17
  mind growing the brain, 137
  motive activation and arousal, 124
  neural architecture, 183
  neural mechanisms, 82
  neurodevelopment process, 129–38, 191–97
  neurodevelopmental process, x
  posterior brain and action regulation, 70–72
  postnatal ontogenetic mechanisms, x
  self-regulation, 120–21
  self-regulation of neural morphogenesis, 129–31
  shaping connections thorugh intention, 131–32

cognition (*Cont.*)
  shaping connections through attention, 132–33
  visceral and somatic domains, 196
  viscerosomatic consolidation, 36
cognitive control
  associative attention, 94–96
  executive attention, 94–96
  processes, 92
cognitive evaluation
  emotional basis of, 107–9
cognitive neuroscience
  development, 91
  executive control, 92–94
  human frontal pole, 79–80
cognitive process
  executive control, 91–94
cognitive processing
  neuromodulators, 181–83
cognitive psychology
  objectivity, 120
collapse of the reptilian telencephalon, x
collothalamic division
  dorsal thalamus, 168–69
collothalamic projections
  brainstem, 155–57
communication
  inter-cortical, 46
communicative process
  language, 136
complementarity
  neurocybernetics of activation and arousal, 114–16
  organization of experience, 128–29
  process and structure, 113
  representation and control, 195–97
complementary opposition
  representation and control, 195–97
component concepts
  notion, 49
computational models
  distributed representation, 26–27
concept, 24
concept components
  association cortex, 134
conditioned stimulus
  animal learning, 5–6
conditioning
  repeated association, 5
configural memory
  hippocampal-dorsal networks, 56
congruent
  complementarity, 196
connectional architecture
  mammalian cortex, 173–76
connectionist
  computational models, 26–27
connectivity
  anterior and posterior association, 38
  brain's, 191–93
  reasoning, 192

conscientiousness
  personality, 107
conscious self-monitoring
  human development, 203–5
consolidation
  dorsal and ventral networks, 75–78
  dorsal and ventral specialization, 66–68
  excitement of, 40–41
  memory, 24–26
  motive control of, 208–11
consolidation process
  subcortical regulation, 42–53
control processes
  organization of action, 90–91
control theory
  learning, 7
coping efforts
  anxiety and tension, 106
core parvalbumin projections
  embryogenesis, 180
core-association-shell organization
  hemispheric structure, 39–40
corollary discharge
  concept, 64
  frontal lobe, 63
corollary discharge proposal
  frontal lobe function, 69
corpus giganticothalamicum, 202
cortex
  anterior-posterior complementarity, 38
  complementary opposition, 196
  dorsal and ventral divisions, 36–38
  dual anlagen, 154–55
  somatic sensory and motor regions, 37
  thalamic regulation, 44
cortical differentiation
  gradients of genetic control, 174
cortical differentiation concept
  dual-trend organization, 169–71
cortical lesions
  human memory, 41–42
cortical levels
  memory consolidation, 54–57
cortical networks, 46
  forebrain cholinergic control, 47–48
cortical processing
  mechanisms, 34
cortical-subcortical organization
  dual configurations, 168
cortical-thalamic networks
  specialization, 143–44
cortico-cortical connectivity
  regulation, 45
corticolimbic divisions
  dorsal and ventral, 199
corticolimbic networks
  global organization, 129–31
  motivation and memory, viii
  representation in, 27–34
corticolimbic pathway, 33
corticolimbic sensory pathways
  architecture, 34

corticolimbic systems
  action regulation, 58
  connectional architecture, 140–42
  cytoarchitectonic specializations, 142–43
  differentiation between dorsal and ventral, 139–40
  dorsal and ventral interaction, 104
  hemispheric specialization, 141–42, 198
  motor control, 216
  subcortical control of inhibitory specialization in ventral, 144–45
  subcortical mechanisms of visceral expression in dorsal, 145–47
corticolimbic traffic
  memory consolidation, 54
corticothalamic network
  circuitry, 46
corticothalamocortical cell assemblies
  mechanisms, 47–48
critical reasoning
  human intelligence, 189
critical thinking, 101–2
crocodilian brain
  dorsal division, 165
crocodilian pallium
  dorsal region, 165
cross-modal processing
  dorsal networks, 184
cybernetic algorithms, x
cybernetic analysis
  Wiener, 7
cybernetic biases
  dorsal and ventral divisions, 178
  neurophysiology, 22
cybernetics, viii
  action control, 113
  action regulation, 63–64, 68
  cognitive, of mammalian brain, 197
  complementarity in representation and control, 195–97
  control modes, 21
  control processes, 90
  dorsal and ventral modes, 90–91
  feedforward and feedback control, 35, 60
  Greek term, 69
  habituation and redundancy, 152
  hedonic context model, 86
  mammalian, inversion, 211–13
  memory in time, 213–15
  modes of action regulation, 76
  motive, and executive functions, 68–70
  motor and cognitive, 123
  motor control, 116
  primitive neural, 132
  projectional and reactive vectors, 83
  redundancy bias, 187
  vectoral, of action, 34–38
  visceromotor control, 90
cytoarchitectonic specialization
  dorsal and ventral divisions of cortex, 176–78

cytoarchitectonic specializations
  dorsal and ventral networks, 66
cytoarchitectonics
  cortical connectivity, 139
  dominance of pyramidal cells, 184
  dorsal and ventral networks, 84, 140–42, 195
  pyramidal vs. granular specialization, 178–79
  representational specialization, 142–43
  specialized thalamic modulation, 143–44
  ventral limbic and cortical networks, 138
cytoarchitecture
  divisions of frontal lobe, 68

deep structure
  language, 135
default mode
  network, 92
delayed alternation
  testing paradigm, 61
dense amnesia
  cognition, 196
dense array electroencephalography (dEEG), vi
depression
  pseudodepression syndrome, 128
diencephalic level
  memory consolidation, 54–57
diencephalon, 42
  interbrain, 8, 9
discrimination learning
  rats, 5–6
discrimination learning paradigms
  animals, 10
disinhibition syndrome, 128
distributed representation
  computational models, 26–27
dopamine
  activation control, 212
  neuromodulation, 52
  neuromodulator, 147, 182–83, 188
  neurotransmitter of basal ganglia, 52
  redundancy bias, 127–28
  ventral corticolimbic neuromodulation, 148–49
dorsal and ventral divisions
  cytoarchitectonic specialization, 176–78
  differentiations, 181
dorsal and ventral networks
  connectivity and cytoarchitectonics, 195
dorsal brain
  neural activity control, 197
dorsal control system
  feedforward, viii
dorsal cortical
  projection pattern, 148
dorsal cortical networks
  evolution, 199
dorsal corticolimbic networks
  neuromodulation by norepinephrine and serotonin, 147–48

Index   263

dorsal corticolimbic networks (Cont.)
    subcortical mechanisms, 145–47
dorsal frontolimbic control system, viii
dorsal frontolimbic networks
    impetus, 215
dorsal hemisphere
    pallial anlagen, 183–86
dorsal limbic networks
    visceromotor control, 122, 175
dorsal neocortex
    matrix projections, 181
    pyramidal dominance, 178
    specialized subcortical circuitry, 179–81
dorsal networks
    consolidation, 77
    cortex, 36–38
dorsal noradrenergic bundle, 148
dorsal pathway
    consolidation processing, 184
dorsal premotor cortex (dPMC)
    executive control, 92–94
dorsal system
    function of frontal components, 96
    functions, 73
    vector of operation, 73
dorsal thalamus
    divisions, 168
dorsal-ventral complementary opposition, 83–85
dorsal-ventral differentiation
    developmental framework, 121
dorsal-ventral specialization
    mammalian, 58
    structural clues, 139–40
dorsolateral prefrontal cortex (DLPFC)
    anatomical connectivity, 61
    executive control, 92–94
    working memory, 94
drug manipulations
    motive controls, 127–28
dual gradients
    neutral factors, 155
dual limbic systems
    complementary opposition, 83–88
    self-regulation, 83–85
dual memory circuits, 54–57, 55, 56
dual-trend organization
    cortical differentiation concept, 169–71
dumb neurons, 26
dynamic reciprocity
    action regulation, 134
dynamism, 107, 110

early stages of learning, 10–11
effortful control, 111
egocentric orientations, 73
egocentric reference frame
    meaning, 75
egocentrism, 125
elation, 108
    clinical mania, 124
    learning, 129

electroconvulsive treatments
    memory, 25
Em2
    genetic transcription factor, 173
    trophic gradient, 211
    trophic gradients, 176
embryogenesis
    core parvalbumin projections, 180
    laminar differentiatioin, 169–71
    laminar stages of cortical, 157–59
    mechanisms, 8
    reptilian foundations, x
    transcription factors and gene expression, 171
embryonic gene expression
    neurotrophic gradients, 168
embryonic mechanisms
    hemispheric specialization, 201
emotion
    cognitive evaluation, 107–9
Emx2
    trophic gradient, 176
Emx-2
    embryogenesis, 167
encephalization
    concept, 48
    visceral function, 49
    viscerosensory and visceromotor functions, 49
energy and vigor
    experience, 106, 107
epilepsies, 40
epileptic seizures, 81
error negativity, 104
error-related negativity (ERN), 104
evolution
    capacity for memory, 120
    context of, 206–18
    cybernetics of memory in time, 213–15
    development of neocortex, 31–32
    general progression, 170
    hemispheric specialization, 198–201
    human frontal pole, 79
    human uniqueness, 197–98
    intelligence, 3
    inverting locus of control, 215–17
    mammalian brain, 160–66
    mammalian cybernetic inversion, 211–13
    mammalian embryogenesis, 152–60
    mammalian learning, 7
    mammalian memory, 151–52
    mind, 137–38
    motive control of consolidation, 207–8
    mutations at hemispheric core, 201–2
    nested frames of, and development, 207–8
    ontogenesis, ix
    self-organizing ontogenesis, 217–18
    working memory in mammals, 115
evolutionary-developmental analysis
    neurodevelopment process, ix
executive attention
    cognitive control, 94–96

executive control
  cognition, 92, 99
  cognitive neuroscience, 92–94
  frontal lobe, 60
  working memory, 63
executive functions
  self-regulation, 78
expectancy
  neural control of, 8–14
expectant
  mammalian memory, 3
experience-dependent
  plasticity, 3
experience-expectant
  plasticity, 3
exteriorization
  process, 36
external control
  artus, 214
external striatum
  anterior dorsal ventricular ridge (ADVR), 156
  disappearance of, 159–60
  reptilian, and mammalian cortex, 163
externalizing
  psychological problems, 110
extraversion
  control influence, 185
  impetus, 215
  mode of, 126
  modes, 217
  personality, 107, 110, 125
  psychological characteristics, 215
extravert
  intentional, 126

factors
  correlations, 106
feedback
  action control, 96, 102
  action regulation, 66, 67, 175
  control, viii
  control constraints, 13
  control mode, 88
  cybernetic control, 10
  cybernetic mode, 90
  cybernetics, 60
  direction of control, 32–34
  guidance in learning, 12
  interaction with feedforward in learning, 102–5
  motor control, 35
  restrictive control, 84
  robotics, 7
  ventral frontal lobe, 186
feedforward
  action control, 96, 102
  action regulation, 66
  control, viii
  cybernetic control, 10
  cybernetic mode, 90
  cybernetics, 60

cybernetics control, 13
direction of control, 32–34, 67
expectant learning, 22
habituation bias, 126
human cognition, 97
interaction with feedback in learning, 102–5
motor control, 35
projectional control, 84, 87
projectional mode of control, 97
robotics, 7
forebrain
  neuromodulators, 147–49, 181–83
forebrain cholinergic projections
  control over thalamus and cortex, 47–48
framing effect
  neural activity, 100
Freud, Sigmund, 4
  motive-memory, 209
frontal cortex
  human and monkey, 62
frontal granular cortex
  dominance of granular layer, 179
frontal lobe
  corollary discharge, 63
  cytoarchitectonic specializations, 142–43
  dorsal and ventral divisions, 88
  dorsal and ventral networks, 66
  executive control, 60
  premotor systems, 65–66
  self-regulation, 64–65
  working memory, 61–63, 61–63
frontal lobe working memory buffer
  concept, 61
frontal networks
  motive cybernetics and executive functions, 68–70
frontal pole
  human evolution, 204
  roles of dorsal and ventral divisions, 204
frontal pole (BA10)
  expansion, 79–80
  humans, 78–79
  thalamic reticular nucleus (TRN), 82
frontal pole (BA9)
  dorsal regions, 180
frontolimbic networks
  consolidation, 31
frontothalamic circuitry
  attnetion vs. intention, 82–83
frontothalamic control
  circuitry, 82
frontothalamic representation
  conscious learning, 205
*function*, v
functional magnetic resonance imaging (fMRI), 91, 192
  brain activity, vi
  humans, 62

GABA interneurons, 171
GABA neuron migration
  human mutation, 203

Index   265

GABAergic inhibitory neurons, 212
ganglionic eminence (GE), 153, 154
  cortical migration from, 159–60
  GABAergic migration, 203
  neuron precursors from, 171
  paleocortex, 154
generation
  internal vs. external, of actions, 98–101
genetic control
  neural activity, 217
genetic regulation
  neural morphogenesis, 167–68
Gestalt theory
  learning, 5
GingerALE v2.0, 98
global synchronization
  neural systems, 129–31
goal concepts
  action plan, 80
goal-directed attention
  control, 95
Goldberg and Costa model
  hemispheric specialization, 116
growth rings
  cortex of primitive mammals, 165
  evolutionary complexity, 31–32
  neocortex, 164

habituation, vii
habituation bias
  action regulation, 210
  attention and working memory, 19
  cognitive cybernetics, 197
  control of vertebrate pallium, 213
  cybernetic mode of sensory systems, 213
  feedforward control, 112
  liking element, 19
  mammalian learning, 193
  neural activity, 20–21, 24, 87, 152
  norepinephrine and serotonin, 149
  primitive cybernetic mode, 215
  primitive learning, 87
  response to change, 185
  sensory control, 116
  sensory memory, 140
  sensory memory control, 175
  slow learning system, 185
  working memory, 50, 115, 215
hallucination constrained by the sensory data
  perception, 33
Hebb, Donald
  cell assembly model, 4
hedonic expectancies
  primitive cognition, 85–86
hedonic expectancy
  approach learning, 184
  learning, 13
hemispheric pallium
  reptilian brain, 161
hemispheric specialization
  asymmetric, 116–19

  dorsal and ventral, 141–42
  human evolution, 198–201, 198–201
  motor control, 144
hemispheric structure
  sensory and motor cortices, 39–40
heterocronic mutation, 201–2
heteromodal
  association cortex, 31, 32
  network level, 32
hidden units, 26
hippocampal-dorsal networks
  spatial or configural memory, 56
hippocampus
  learning, 10
  mammalian, 159
  mediodorsal system, 96
  memory consolidation, 55
homeostasis, 70
hominoid cognition, 201
*homo erectus*
  frontal pole, 79
*homo floresiansis*, 79
hostility, 127
Hughlings, John, 60
human brain
  conscious learning, 203–5
  development, 2
  development strategies, x
  hemispheric specialization, 116–19, 141–42
  patterns of mammalian and reptilian brains, 207
  spindle cells, 201–2
human cognition
  mechanisms, x
  neurodevelopment process, 191–97
  neurophysiological learning model, 11
human consciousness
  dual frontopolar networks, 180
human development
  context of evolution, 206–18
  radical neoteny, 205–6
human dorsal neocortex
  embryogenesis, 179
human emotion
  cognitive evaluation, 107–9
human evolution
  specialized thalamic modulation, 178–79
  temperamental dispositions, 139
human frontal lobe
  self-regulation, 78–79
human frontal pole
  evolution, 79
human intelligence
  complex capacities, 189
  dorsal and ventral specialization, 149
  hemispheric specialization, 198–201
human learning
  phenomenon, 1
human limbic system, 9
human maturation
  learning capacity, 209

human mutation
  diencephalic counterpoint, 202–3
human neurodevelopmental
  thalamus and cortex, 202
human uniqueness, 197–98
humans
  frontal cortex, 62
  frontal pole (BA10), 79–80
hypothalamic control
  sleep, 50
hypothalamus
  diencephalon, 8
hypotheses
  biases, 6
hypothesis
  mediodorsal system, 96
hypothesis generation
  action control, 102
  actions, 98

ideational apraxia, 74
ideomotor apraxia, 74
imperception
  right hemisphere lesions, 200
impetus
  cognition, 206
  control mode, 129
  dorsal frontolimbic networks, 215
  dorsal projectional mode, 91
  elation, 129
  impulse, viii
  intentionality, 133
  internal control, 214, 215
  mechanisms, ix
  motivated learning, 210
  motivational control, 119
impulsiveness
  self-regulation, 111
incentive
  wanting vs. liking, 19
individual-specific plasticity, 3
infants
  learning new words, 137
inferior frontal junction (IFJ)
  executive control, 92–94
inhibitory capacity
  thalamus, 202
inhibitory control
  neural networks, 203
inhibitory specification
  basal ganglia, 52–53
  speech, 135
  subpallial mechanisms, 187
  ventral corticolimbic networks, 144–45
inhibitory suppression
  actions, 188
insula
  viscerosensory function, 49
  viscerosensory functions, 146
integral mechanisms
  primitive cognition, 85–86

intelligence, 207
  evolution of, 3
  human cognition, 205
  neural evolution, 200
intention
  differentiated attention vs. integrative, 82–83
  hemispheric asymmetries, 119
  parallel architectures for attention and, 78–83
  regulation in mammalian brain, 216
  shaping connections through, 131–32
intentional
  control, 96
  direction of behavior, 119
  learning mode, 131
intentionality, 122
  autism, 123
interbrain
  diencephalon, 8
interconnectivity
  limbic networks, 140–42
inter-cortical communication
  thalamocortical and corticothalamic projections, 46
internal control
  impetus, 214
internalizing
  psychological problems, 110
internalizing-externalizing dimension
  psychological disorders, 109
International Affective Picture System (IAPS), 108
interoceptive responses
  central nervous system, 64
intersubjectivity
  autism, 123
intralaminar nuclei
  thalamic projections, 53
introversion, 216
  mode of, 126
  modes, 217
  personality, 125
introvert
  constraint, 126
inversion
  mammalian cybernetics, 211–13
  primordial redirection, 176
Iowa Gambling task, 97
item memory, 67

juvenile mammals
  learning, 4
juvenilization
  human development, 205–6

kindling
  phenomenon, 40, 55

laminar projection
  feedforward, 34

language
  child in social context, 136
  deep structure, 135
  hemispheric specialization, 198, 201
  intentionality, 122
  network representations, 196
  opponent complementarity in developing, 133–37
  self-regulation of cognition, 196
late stages of learning, 10
learning
  brain, ix
  expectant and reactive forms, 194
  feedforward and feedback in, 102–5
  Gestalt theory, 5
  habituation bias, 185
  inhibitory specification, 187
  mammalian, vi
  modern theory, 5–8
  motivated, 11–14
  neural mechanisms, 82
  neural morphogenesis, 129–31
  neurodevelopment, vii
  neurodevelopmental mechanisms of, 1–2
  neurophysiological mechanisms, viii
  plasticity mechanisms, 2–5
  shaping connections through attention, 132–33
  shaping connections through expectant intention, 131–32
  vertical integration of, 14–15
learning/expectancy bias
  cognition, 85–86
lemnothalamic divisions
  dorsal thalamus, 168–69
lemnothalamic projections
  brainstem, 155–57
  dorsal thalamus, 180
lesions
  amnesia and brain, 196
  approach learning, 103
  dorsal and ventral brain, 128
  frontal lobe, 63
  hemispheric effects on sensory and motor skills, 200
  human brain, and cognition, 208
  mediodorsal frontal, and speech deficit, 136
liking
  incentive, 19
limbic
  memory consolidation, 54–57
  network level, 32
  subcortical system, 60
limbic control
  right hemisphere specialization, 41–42
limbic cortex, 28, 31
  dorsal and ventral regions of, 75–78
  memory and adaptive controls, 48–50
  visceral influence, 49
limbic networks. See corticolimbic systems
  action, 64–70
  consolidating memory, 24–26

cybernetics and executive functions, 68–70
  dorsal and ventral, 83–85
  dual routes to, 76
  regulating memory, 48–50
limbic set points
  evaluative guides, 64
limbic system
  diencephalon, 8
  dual organization, 196
limbic-thalamic-hypothalamic circuits, 40
limbifugal processing
  consolidation, 197
  feedforward, 33
limbifugal traffic
  consolidation, 177
limbipetal processing
  consolidation, 66, 197
  corticolimbic traffic, 177
  motive-action plan, 35
locus of control
  mammalian brain, 215–17
locus-function specificity, 141
long-term potentiation
  hippocampal phenomena, 55

macaque
  association cortex, 140
  frontothalamic connectivity, 82
machine learning
  animals and, 6–8
  stability-plasticity dilemma, 7–8
machine learning model
  adaptive resonance, 41
mammalian brain
  balanced representation and control, 195–97
  cognitive cybernetics, 197
  development, vii
  dorsal, archicortical division, 185
  evolution, 160
  human uniqueness, 197–98
  locus of control, 215–17
  management of learning, 27
  memory, vii
mammalian cognition
  ventral hemisphere, 186
mammalian cortex
  connectional architecture, 173–76
  inversion of functional control, 213
  laminar differentiation, 169–71
  pallial and subpallial self-regulation, 191
  subcortical control, 140
mammalian cybernetic inversion, 213, 214
  complementary opposition, 211–13
mammalian embryogenesis
  cortical migration from ganglionic eminence (GE), 159–60
  evolution of, 152–60
  genetic regulation of neural morphogenesis, 167–68
  laminar stages of cortical, 157–59

lemnothalamic and collothalamic
  brainstem primordia, 155–57
mammalian evolution
  reptilian telencephalon, x
mammalian learning
  cognitive expectancy, 22
  expectancy, 41
  neurophysiology, vi
mammalian memory
  collapse of reptilian telencephalon,
    171–73
  consolidating, 24–26
  mechanisms, 3
mammalian neocortex
  dorsal and ventral divisions, x, 183
  dorsal hemisphere, 183–86
  dual origins of, 166–67
  elaboration and expansion, 42
  evolutionary-development order,
    31–32
  memory consolidation, 172
  mutations, 188, 201–2
  organization, vii, 14–15
  ventricular/subventricular zone (VZ/SVZ),
    155
mammalian neural systems
  evolution, 138
mammalian neuroembryology
  evolutionary-developmental analysis, ix
mammalian variation
  dual origins of thalamocortical traffic,
    168–69
  embryonic patterns of laminar
    differentiation, 169–71
  emergence of mammalian memory,
    171–73
  genetic regulation of neural
    morphogenesis, 167–68
  interpreting dual origins of neocortex,
    166–67
  neocortical evolution, 160–66
  vertebrate theme, 160–73
mammals
  cortex organization, 77
  learning, 5
marsupials, 165
matrix projections
  thalamic nuclei, 143
maturation
  radical neotony, 205–6
medial frontal cortex
  internal vs. external generation of
    actions, 98–101
  learning studies, 97
  probability tracking studies, 97
  social economic game studies, 97–98
  studies of functions, 97–101
mediodorsal (MD) nuclei
  learning, 10
mediodorsal frontal cortex
  control of actions, 213
  sensory inputs, 67
mediodorsal frontal lesions
  speech deficit, 136
mediodorsal networks
  cortex, 36–38
memory
  asymmetric structure of, 38–42
  consolidating, 24–26
  cortical lesions and limbic disinhibition,
    41–42
  excitement of consolidation, 40–41
  Freud's reasoning, 4
  hemispheric specialization, 39–40
  mammalian learning, 193
  neurophysiology, vii
  postnatal ontogenetic mechanisms, x
  rabbits, 10–11
  self-regulation, 120–21
  spontaneous consolidation, 204
  visceral vs. somatic boundaries, 39–40
  working of, 61–63
memory capacity
  executive control, 63
memory consolidation
  cortical, limbic and diencephalic levels,
    54–57
  cybernetics, 191
  dorsal and ventral circuits, 194
  dorsal and ventral specializations,
    66–68
  mammalian brain regulating, 194
  mammalian neocortex, 172
  striatum and pallidum, 50–52
  vertical integration, 53–57
memory disorders
  microgenetic analysis, 196
memory integration
  allostatic process, 70
memory of the future
  extension, 63
memory structure
  cognitive representation, vii
mesolimbic dopamine system
  neuroscience, 19
microgenesis
  nested frames, 208
  Werner's concept of, 208
microgenetic analysis
  action, 68
microgenetic theory
  Brown, 36
  language, 134
mind
  brain, v
  evolution, 137–38
mirror neurons, 38
modern learning theory, 5–8
monkeys
  frontal cortex, 62
  neuromodulators in, 148
  working memory, 61
monotremes, 165
Montreal Neurologic Institute (MNI), 98

morphogenesis
　mutation and selection, 152
　program of genome, 217
　transition of vertebrate speciation, 207
motility control
　Yakovlev's notion, 36
motivated learning, 11–14
motivational biases
　activity controls, 210
motivational control
　limbic base, 69
motive controls
　drug manipulations, 127–28
　self-regulation, 129
motive expansion
　working memory, 19–20
motive restriction
　working memory, 18–19
motive vectors
　approach and avoidance, 137
motive-memory
　Freud, 209
　learning, 4
　psychoanalysis, viii
motor control
　action plans, 38
　basal ganglia, 52–53
　feedforward, 176
　surround inhibition, 144
　thalamo-cortical projections, 53
motor memory, 31
motor readiness
　action regulation, 115
mutation
　humans, 197–206

Negative Affect, 108, 125
negative arousal dimension, 106
negative contrast effect
　learning, 41
Negative Emotionality, 107, 110
neocortex
　consolidating memory, 24–26
　consolidation process, 180
　dorsal and ventral divisions, 21, 75–78
　dual origins of, 166–67
　evolution, 160–66
　hybrid organization, 167
　mammals, 54
　organization, 176
　pallial anlagen of dorsal hemisphere, 183–86
　points of origin, 152
　ring of, 28
　sensory and motor maps of, 57
　sensory-motor specializations, 58
neostriatum, 50
neotony
　human development and radical, 205–6
nerve fiber tractography, vi
nervous system
　somatic, vii

visceral, vii
nested frames
　neural development, 207–8
network architecture
　self-organizing ontogenesis, 217–18
network connectivity
　expectancies, 194
network structure
　neurodevelopmental theory, 4–5
　sensory and memory levels, 32–34
neural activity
　control of, vii
　dorsal corticolimbic neuromodulation, 147–48
　dorsal networks, 87
　framing effect, 100
　mind growing the brain, 137
　motivational vectors, 209
　motive mechanisms, 124
　primitive controls, 216
　redundancy and habituation, 20
　self-regulation, vii
　subcortical circuits, 87
　ventral corticolimbic neuromodulation, 148–49
neural arousal
　mehcanistic theory, 114
neural connectivity
　habituation and redundancy biases, 193
neural control
　expectancy, 8–14
　thalamic mechanisms, 53
　theory of intentions, 131
neural differentiation
　mechanisms, 1
neural embryogenesis
　stages, 1
neural mechanisms
　dorsal and ventral corticolimbic networks, 139–40
neural morphogenesis
　self-regulation, 129–31
neural networks
　computational models, 201
　viscerosomatic consolidation, 29–31
neural plasticity
　learning, 2–5
　learning and memory, vi
　neurodevelopmental theory, 193–95
　regulation, 130
neuraxis
　vertical integration, 14–15
neuroanatomy
　neocortical evolution, 160–66
neuroarchaeology, x
neurocybernetic mechanisms
　pallial and subpallial, x
neurocybernetics
　activation and arousal, 114–16
　representation and control, 195–97
　cognition, 129–38

neurodevelopment process
    evolution of mind, 137–38
    shaping connections through attention, 132–33
    shaping connections through intention, 131–32
neurodevelopmental approach
    cognition, 120–21
neurodevelopmental mechanism
    corticol networks, 185
neurodevelopmental mechanisms
    learning, 1–2
neurodevelopmental process
    brainstem and forebrain neuromodulators, 181–83
    cognition, 129–38, 191–97
    connectivity of brain, 191–93
    continuity of ontogenesis, 218
    evolution of mind, 137–38
    mammalian memory, 151–52
    mind growing the brain, 137
    motivated anatomy, 149–50
    representation and control complementarity, 195–97
    self regulation of neural morphogenesis, 129–31
    shaping connections through attention, 132–33
    shaping connections through intention, 131–32
    motivated anatomy, 149–50
neurodevelopmental theory
    challenge, 193
    plasticity and network structure, 4–5
    principles, 193–95
    psychological self-regulation, 133
    vertical integration of learning, 14–15
neuroembryogenesis
    dual trophic gradients, 167–68
    mammalian memory, 151–52
    neuromodulation, 20
    stages, 1
neuroembryology
    cortical-subcortical organization, 168–69
    mammalian, x
    mammalian cortex, 149
    mammalian neural evolution, ix
neuroimaging literature, 92
neuroimaging tools
    human brain, 192
neuromodulator projection systems, 17
    neural plasticity, 129–31
neuromodulators
    brainstem, 16–18
    brainstem and forebrain, 147–49, 181–83
    dopamine and acetylcholine, 148–49
    motivated anatomy, 149–50
    norepinephrine, 185
    norepinephrine and serotonin, 147–48
    self-regulation, 190

neurophysiological learning model
    human cognition, 11
neuropsychological theory
    action regulation, 89
    asymmetric hemispheric specialization, 116–19
    control process and cognitive structure, 112–14
    intention and attention, 119
    structure and process, 111–19
neuroscience
    development, v
    feedforward and feedback projections, 35
neurotic styles
    cognition, 112
neuroticism
    personality, 107, 110
neurotrophic factors, 156
    dual gradients of, 155
non-specific binding circuit, 44
noradrenergic brainstem projection
    arousal, 115
norepinephrine
    dorsal corticolimbic neuromodulation, 147–48
    habituation bias, 127–28, 127–28
    neural plasticity, 181–82
    neurodevelopmental role, 170
    neuromodulator, 147, 185
    working memory, 50

object
    processing of, 72
object memory, 67
object relations theory
    self, 111
object specification
    perception, 86
obsessive-compulsive disorders, 125, 127
ontogenesis
    developmental process, 89
    evolution, ix
    nested frames, 208
    neural, vii
    reasoning, 192
    self-organizing, 217–18
    self-organizing, on phyletic frame, 190–91
ontogeny, ix
openness
    personality, 107
optic ataxia
    parietal lobe, 74
orthogenesis
    concept, 208
output
    projections, 33

paleocortex
    ganglionic eminence (GE), 154
    neocortex evolution, 152
    ventral neocortex, 166
    ventral system, 140, 145

paleocortical base
   learning and consolidation, 43
paleocortical networks
   ventral base of cortex, 174
paleostriatum, 50–52
pallial and subpallial
   neural algorithms, 167
   neural architecture, 166
pallial anlagen
   dorsal hemisphere, 183–86
pallial architecture
   advantages, 172
pallial mound, 156
pallidum
   memory consolidation, 50–52
pallium
   amphibians and reptiles, 159
   origin of neocortex, 166–67
   vertebrate brain, 161
Papez circuit, 204
   learning, 10
   limbic system, 146
paradoxes
   internal and external control, 121–23
paralimbic, 28
paranoid personality disorders, 127
parietal lobe
   dorsal division, 73
   lesions to inferior, 74
   object processing, 71
parietal lobe damage
   deficits, 72–74
parietal lobes
   object meaning, 74
partial patterns
   alligator, 165
pattern of connections
   neurons, 26
Pax6
   developing cortex, 167
   genetic transcription factor, 173
   trophic gradient, 176, 211
Pax6 mutants, 169
perception
   sensory data, 134
perceptron
   connectionist model, 7
personal expectancy
   action, 119
personality
   dimensions of, 107
   disorders, 125
   representation and control, 216
personality disorders
   psychological development, 109–11
   psychological testing, 112
   self-regulation, 126
petit mal epilepsy, 81
phasic arousal system
   sensory vs. motor bias, 115
phasic fashion
   attention, 16

phylogenesis
   nested frames, 208
phylogeny, ix
plasticity
   neurodevelopmental theory, 4–5
   spatiotemporal trade-offs, 20–21
Positive Affect, 108, 124
positive arousal dimension, 106
Positive Emotionality, 107, 110
positive expectancy
   learning, 103
posterior attention system
   primates and humans, 11
posterior brain
   action regulation, 70–72
   modes of controlling action, 75–78
posterior cingulate cortex (PCC)
   mediodorsal system, 96
   perceptual information, 75
posterior cortical networks
   complementary of divisions, 75–78
posterior networks
   dorsal and ventral, 113
postural-affective matrix
   Werner, 36
pragmatic
   cognition, 76
   processing, 72
pragmatics
   use of language, 77
preconceptions
   psychological theory, 190
prestarting synthesis
   action regulation, 63–64
primary
   network level, 32
primary sensory
   cortex, 31, 32
primate cortex
   networks, 28
primitive
   spindle neurons, 201
primitive cybernetics
   action regulation, 86
primordial pallii dorsalis or P.P.D., 161
probability tracking, 97
*Project for a Scientific Psychology*
   Freud, 4
projectional vectors
   action, 65–66
protomaps, 153
pseudodepression syndrome, 128, 215
pseudopsychopathic syndrome, 128, 216
psychiatric depression, 128
psychological controls
   self-regulation, 90–91
psychological organization
   motive process, 190
psychological process
   dorsal and ventral modes, 116
psychological self-regulation
   affect and arousal, 105–7

dimensions, 105–11
psychological theory
  human development, viii
  preconceptions, 190
  self-regulation, 109–11
psychology
  development, v
  evolution of mind, 137–38
  scientific analysis, 191
  self-regulation, 105
psychometric methodology
  emotion, 105–7
psychomotor retardation
  depression, 128
psychopathology
  development approach, 110
purposeful anticipation
  cybernetic concept, 69
pyramidal neurons
  dorsal division, 142

qualitative controls
  neural activity, 20
quickening
  beginning of life, 206

rabbits
  approach and avoidance, 22
  avoidance action, 12
  elementary learning, 86
  learning challenges, 10–11
radial migration
  embryogenesis, 153
  mammals, 170
  neuronal, 170
radial unit hypothesis
  mammalian neocortex, 153
radical neoteny
  developmental mechanisms, x
  human development, 205–6
  human evolution, 205–6
  limbic networks, 209
rats
  discrimination learning, 5–6
  feedback training, 103
  intentions, 131
  memory deficits, 48
reactive vectors
  action, 65–66
redundancy, vii
redundancy bias
  action pattern, 212
  artus, 214
  cognitive consolidation, 210
  cognitive cybernetics, 197
  concept, 212
  cybernetics, 187
  dopamine, 128
  dopamine modulation, 183
  feedback control, 112
  language, 135
  learning, 132

mammalian learning, 193
motor control, 140, 175, 187
motor sequencing, 116
neural activity, 20–21, 24, 152
neuromodulator, 169
neuromodulators, 188
premammalian telencephalon, 214
primitive cybernetics, 87
primitive memory, 209
reactive attention, 133
viscerosensory function, 176
working memory, 18, 52, 114, 125, 216
representation of the regulatory function, 205
  cognition, 132
  integration, 60
representational levels
  learning, 26
representations
  object-response, 93
reptiles
  external striatum, 155–57
  motor capacities, 213
  pallium, 159
  primitive general cortex, 164
  pyramidal cells, 159
reptilian brain
  organization, 165
reptilian pallium
  mammalian memory, 171–73
reptilian telencephalon
  activation control, 211
reset signal, 48
resting state, 92
retrograde amnesia, 25
reward circuits
  concept, 53
*Riddle of Frontal Lobe Function in Man*, Teuber, 63
robotics
  control theory, 7
rostral bias
  viscerosensory function, 175
  growth ring interpretation, 34

Sanides hypothesis
  connectivity studies, 173
  dorsal-ventral separation, 37
  growth ring interpretation, 34
  limbic cortex, 31
  neocortex, 154
  primate cortex, 164
schemas
  selection of, 92
schizophrenia, 110, 124
  delusions, 123–24
scientific analysis
  objectivity for flexible, v
scientific theory
  brain and mind, 192
seizures
  absence spells, 68

seizures (*Cont.*)
  epileptic, 81
  limbic, 82
selective attention
  thalamic gating, 204
self-generated activity
  embryonic, 4
self-organization
  adaptive learning, 5
  dual modes of neurodevelopmental, 194
  embryo, ix
  mammalian neural connections, 3
  memory, x
  ontogenesis, 217–18
  ontogenesis on phyletic frame, 190–91
  spontaneous acts, 4
self-regulation
  activation and arousal, 15–21, 86–88
  algorithms in cognition, 207
  conscious learning, 203–5
  control and cognitive structure, 124–27
  deficits in social, 128
  developmental literature, 110
  dopamine and acetylcholine, 182–83
  dorsal and ventral brain lesions, 128
  dual limbic systems, 83–85
  evolved roots of, 183–89
  feedforward and feedback, 7
  human frontal lobe, 78–79
  internal and external control in social context, 121–23
  losing locus of control, 123–24
  mammalian inversion, 214
  memory, 120–21
  neural morphogenesis, 129–31
  neuropsychological theory, 111–12
  organization of experience, 128–29
  popular culture, 112
  psychological controls, 90–91
  psychological development, 109–11
  psychology, 105
  visceral and social contexts of human, 119–29
  visceral and somatic modes, 167
self-regulatory circuitry
  mammals, 54
semantic
  cognition, 76
  processing, 72
semantics
  meanings of words, 77
sense of agency
  loss of, 123–24
sensory cortex
  input projections, 33
sensory feedback
  inter-cortical communication, 34
sensory input
  action regulation, 176
sensory representation
  right hemisphere, 200

sensory systems
  control bias, 115
sensory-to-association
  limbipetal, 34
serotonin
  dorsal corticolimbic neuromodulation, 147–48
  neural plasticity, 181–82
  neuromodulator, 147
  working memory, 50
shock treatments
  memory, 25
short-latency afferent inhibition
  phenomena, 144
simple structure
  factor analysis, 106
*Simultaneous Matching Adaptive Resonance Theory* or SMART model, 47
sleep spindles, 81
sleep stages, 50
social adaptation
  self-regulation, 123
social attachment
  intentionality, 122
social economic game studies, 97–98
social propriety, 107, 110
somatic
  nervous system, vii
  term, 64
somatic constraints
  neural networks, 29–31
somatic motor network
  specialization, 90
somatic networks
  neocortex, 58
somatic sensory network
  specialization, 90
somatic set-points
  action, 64
somatic-marker hypothesis
  action, 64
sources
  executive attention, 94
spatial memory
  differentiation, 117
  hippocampal-dorsal network, 56
spatial relations
  processing, 72
spatiotemporal complementarity
  activation and arousal, 114–16
spatiotemporal trade-offs
  activity-dependent plasticity, 20–21
specialization
  asymmetric hemispheric, 116–19
  limbic and neocortical networks, 41–42
  right and left hemispheres, 39–40
  somatic input/output, 195
specialized connectivity
  thalamus and cortex, 143–44
species-specific plasticity, 3

speech
  complementarity in developing language, 133–37
speech deficit
  mediodorsal frontal lesions, 136
sphere of visceration, 36
spike-wave discharges
  animal models, 81
spindle cells
  functional role, 201–2
stability-plasticity dilemma
  learning, 26, 209
  machine learning, 7–8
stimulus-response mappings
  hypotheses, 100
  learning, 97
striatum
  memory consolidation, 50–52
Stroop task
  representation, 93
structure-process relations
  action regulation, 112–14
subcortical circuitry
  dorsal corticolimbic networks, 145–47
  specialization, 144–47
  specialized, 179–81
  ventral corticolimbic networks, 144–45
subcortical circuits
  neural activity, 87
subcortical regulation
  basal ganglia, 52–53
  consolidation process, 42–53
  forebrain cholinergic control, 47–48
  limbic elaboration of controls, 48–50
  thalamic control of cortical traffic, 43–47
  ventral striatum and paleostriatum, 50–52
subcortical systems
  vertical integration, 60
subpallial
  basal ganglia differentiation, 155
subpallial anlagen
  ventral hemisphere, 186–89
subpallial thalamic projections
  incorporation in ventral division, 186
subventricular zone (SVZ), 153, 154
successive negative contrast effect
  expectancy, 13
supplementary motor area (SMA)
  basal ganglia circuits, 69
  cytoarchitectonics, 66
  mediodorsal frontal lobe, 65
surround inhibition
  motor control, 144
  striatal circuits, 178
swellings
  lateral frontal, 79
swimming movements
  dorsal cortex, 164
synaptic plasticity
  learning and, 2–5
synaptogenesis, vi

Talairach coordinates, 98
tangential migration
  embryogenesis, 153
  mammalian neocortex, 154
telencephalon
  activation control in reptilian, 211
  architecture, 161
  dorsal and ventral divisions, 195
  human brain, 9
  reptilian to mammalian form, 164
  subcortical, 42
temperament
  self-regulation, 110
tension
  experience, 106, 107
thalamic
  subcortical system, 60
thalamic mechanisms
  attention, 78
thalamic modulation
  specialized, 178–79
thalamic regulation
  concepts of, 80–82
  cortical connectivity, 43
  cortical traffic, 43–47
thalamic reticular nucleus (TRN), 180
  frontal pole, 82
  frontal pole connectivity, 203
thalamocortical connections
  specialization, 143–44
thalamocortical matrix binding circuit
  cortex, 45
thalamocortical networks
  memory control, 43
thalamocortical projections
  regulating corticocortical connections, 44
thalamocortical regulation
  dorsal neocortex, 169
thalamus
  core projections, 45
  diencephalon, 8
  human neurodevelopment, 202–3
theory of action regulation, viii
tiredness
  experience of, 106
tone
  conditioned stimulus, 5–6
tonic activation, 16
  control, 16
tonic activation system, 114, 115, 116, 147, 212
training-induced activity
  learning, 10
transcortical motor aphasia
  syndrome, 136
transcranial magnetic stimulation, 43, 143
transference
  prior experience, 209
transformational
  cognitive process, 209
triangular circuit
  consolidation mechanism, 189

triangular circuit (*Cont.*)
  sensory control, 67
Trust Game, 98
Tucker and Williamson model
  cognition control, 115
  sensory systems, 115
  tonic activation and phasic arousal, 116

unimodal
  association cortex, 31, 32
  network level, 32
utilization behavior
  frontal lobe, 64

valence, 106, 147
ventral brain
  neural activity, 197
ventral control system
  feedback, viii
ventral corticolimbic networks
  inhibitory specialization, 144–45
  viscerosensory response, 194
ventral frontolimbic control system, viii
ventral hemisphere
  subpallial anlagen of, 186–89
ventral limbic networks
  amygdalar connections, 180
  feedback guidance, 12
  viscerosensory control, 175
ventral neocortex
  complexity, 187
  incorporation of subpallial elements, 186–89
  neuromodulators, 188
ventral networks
  consolidation, 77
  cortex, 36–38
ventral premotor cortex (PMv)
  action control, 65
ventral striatum
  activity, 50–52
ventral system
  functions, 72
  vector of operation, 73
ventricular zone (VZ), 153, 154
ventrolateral frontal networks
  sensory feedback, 67
ventrolateral networks
  cortex, 36–38
vertebrate brain
  cortical architecture, 161
  specialized thalamic modulation, 178–79
vertebrate organization
  input/output circuits, 161
  reptiles, 166
vertical integration
  consolidation of experience, 53–57
  learning, 14–15

neural activity, 22
visceral
  nervous system, vii
visceral constraints
  neural networks, 29–31
  visceromotor cingulate limbic networks, 199
visceromotor control
  dorsal limbic networks, 175
visceromotor function
  cingulate cortex, 146
  limbic cortex, 37
visceromotor functions
  action regulation, 86
  regulation, 49
visceromotor networks, 199
visceromotor vs viscerosensory
  complementarity, 84
viscerosensory control
  motive bias, 123
  ventral limbic networks, 175
viscerosensory function
  limbic cortex, 37
viscerosensory functions
  feedback control, 86
  insula, 49, 146
viscerosensory insular-amygdalar
  networks, 199
viscerosensory networks, 199
vision
  cortical pathways, 71
visual form agnosia
  processing, 72
visual processing
  recognition, 72–74
Von Economo neurons, 201

waking state stimulation, 143
wanting
  anticipation, 53
  incentive, 19
Werner, Heinz, 36
Wernicke's
  comprehension, 135
Wernicke's aphasia, 196
working
  component of working memory, 63
working memory, 61–63
  evolution in mammals, 115
  executive control, 93
  frontal lobe, 88
  limbic networks, 48–50
  mechanisms, 14
  motive biases, 59
  motive expansion, 19–20
  motive restriction, 18–19
  norepinephrine and serotonin, 50